Methods in Molecular Biology

Series Editor
John M. Walker
School of Life and Medical Sciences
University of Hertfordshire
Hatfield, Hertfordshire, AL10 9AB, UK

For further volumes:
http://www.springer.com/series/7651

Germinal Centers

Methods and Protocols

Edited by

Dinis Pedro Calado

Immunity and Cancer Laboratory
The Francis Crick Institute, and King's College London
London, UK

Humana Press

Editor
Dinis Pedro Calado
Immunity and Cancer Laboratory
The Francis Crick Institute, and King's College London
London, UK

ISSN 1064-3745 ISSN 1940-6029 (electronic)
Methods in Molecular Biology
ISBN 978-1-4939-7094-0 ISBN 978-1-4939-7095-7 (eBook)
DOI 10.1007/978-1-4939-7095-7

Library of Congress Control Number: 2017941511

Cover Image: Murine GC: GCR induced by IC retention on FDCs in a mouse LN. AMCA-OVA ICs (blue) are retained in the FDC reticulum. GC B cells (green) are surrounded by a resting B cell mantle (red)

Printed on acid-free paper

This Humana Press imprint is published by Springer Nature
The registered company is Springer Science+Business Media LLC
The registered company address is: 233 Spring Street, New York, NY 10013, U.S.A.

Preface

In 1884 Walther Flemming described histologically defined sites, within the follicles of secondary lymphoid organs, where large cells undergoing mitosis were present. Flemming called these sites germinal centers (GCs), in accordance with his proposal of these sites functioning as the places where cells originate in the body. Although the name of these sites remained, Flemming's hypothesis has since been disproven. Currently it is accepted that GCs are formed during immune responses and are the sites of B cell clonal expansion in T cell-dependent antibody responses to infection. In the GC, B cells undergo processes of DNA recombination, mutation, and differentiation by which long-lived high affinity antibody secreting cells (plasma cells) and memory B cells develop, assuring long-term protection against the infectious agents.

GCs are dynamic structures formed within B cell follicles and are composed of a diverse set of specialized cells. These include primarily highly proliferative GC B cells; follicular dendritic cells (FDCs), which provide migration and survival cues to GC B cells, and that retain antigen in their surfaces, important for affinity maturation under antigen-limiting conditions; GC T cells composed mainly of $CD4^{pos}$ T follicular helper T cells (Tfh), key for the formation of GCs and for processes of GC B cell selection; and a smaller subset of $CD4^{pos}$ $Foxp3^{pos}$ regulatory T cells (Tfr), thought to limit the extent of the GC reaction. GC cellular populations also include tingible body macrophages (TBM) that phagocyte and eliminate dying B cells arising in the process of GC B cell selection, as well as other less well-characterized cellular subsets including $CD8^{pos}$ T cells and bone marrow-derived dendritic cells for which a function remains to be determined.

Initial studies on GCs relied on tissue light microscopy, and despite the rudimentary technology available, key concepts on GC biology were established early on leading to highly active areas of research at the present time. An example of such is the observation of GC compartmentalization in two morphologically distinct areas: a dark zone (DZ) adjacent to the T cell zone and a light zone (LZ) contiguous to the splenic marginal zone or to the lymph node capsule. The lighter and darker appearances of the GC LZ and DZ resulted from the fact that B cells in the LZ are scattered among a network of FDCs and that these are largely absent from the DZ. GC T cells as well as FDCs localize in the LZ, where B cell selection takes place; TBM on the other hand are scattered among both areas.

The advent of cellular analysis by fluorescence-activated cell sorting, intravital microscopy, and gene expression profiling allows today the study GC biological processes to a high level of sophistication. Some of the initial concepts have since been solidified while others having to be reanalyzed. Initial studies using pulse labeling with radioactive nucleotides demonstrated the migration of cells in the DZ to the LZ, suggesting a precursor–progeny relationship between the two GC areas. Recently, studies using intravital microscopy showed a more complex migratory pattern of GC B cells, where bidirectional migration takes place, in accordance with the model of cyclic re-entry in which stepwise antibody affinity maturation takes place through multiple rounds of selection.

In vitro culture systems can mimic some of the features of GC B cell biology, including high proliferative index and class-switch recombination. Still studies in living organisms, and particularly in the mouse, have been crucial for an in-depth understanding of GCs. In addition to the use of wild-type animals, the field has been prolific in the generation of genetically modified mice that allow the analysis of processes such as GC B cell selection, and affinity maturation. GC B cells arise from antigen-activated naïve B cells in a T cell-dependent manner through the course of an immune response. Therefore animals displaying impaired early development of these populations are improper for the study of GC biology. The generation of tools allowing targeted genetic mutational studies specifically to GC B cells, using primarily the Cre-Loxp system, has provided unique experimental approaches to tackle the relevance of genes, noncoding RNAs, and pathways for this stage of B cell development.

Animal model systems have also been used to address the concept of oncogenicity of the GC reaction. In the GC microenvironment B cells undergo processes that involve DNA recombination and mutation while rapidly dividing, namely class-switching and somatic hypermutation. These are essential for the development of high affinity antibodies; however infidelity in them may lead to oncogenic DNA lesions and be at the origin of cancer. In fact, and although human cancers may arise from B cells at several stages of development, cancers derived from GC B cells or from B cells that have passed the GC reaction are the most common hematological malignancy in adults overall.

GCs are key for antibody immune responses, vaccination, and are strongly implicated in cancer. GC biological processes involve cell-to-cell interaction; cellular activation, division, death, migration, selection, and differentiation; as well as DNA damage and repair. It is thus not surprising that GCs are a highly active research field and that its community includes scientist from the most diverse backgrounds. This MiMB volume provides key methods and protocols from those laboratories with the expectation of stimulating further research and to aid scientists in the study of GC biology and pathology.

London, UK *Dinis Pedro Calado*

Contents

Contributors

DAVIDE ANGELETTI • *Laboratory of Viral Diseases, National Institutes of Allergy and Infectious Diseases, National Institutes of Health, Bethesda, MD, USA*

SEBASTIAN C. BINDER • *Systems Immunology Department and Braunschweig Integrated Centre of Systems Biology, Helmholtz Centre for Infection Research, Braunschweig, Germany*

GEOFFREY BROWN • *Institute for Immunology and Immunotherapy, Institute for Biomedical Research, University of Birmingham Medical School, Birmingham, UK*

MICHAEL C. CARROLL • *Program in Cellular and Molecular Medicine, Boston Children's Hospital, Boston, MA, USA*

STEFANO CASOLA • *The FIRC Institute of Molecular Oncology Foundation, Milan, Italy*

GIORGIO CATTORETTI • *Dipartimento di Medicina e Chirurgia, Universitá degli Studi di Milano-Bicocca (UNIMIB), Monza, Italy; Anatomia Patologica, Azienda Ospedaliera San Gerardo, Monza, Italy*

VAN TRUNG CHU • *Max-Delbrück-Center for Molecular Medicine, Berlin, Germany*

GRAEME COWAN • *The King's Buildings, Ashworth Laboratories, Charlotte Auerbach Road, Edinburgh, UK*

MANUEL D. DÍAZ-MUÑOZ • *Laboratory of Lymphocyte Signalling and Development, The Babraham Institute, Cambridge, UK; Division of Immunology, Infection and Inflammatory Disease, Department of Immunobiology, King's College London, London, UK*

DAVID DOMINGUEZ-SOLA • *Department of Oncological Sciences, Icahn School of Medicine at Mount Sinai, New York, NY, USA; The Tisch Cancer Institute & Precision Immunology Institute, Icahn School of Medicine at Mount Sinai, New York, NY, USA*

GREGORY M. FRANK • *Laboratory of Viral Diseases, National Institutes of Allergy and Infectious Diseases, National Institutes of Health, Bethesda, MD, USA*

SIMON LE GALLOU • *Institut Necker-Enfants Malades, INSERM U1151-CNRS UMR 8253, Université Paris Descartes, Sorbonne Paris Cité, Faculté de Médecine-Site Broussais, Paris Cedex, France; INSERM U1236, Faculté de Médecine, Université Rennes 1, Rennes Cedex, France*

LAURA GARCIA-IBANEZ • *Institute for Immunology and Immunotherapy, Institute for Biomedical Research, University of Birmingham Medical School, Birmingham, UK*

ROBIN GRAF • *Max-Delbrück-Center for Molecular Medicine, Berlin, Germany*

KEI HANIUDA • *Division of Molecular Biology, Research Institute for Biomedical Sciences (RIBS), Tokyo University of Science, Noda, Chiba, Japan*

ANJA E. HAUSER • *Immune Dynamics and Intravital Microscopy, Charité Universitätsmedizin, Berlin, Germany; Deutsches Rheumaforschungszentrum, Berlin, Germany*

BALTHASAR A. HEESTERS • *Program in Cellular and Molecular Medicine, Boston Children's Hospital, Boston, MA, USA*

NICOLE HEISE • *Herbert Irving Comprehensive Cancer Center (adjunct), Columbia University (adjunct), New York, NY, USA*

JOHANNE T. JACOBSEN • *Laboratory of Lymphocyte Dynamics, Rockefeller University, New York, NY, USA*

CHRISTOPH JANDL • *Department of Immunology, Garvan Institute of Medical Research, Darlinghurst, NSW, Australia; Department of Medicine, St. Vincent's Clinical School, University of NSW, Sydney, NSW, Australia*

ELENI KABRANI • *Max-Delbrück-Center for Molecular Medicine, Berlin, Germany*

CECILE KING • *Department of Immunology, Garvan Institute of Medical Research, Darlinghurst, NSW, Australia; Department of Medicine, St. Vincent's Clinical School, University of NSW, Sydney, NSW, Australia*

DAISUKE KITAMURA • *Division of Molecular Biology, Research Institute for Biomedical Sciences (RIBS), Tokyo University of Science, Noda, Chiba, Japan*

ULF KLEIN • *Herbert Irving Comprehensive Cancer Center (adjunct), Columbia University (adjunct), New York, NY, USA; Leeds Institute of Cancer and Pathology, University of Leeds, Leeds, NY, UK*

RALF KÜPPERS • *Institute of Cell Biology (Cancer Research), Medical Faculty, University of Duisburg-Essen, Essen, Germany*

RANDALL L. LINDQUIST • *Deutsches Rheumaforschungszentrum, Berlin, Germany*

MICHELLE A. LINTERMAN • *Lymphocyte Signalling and Development, Babraham Institute, Cambridge, UK*

CLAUDIA LOETSCH • *Department of Immunology, Garvan Institute of Medical Research, Darlinghurst, NSW, Australia; Department of Medicine, St. Vincent's Clinical School, University of NSW, Sydney, NSW, Australia*

MICHAEL MEYER-HERMANN • *Systems Immunology Department and Braunschweig Integrated Centre of Systems Biology, Helmholtz Centre for Infection Research, Braunschweig, Germany*

ELISA MONZÓN-CASANOVA • *Laboratory of Lymphocyte Signalling and Development, The Babraham Institute, Cambridge, UK; Department of Biochemistry, University of Cambridge, Cambridge, UK*

IMOGEN MORAN • *Immunology Division, Garvan Institute of Medical Research, Darlinghurst, NSW, Australia; Faculty of Medicine, St Vincent's Clinical School, UNSW Australia, Darlinghurst, NSW, Australia*

RINAKO NAKAGAWA • *Immunity and Cancer Laboratory, The Francis Crick Institute, London, UK*

TAKUYA NOJIMA • *Division of Molecular Biology, Research Institute for Biomedical Sciences (RIBS), Tokyo University of Science, Noda, Chiba, Japan*

VALENTINA PETROCELLI • *The FIRC Institute of Molecular Oncology Foundation, Milan, Italy*

TRI GIANG PHAN • *Immunology Division, Garvan Institute of Medical Research, Darlinghurst, NSW, Australia; Faculty of Medicine, St Vincent's Clinical School, UNSW Australia, Darlinghurst, NSW, Australia*

COSTANTINO PITZALIS • *Centre for Experimental Medicine and Rheumatology, William Harvey Research Institute, Barts and the London School of Medicine and Dentistry, Queen Mary University of London, London, UK*

CEES E. VAN DER POEL • *Program in Cellular and Molecular Medicine, Boston Children's Hospital, Boston, MA, USA*

HAI QI • *Laboratory of Dynamic Immunobiology, Department of Basic Medical Sciences, Tsinghua-Peking Center for Life Sciences, Institute for Immunology, School of Medicine, Tsinghua University, Beijing 100084, China*

KLAUS RAJEWSKY • *Max-Delbrück-Center for Molecular Medicine, Berlin, Germany*

Ananya Rastogi • *Systems Immunology Department and Braunschweig Integrated Centre of Systems Biology, Helmholtz Centre for Infection Research, Braunschweig, Germany*

Claude-Agnès Reynaud • *Institut Necker-Enfants Malades, INSERM U1151-CNRS UMR 8253, Université Paris Descartes, Sorbonne Paris Cité, Faculté de Médecine-Site Broussais, Paris Cedex, France*

Philippe A. Robert • *Systems Immunology Department and Braunschweig Integrated Centre of Systems Biology, Helmholtz Centre for Infection Research, Braunschweig, Germany; Institut de Génétique Moléculaire de Montpellier, CNRS, UMR 5535, Université de Montpellier, Montpellier, France*

Surinder S. Sahota • *Tumour Immunogenetics Group, Cancer Sciences Academic Unit, Faculty of Medicine, University of Southampton, Southampton, UK*

Sandrine Sander • *Adaptive Immunity and Lymphoma, German Cancer Research Center and National Center for Tumor Diseases, Heidelberg, Germany*

Riham El Sayed • *Centre for Experimental Medicine and Rheumatology, William Harvey Research Institute, Barts and the London School of Medicine and Dentistry, Queen Mary University of London, London, UK; Department of Clinical and Chemical Pathology, Kasr Al-Ainy Faculty of Medicine, Cairo University, Egypt*

Marc Seifert • *Institute of Cell Biology (Cancer Research), Medical Faculty, University of Duisburg-Essen, Essen, Germany*

Changming Shih • *Laboratory of Dynamic Immunobiology, Department of Basic Medical Sciences, Tsinghua-Peking Center for Life Sciences, Institute for Immunology, School of Medicine, Tsinghua University, Beijing 100084, China*

Mohey Eldin M. El Shikh • *Centre for Experimental Medicine and Rheumatology, William Harvey Research Institute, Barts and the London School of Medicine and Dentistry, Queen Mary University of London, London, UK*

Laura Tech • *Deutsches Rheumaforschungszentrum, Berlin, Germany*

Kai-Michael Toellner • *Institute for Immunology and Immunotherapy, Institute for Biomedical Research, University of Birmingham Medical School, Birmingham, UK*

Martin Turner • *Laboratory of Lymphocyte Signalling and Development, The Babraham Institute, Cambridge, UK*

Carolin Ulbricht • *Immune Dynamics and Intravital Microscopy, Charité Universitätsmedizin, Berlin, Germany*

Ine Vanderleyden • *Lymphocyte Signalling and Development, Babraham Institute, Cambridge, UK*

Gabriel D. Victora • *Laboratory of Lymphocyte Dynamics, Rockefeller University, New York, NY, USA*

Jean-Claude Weill • *Institut Necker-Enfants Malades, INSERM U1151-CNRS UMR 8253, Université Paris Descartes, Sorbonne Paris Cité, Faculté de Médecine-Site Broussais, Paris Cedex, France*

Nicola J. Weston-Bell • *Tumour Immunogenetics Group, Cancer Sciences Academic Unit, Faculty of Medicine, University of Southampton, Southampton, UK*

Jonathan W. Yewdell • *Laboratory of Viral Diseases, National Institutes of Health, National Institutes of Allergy and Infectious Diseases, Bethesda, MD, USA*

Yang Zhang • *Institute for Immunology and Immunotherapy, Institute for Biomedical Research, University of Birmingham Medical School, Birmingham, UK*

Chapter 1

Analysis of the Germinal Center Reaction in Tissue Sections

David Dominguez-Sola and Giorgio Cattoretti

Abstract

Germinal centers are short-lived microanatomical compartments with essential roles in adaptive immunity. These lymphoid structures can be identified in secondary lymphoid organs using both flow cytometry and immunohistological analyses, but only the latter provides useful architectural and spatial information. Here we describe how to use immunofluorescence and immunohistochemistry with specific antibodies to precisely highlight the cellular and architectural features of germinal centers, both in human and mouse secondary lymphoid organs, and to study their normal development and disturbance in disease.

Key words Immunohistochemistry, Immunofluorescence, Frozen sections, Formalin-fixed paraffin-embedded sections, Epitope, Antigen retrieval

1 Introduction

The close connection between tissue architecture and tissue function is routinely used in the diagnostic assessment of diseased states. It is also useful to better understand how tissues and organisms develop, and as such it has provided key conceptual clues on how diseases arise and evolve. Recent advances in antibody generation, and the subsequent development of spatial pathology-based annotations of protein expression, have rapidly advanced this idea and allowed for the analysis of tissue-based proteomes in any organ and disease [1, 2]. These technical efforts have for instance been compiled in multi-data resources like the Human Protein Atlas (www.proteinatlas.org) [1–3], and make feasible mapping complex biological processes and phenotypic changes within any tissue at the cellular and molecular level, by using immunohistochemistry and immunofluorescence with high-affinity antibodies as main tools.

Immunohistochemistry and immunofluorescence were first conceptualized and developed by Albert Coons and colleagues as a method to reveal the presence of specific antigens in tissue sections, in an effort to understand the topographical relationship between diseases and their biological basis. These methods quickly

Dinis Pedro Calado (ed.), *Germinal Centers: Methods and Protocols*, Methods in Molecular Biology, vol. 1623,
DOI 10.1007/978-1-4939-7095-7_1, © Springer Science+Business Media LLC 2017

evolved into fundamental strategies for the detection of specific markers in cells or tissue structures [4, 5]. Immunohistochemistry and immunofluorescence use antibodies to specifically detect the presence of epitopes in tissue sections. Conjugation of antibodies to certain moieties, e.g., enzymes or fluorochromes, allows for mapping the distribution of these epitopes in cells, structures, and microanatomical compartments at the microscopic level and at high resolution. With this strategy it is possible to define the phenotypic features of cells and precisely determine the cellular composition of any tissue under study, generating functional architectural maps of normal and diseased states—by combining for example the detection of phenotypic markers and signaling cues through antibodies specific for particular posttranslational modifications—or determining the cellular lineage of cancerous cells of uncertain origin. The powerful combination of morphology and immunohistological methods has proven useful in both diagnostic and research settings [6], and has helped to decipher the roles of gene programs and signaling pathways in controlling cellular dynamics, tissue organization and function. Findings like the existence of cellular hierarchies in epithelial cancers [7], the activation of senescence and DNA damage responses in precancerous lesions [8, 9], or the disruption of germinal center polarity upon loss of certain signaling pathways [10–12], exemplify well this notion.

Germinal centers can be easily identified in secondary lymphoid tissues in response to T-cell dependent antigens. These lymphoid structures develop within few days after antigen exposure and sustain the rapid expansion and selection of B cell clones producing antibodies (antigen receptors) of increasing affinity, key elements of the immune adaptive system [13]. The phenotypic analysis of germinal centers via flow cytometry and immunohistology has revealed a high degree of architectural organization within these microscopic structures, mainly composed by B cells, but also T cells, macrophages, and the so-called follicular dendritic cells (FDCs). The distribution of these cell types in the germinal center follows a strict pattern directly correlated with the two known microanatomical compartments classically defined within this structure: the so-called Dark and Light zones. Specifically, FDCs and T cells are located in the light zone (as opposed to the dark zone), while B cells distribute and transit between both zones [13, 14] (Fig. 1). This organization underlies critical biological differences between Light and Dark zone, and is required for the functional success of the germinal center reaction during adaptive immune responses [12, 15].

The architectural and cellular dynamics of the germinal center reaction can be studied in mouse and human lymphoid tissues using flow cytometry, immunohistochemistry, or immunofluorescence [16]. Germinal center B cells can be easily distinguished from other B cell subsets according to the expression of specific phenotypic markers, like cell-specific enzymes (activation-induced

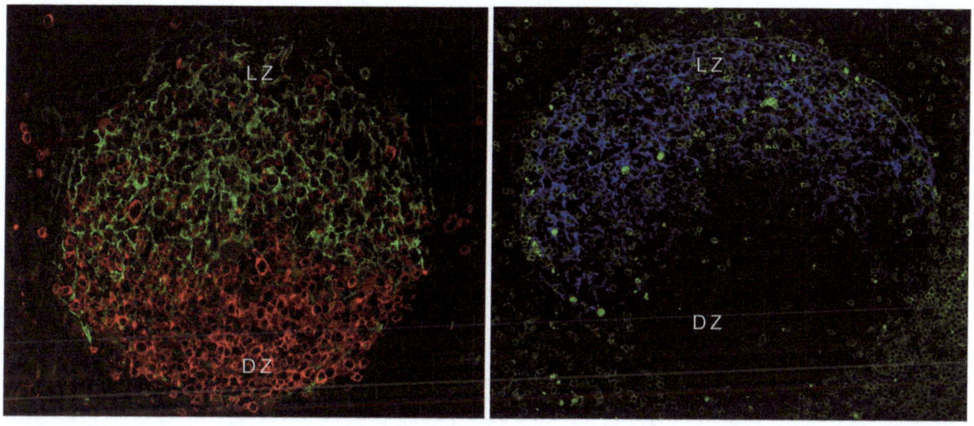

Fig. 1 Light/Dark Zone division in a germinal center, as seen using immunofluorescence. Germinal centers (GCs) identified in sections of human tonsil (FFPE), are here stained for CD23 (*green*), AID (*red*) (*left panel*); or CD3 (*green*), CD23 (*blue*) (*right panel*). AID (activation-induced cytidine deaminase) is expressed in GC B cells, most abundantly in B cells located in the Dark Zone of GC, highlighted in *red* in the *left panel*. This distribution is in contrast to that of CD23 (*left panel* = *green*, *right panel* = *blue*), that highlights the population of follicular dendritic cells (FDC). As exemplified by these images, FDCs (CD23+) and T cells (CD3+) are preferentially found in the GC Light Zone. Original magnification, 20X

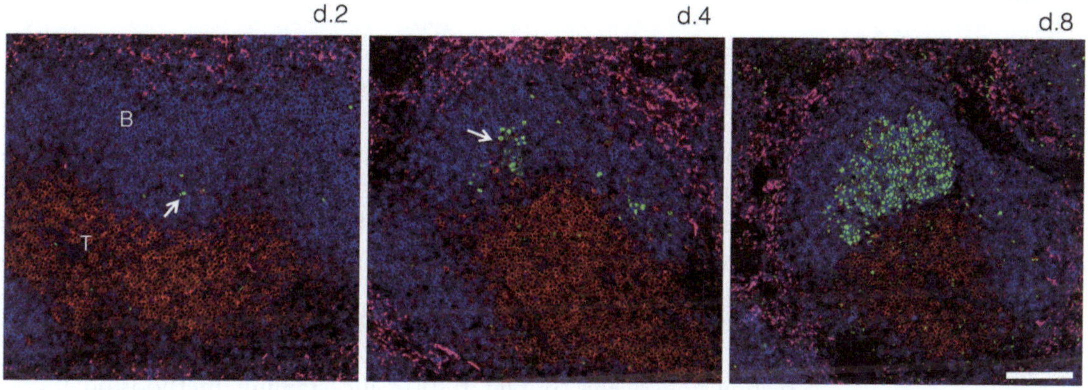

Fig. 2 Sequential analysis of the Germinal Center reaction via immunofluorescence. Immunofluorescence analysis of mouse spleens, harvested at different times post-immunization with T-dependent antigens (intraperitoneal injection of sheep red blood cells; C57BL/6 mice). B cells are identified by their surface expression of B220 (*blue*). T cells by their surface expression of CD3 (*red*). B cells committed to the germinal center B cell fate are identified by their expression of Bcl6 (*green*). Note how the B220+/Bcl6+ population appears close to the B-T border (*arrow*, d.2), travels to the B cell follicle (d.4, *arrow*) and expands in the days after immunization (d.8). Original magnification, 10X

cytidine deaminase, AID) [17] or transcription factors like Bcl6 [18], which are not expressed in any other mature B cell subsets [16, 19]. Using Bcl6 as a marker of germinal center B cell differentiation we can accurately follow the germinal center response over time in tissue sections from immunized mice, and obtain valuable spatiotemporal information of this process (Fig. 2). The immunohistological analysis of the germinal center reaction has thus been instrumental to understand important aspects of its organization and physiology in connection to the identification of specific B cell subsets, based on the expression of specific proteins [11, 16, 20, 21].

In this chapter we describe in detail how to use immunofluorescence/immunohistochemistry to detect multiple epitopes and reveal the cellular milieu and architectural organization of germinal centers in secondary lymphoid tissues. We further discuss the technical nuances of these methodologies, with special emphasis on the particularities of this tissue type. Since the use of antibodies as detection tools in secondary lymphoid tissues can be technically challenging due to the presence of similar molecules and antibody receptors in the surface of specific cell populations, we also provide different strategies to circumvent critical technical pitfalls. Finally, we briefly discuss some resources and more sophisticated technical strategies for antibody-based tissue analysis developed in the last few years, which might be of particular interest in a research laboratory setting.

2 Materials

Germinal centers have been described in vertebrates evolutionary younger than amphibians [22]. It is worth noting, however, that using lymphoid tissues from species that are also the source of the specific immunoglobulins used as reagents may represent a technical challenge [23, 24]. Thus, antibody-based detection of antigens in human and vertebrate tissues other than rabbit, mouse, rat, swine, goat, and sheep will be treated differently (see below). An excellent immunohistochemistry handbook is published and updated regularly by Agilent/Dako (Glostrup, Denmark) and can be found in their Knowledge Center: http://www.agilent.com/cs/library/technicaloverviews/public/08002_ihc_staining_methods.pdf).

2.1 Deparaffinization

Xylene or limonene (*see* **Note 13**). 100% ethanol, 95% ethanol, 70% ethanol, 50% ethanol, distilled water.

2.2 Fixative

Buffered formaldehyde fixation solution (FA): 4% formalin, 0.05 M sodium phosphate buffered solution, 0.01% v/v methanol (*see* **Notes 1** and **2**).

2.3 Buffers

1. Antibody dilution and storage buffer: Tris-buffered saline solutions are preferred to phosphate buffered solutions for the dilution of antibodies or proteins, both unconjugated or conjugated to an enzyme or fluorochrome [25] (*see* **Notes 3** and **4**).
2. Antigen retrieval buffers: retrieve immunoreactive epitopes in FFPE tissues using a combination of heat and chelating agents,

commonly known as "Antigen Retrieval" (AR) [26, 27] (*see* **Note 5**). *Sodium citrate buffer*: 10 mM sodium citrate, 0.05% Tween 20, pH 6.0. Weight 2.94 g of tri-sodium citrate (dihydrate), mix and dissolve in 1 L of distilled water, and adjust pH to 6.0 with 1 N HCL. Add 0.5 mL of Tween 20 (optional) and mix well. Freshly prepared buffer can be stored at room temperature for 3 months, or at 4C for longer storage. *1 mM EDTA, pH 8.0*: Weight 0.37 g of EDTA, mix and dissolve in 1 L of distilled water. Store at room temperature for 3 months. *Tris–EDTA Buffer, pH 9.0*: 10 mM Tris base, 1 mM EDTA Solution, 0.05% Tween 20, pH 9.0. Weight 1.21 g of Tris base, and 0.37 g of EDTA, mix and dissolve in 100 mL to make 10×, or 50 mL to make 20×. pH will be very close to 9.0. Add 0.5 mL of Tween 20 and mix well. Store at room temperature for 3 months or at 4C for longer storage.

3. Washing buffers: TBS: 50 mM Tris–HCl, 150 mM NaCl, pH 7.5; TBS-T: 50 mM Tris–HCl, 150 mM NaCl, 0.05% Tween 20, pH 7.5 (*see* **Note 6**).
4. Stripping buffer: mix 20 mL of 10% w/v sodium dodecyl sulfate (SDS) with 12.5 mL 0.5 M Tris–HCl, pH 6.8, and 67.5 mL ultrapure water (=100 mL of *stripping buffer*). Under a fume hood, add 0.8 mL of beta-mercaptoethanol (2-ME). Pour in a vertical staining jar. Scale the volume according to the number of slides to treat.

2.4 Chromogenic Dyes

1. Horseradish peroxidase (HRP) detection: diaminobenzidine (DAB); amino-ethyl carbazole (AEC) (*see* **Note 7**).
2. Alkaline phosphatase (AP) detection: 5-bromo-4-chloro-3-indolyl phosphate coupled to nitroblue tetrazolium (NBT/BCIP) (*see* **Note 8**).

2.5 Fluorochromes

Directly fluorochrome-conjugated antibodies; or unconjugated primary antibodies and species specific (against primary) fluorochrome conjugated antibodies (*see* **Notes 9** and **10**).

2.6 Mounting Media

1. For insoluble immunohistochemistry substrates such as DAB and AP, preparations may be coverslipped with permanent solvent-based hardening media or an appropriate transparent film.
2. For soluble imunohistochemistry chromogens use a water-based media for coverslipping (*see* **Note 11**).
3. For immunofluorescence use an antifading compound, a hardening substance and a nuclear counterstain (commonly DAPI, a facultative DNA dye) (*see* **Note 12**).

3 Methods

The following methods refer to manually performed techniques, often the only fashion by which immunolabeling is done in a research laboratory. However, automated immunostainers have also been successfully used in research settings.

3.1 Deparaffinization and Antigen Retrievel

1. Remove the paraffin with 2 × 10 min immersion of the tissue sections in xylene or limonene (*see* **Note 13**).
2. Transfer to a sequence of 3-min incubations in 100% ethanol (2×), 95% ethanol (1–2×), 70% ethanol (2×), 50% ethanol (1×), distilled water. Slides can be stored in water until antigen retrieval is performed.
3. Transfer the section/s to a radiotrasparent rack, inside a 1 L heat-resistant container, filled with 800 mL of AR (antigen retrieval) solution of desired pH (*see* **Note 14**).
4. Place the container in a microwave oven, set the device to "high" until the liquid begins to boil.
5. Set the microwave oven to an intermediate power, so that the liquid keeps simmering. Boil for 20 min (*see* **Note 15**).
6. Cool down the container at RT. Transfer the sections to TBS-T. Wash once.

3.2 General Immunostaining

1. Remove most of the buffer covering the section by gently blotting one end of the slide against absorbent paper. Do not let dry the section at any time (*see* **Note 16**).
2. Block the tissue sections with by incubating the slides in TBS-T with an addition of protein (usually 3% bovine serum albumin—BSA—or 5% goat/horse serum). If using biotin-conjugated antibodies, or HRP conjugated antibodies, it is necessary to block these molecules in the tissue prior to staining (*see* **Note 17**).
3. Pour out the blocking solution. Apply a prediluted primary antibody solution to the section, usually 100–200 μL (make sure it covers the entire tissue surface). Antibody concentration may range between 1.0 and 0.1 μg/mL, and needs to be empirically tested. The presence of Tween in the TBS-T buffer allows for an even, smooth distribution of the liquid over the section. Alternatively, 0.025% Triton X-100 can be added (*see* **Note 18**).
4. Incubate in a humid chamber (Fig. 3) for the time experimentally defined, typically a multiple of 15 min or overnight (*see* **Note 19**).

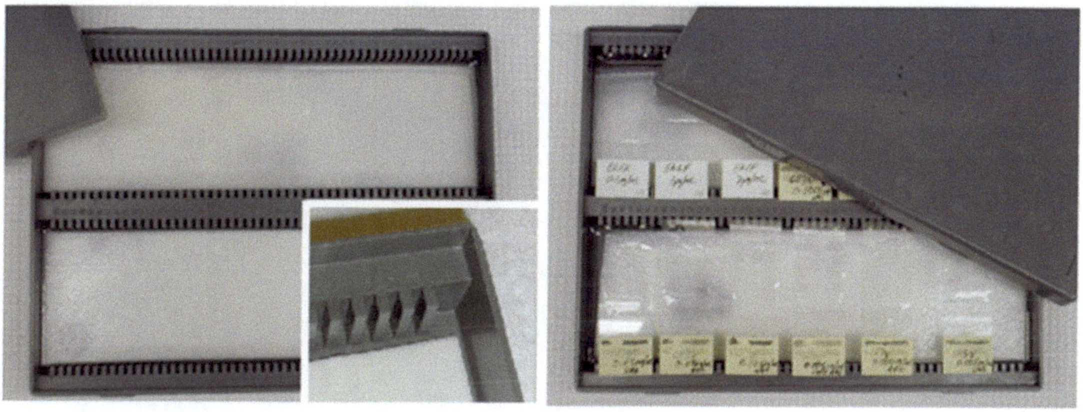

Fig. 3 Humid chamber incubation box. A quick and cheap humid chamber incubation box can be prepared using a slide box, to which strips of blotting paper moistened in water are applied at the bottom (*left*). The slides are placed and incubated at the desired temperature (*right*). A horizontal placement of the incubating slides is checked with a level before use. The box shown is produced by Kartell (Italy). Alternatively, commercial versions of humid chambers are sold by different companies

5. Remove the antibody by either adding an excess buffer to the slide still in the incubation chamber, decanting onto filter paper, or sliding the slide in a Hellendahl vertical 16 slides jar. In the latter case, each slide which follows the previous will be placed behind, so that it does not contact the antibody mixed used on the previous slide. Perform at least 2 × 5 min washes in abundant TBS-T.
6. Remove most of the buffer covering the section by gently blotting one end of the slide against absorbent paper.
7. Cover the tissue sections with the secondary antibody mix, at the desired dilution (polymer enzyme conjugate, fluorochrome-conjugated, or biotinylated). Incubate at room temperature for 45–60 min in a humid chamber.
8. Wash as in **step 5**, and proceed to develop the reaction by incubating the slides with the appropiate substrate (HRP or AP), or with streptavidin particles in the case of biotinylated secondary antibodies. For direct or indirect immunofluorescences, skip to next step.
9. Mount in aqueous medium (glycerol based) (*see* **Note 20**).

3.3 Double Immunohistochemistry

Double IHC staining often employs HRP- and AP-conjugated secondary antibodies. Due to the masking ability of the DAB, and depending to the selected sequence of markers to be stained, it is preferable to perform sequentially immunoperoxidase first, followed by immunoalkaline phosphatase stain (Fig. 4).

1. Perform the first stain as in **steps 1–5** as in Subheading 3.2 above.
2. Develop the HRP with DAB or AEC.

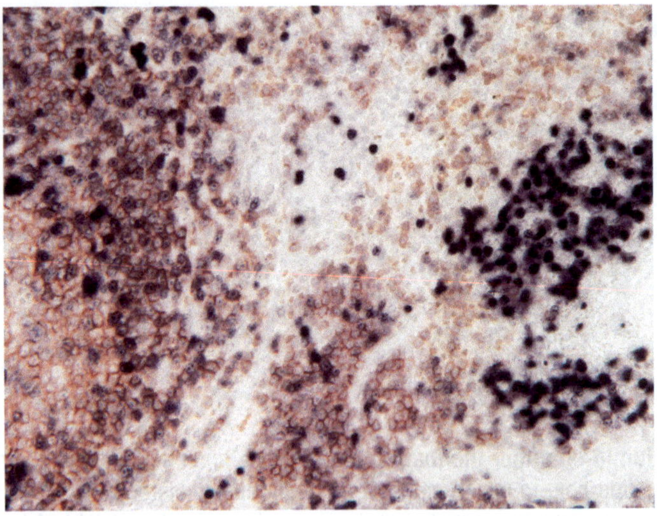

Fig. 4 Double immunohistochemistry for nonoverlapping antigens. Mouse normal spleen stained for the B-cell antigen B220 (rat anti-mouse MAb; HRP; *brown*) and the transcription factor IRF4 (goat anti-human/mouse polyclonal; AP; *purple/black*). Follicular B cells (*left*) lightly stain for IRF4, while red pulp plasma cells (*right*) are strongly positive. Original magnification, 10×

3. After a thorough washing, repeat **steps 1–5** as in Subheading 3.2. above with the second stain.
4. Develop the AP.
5. Counterstain lightly with hematoxylin (optional).
6. Coverslip with aqueous or solvent-based mounting media as appropriate (*see* **Note 21**).

3.4 Double or Tripe Immunofluorescent Staining

Multiple immunofluorescent staining requires the use of primary antibodies directed against nonoverlapping epitopes, coupled to either directly conjugated primary antibodies or a combination of three Ig isotypes of the same species (e.g., IgG1, IgG2a, and IgG2b); three Ig's from three different species (e.g., mouse, rabbit, goat); or a combination of these (e.g., two mouse isotypes and one rabbit Ig), (Fig. 5) (*see* **Notes 22** and **23**).

1. Perform **steps 1–5** of Subheading 3.2, however this time with a pool of the desired antibodies, each one at the final dilution in the pool.
2. Perform **steps 1–3** of the general immunostaining protocol, this time with a pool of appropriately diluted secondary antibody mix.
3. Wash as in **step 5** of the general immunostaining protocol.
4. A way to enhance the signal over the background, is to repeat **steps 1** and **2**, this time for half of the incubation time. This will

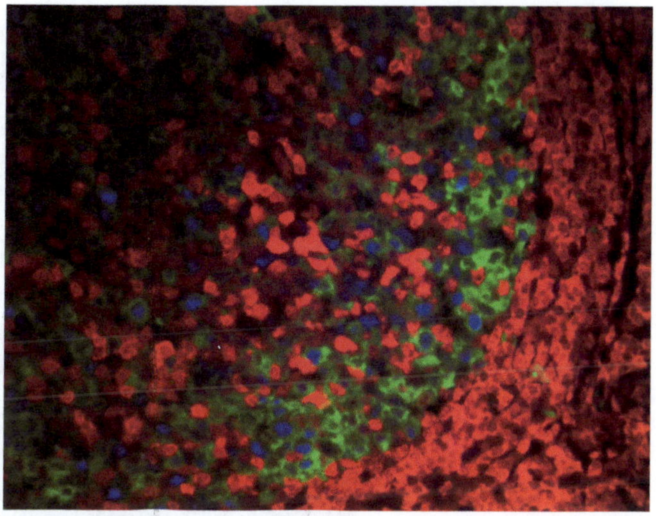

Fig. 5 Triple immunofluorescence on human tonsil. Section of human tonsil (FFPE) stained for CDKN1B (p27) in *red* (mouse MAb, TRITC-labeled secondary antibody), AID in *green* (rat MAb, Alexa 488-labeled secondary antibody) and BCL6 in *blue* (rabbit polyclonal, biotin-conjugated secondary Ab, Avidin-HRP, and *blue* tyramide deposition). Original magnification 40×, no nuclear counterstain

double the amount of fluorescent labeling of the antigens in the section (this increases significantly the signal-to-noise ratio without a major effect on the background).

5. Wash as in **step 5** of the general immunostaining protocol and proceed to coverslipping with antifade containing medium.

3.5 Antibody Stripping for Multiplexing

Multiplex staining is accomplished by combining serial rounds of staining, image capture and antibody stripping on the same slide. This approach results in the compilation of antigen expression maps of a same tissue section that can later be integrated in a single, complex digital image (Fig. 6). It is therefore necessary to capture an image of the previous immunostaining before erasing that stain to apply the next one. This process is best achieved by capturing a digital image of the whole slide (whole slide image, WSI) with a digital scanner. In immunofluorescence multiplexing, and before proceeding with any specific immunostaining, it is also important to obtain a WSI of the non-stained section capturing autofluorescence signals in all channels. This WSI is necessary for digital subtraction from the specific signal.

1. Remove the coverslip of a previously immunostained section; the immunostains should be either immunofluorescent (excluding tyramide-precipitation based labeling, which is insoluble); or immunohistochemical with an alcohol-soluble precipitating chromogen (AEC). When using the latter (AEC), dissolve the chromogen by several dips in absolute

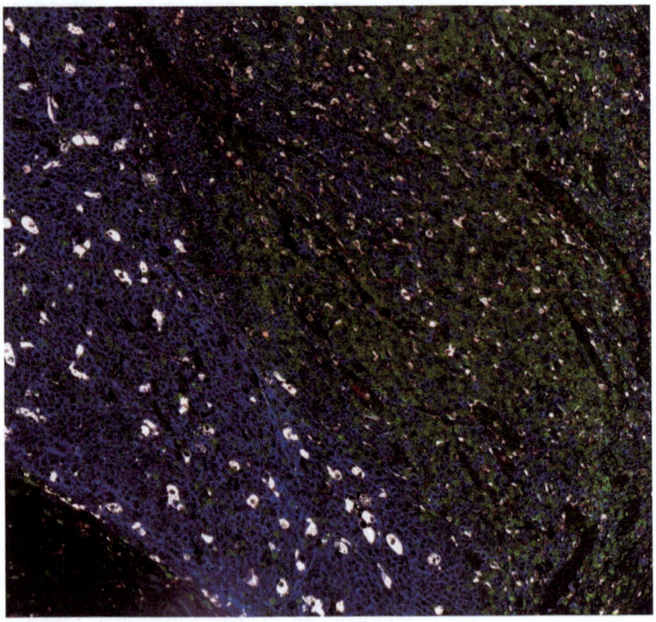

Fig. 6 Multiplexed image of a Germinal Center. A single section of a human tonsil (FFPE) was stained via multiplexing for CD20, CD3, CD68 KP1, and CD68 PGM1 with one round of stripping and restaining. A Germinal Center on the left is populated by B cells (CD20, *blue*) and contains scattered CD3+ T cells (*green*) and GM macrophages that coexpress both epitopes of CD68 (KP1 *white*, PGM1 *red*). Original magnification, 20×. DAPI counterstain (not shown) was used for image registration. Autofluorescence in the *green* and *orange* channels was digitally subtracted

ethanol, before rehydrating the section by sequential incubation in ethanol dilutions of lower percentage until moving the slides to water.

2. Preheat a shaking water bath to 56 °C.
3. Incubate the slides in this stripping buffer for 30 min while shaking at 56 °C. (The recipe for the Stripping Buffer is provided in section 2.3, above).
4. Wash thoroughly the sections in distilled water (at least four washes, one every 15 min).
5. Apply a second round of immunostaining, following the General Immunostaining Protocol described above.

4 Notes

1. Fixation is dependent on time, temperature and thickness of the specimen [28]. When fixing, it is customary to reduce the thickness of the surgical tissue to 2 mm or less, so that in less than 24 h at room temperature (RT) the tissue is considered

"fixed." This is important to maintain optimal morphology and antigen preservation. Note that the formaldehyde (FA) fixation requirement for any given protein may vary, both in whole tissue samples [29] or in sections, usually frozen sections [30]. A shorter fixation time can be accomplished by a combination of heat, molecular agitation, and time [31].

2. The use of non-formalin based fixatives have been proposed to avoid a harmful chemical component and/or better preserve antigens and molecules for detection and extraction [32–34]. However, fixatives based on precipitants (Carnoy, Metacarnoy [35, 36]) may fail to stabilize some proteins in tissues [30, 37, 38]. Tissue sections obtained from a frozen specimen in a microtome (frozen sections) have been traditionally fixed in acetone or mixture of acetone, alcohols (ethanol, methanol), or chloroform [39]. Acetone needs to be kept anhydrous, and thus should be kept in the original bottle. Since the most commonly used and accepted standard fixative is FA, it is advisable to have both a FA fixed specimen and another tissue portion flash frozen for extraction and/or frozen section analysis. Please note that formaldehyde has been classified as a Class I carcinogen and acetone is toxic.
3. In antibody dilution and storage buffers, the sodium chloride contents (NaCl) provide the necessary ionic strength to lessen nonspecific ionic protein–protein interactions, while the Tris salt stabilizes the pH. Other unwanted interactions, such as the adherence of antibody molecules to the wall of the tube or vial, can be prevented by adding an excess of inert proteins such as 0.2–5% bovine serum albumin or 0.1% gelatin. A 5% bovine serum albumin Tris-buffered solution is useful to dilute and store intermediate dilutions of antibodies. This type of solution can maintain antibody solutions stable for over 25 years [40]. Sigma-Aldrich maintains an useful online Buffer Reference Center section. http://www.sigmaaldrich.com/life-science/core-bioreagents/biological-buffers/learning-center/buffer-reference-center.html
4. In order to avoid bacterial or fungal contamination in these protein-containing solutions (antibodies, dilution buffers, etc.), 15 mM sodium azide may be added. Note that sodium azide inhibits horseradish peroxidase and other peroxidases, and thus should be excluded from buffers used in the dilution of secondary antibodies.
5. The development of a method of AR in FFPE tissues was a landmark discovery critical to the role of immunohistochemistry in diagnostic pathology and research. Different improvements to the original protocols for AR have been described ever since (reviewed in [41]). Our current understanding is that

epitopes are better unmasked in buffers with a pH opposite to the isoelectric point of the antigen [42, 43]. Heat is a crucial component for unmasking. Temperatures greater than 97 °C have been attained by using AR buffers with >80% of organic compounds [44]. Notably, automated immunostainers use special buffers with the ability to dewax and retrieve antigens in one single step [44, 45]. In some instances, greater detection sensitivity can be obtained by pretreating the sections with a modified Laemmli extraction buffer [46].

6. Because of its buffering capacity, and its ability to reduce nonspecific binding of antibodies to the tissues as compared to phosphate-based buffers; Tris-buffered saline buffers (*see* http://cshprotocols.cshlp.org/content/2009/6/pdb.rec11830) are recommended for all the washing steps in regular IHC protocols. The addition of Tween-20 to the buffer reduces surface tension. If planning to perform multiplex staining, it is essential to include 10% disaccharide (sucrose, lactose) as a water substitute to avoid drying-associated artifacts, prevent antigen remasking and facilitate stripping the section of predeposited antibodies [47].
7. Two chromogenic substrates are commonly used in immunohistochemistry (IHC) protocols based on *horseradish peroxidase (HRP)* detection methods: diaminobenzidine (DAB) and amino-ethyl carbazole (AEC). Both compounds precipitate at the site of peroxidase activity with a golden brown (DAB) or reddish brown color (AEC). Other chromogenic dyes used for HRP-based IHC exist that are marketed by different manufacturers. The differences between DAB and AEC are the following: sensitivity (DAB > AEC); solubility (DAB: insoluble; AEC: alcohol soluble); ability to block access of antibodies and other reagents to the subcellular structures immunostained with the compound (DAB > AEC); and spectral emission (which can be distinguished using spectral microscopy). Endogenous peroxidases and pseudoperoxidases should be inhibited or blocked before applying an HRP-based stain [48]. This is commonly achieved by incubating the tissues in hydrogen peroxide solution. HRP is also inhibited by sodium azide.
8. Various, soluble and insoluble chromogenic susbtrates are used to develop alkaline phosphatase-based detection methods (usually from calf intestine, AP). These susbtrates greatly differ in sensitivity and color (from red to blue, depending on the substrate). Newfuchsin was the first to be employed, producing a fine, insoluble purple deposition. More recently, naphtol-based compounds are used, often with brand names. The most sensitive dye is 5-bromo-4-chloro-3-indolyl phosphate coupled to nitroblue tetrazolium (NBT/BCIP). This is a deep purple, insoluble dye that cannot block itself and thus can be developed over days. This particular property has for

example been exploited for in-situ hybridization. The disadvantage of NBT/BCIP is that is solvent-soluble, and thus the resulting tissue preparations cannot be counterstained with a nuclear dye of contrasting color. Please note that endogenous alkaline phosphatase is inhibited by high heat during antigen retrieval. AP-conjugates can be stored and diluted in sodium azide-containing Tris buffer.

9. Although not commonly favored by pathologists because tissue morphology and IHC results are not simultaneously displayed; fluorochrome staining of tissue sections has the great advantage of multiplex staining without interference, providing that autofluorescence is controlled and species- or isotype-specific fluorochrome conjugated secondary antibodies are used (unless directly conjugated antibodies are used). The spectrum of visible colors observable with a fluorescence microscope equipped with paired excitation and emission filters ranges from blue to red (390–700 nm). However, due to the large "shoulder" of most available fluorochromes, only three wavelenghts can be accomodated within the visible spectrum: blue, green, and red [49]. More spectra can be accomodated by using quantum dot technology [50], because of the narrower emission of these reagents. The use of detectors instead of eyepieces allows the extension of the sensitivity to the far-red, which allows to accomodate a fourth detection channel [51]. Confocal microscopes and whole slide linear or tiling scanners detect four fluorescence channels or more because are these devices are based on electronic light detectors. The fluorochromes currently commercially available are more photostable and brighter than older ones, thus allowing better performances. To obtain the best performance and results in a fluorescent staining experiment, it is important to match the spectral characteristics and the efficiency of the fluorochromes with the filter set of the instrument used for detection. This can be done using publicly available software such as: searchLight (Semrock: http://searchlight.semrock.com/); fluoScout (Leica Microsystems: http://www.leica-microsystems.com/fluoscout/), microscopyU (Nikon: http://www.microscopyu.com/tutorials/flash/spectralprofiles/index.html). Avoiding tissue autofluorescence is difficult, specially in secondary lymphoid tissues like spleen. This can however be overcome most efficiently by digital subtraction [52–55]. Other alternative methods that use inactivation or obfuscation as means to reduce autofluorescence have also been described (described in detail in [56]).

10. Multiplex staining refers to protocols or techniques designed to accomplish more than two IHC stainings or more than four

IF stainings on the very same tissue section. Double IHC staining with contrasting colors or four color IF are considered as routine "double staining" and thus should not be regarded as "multiplexing." There are less than 20 publications to date dealing with multiplex approaches. Briefly, three methods or strategies can be used: inactivation of fluorochromes [57]; spectral imaging [58]; sequential deposition and removal of antibodies or stains [59–61]. Alkaline-based inactivation of fluorochromes (cyanine dyes) is a proprietary patented solution [57], and thus not commercially available. In addition, the cost of directly conjugated primary antibodies and the reduced sensitivity compared to indirect immunostaining procedures (unconjugated primary + conjugated secondary) make this technique less affordable than others. Spectral imaging is a technique developed after an original publication [62] in which the insensitivity of the tyramide amplification product to high temperatures and quenching of the previously deposited antibody–HRP complexes are exploited for the deposition of a subsequent tyramide product in a different color on a separate cell or cellular structure [58]. With a costly investment in hardware and software, this technique is amenable to the use of standard commercially available antibodies. The sequential deposition and removal of IHC or IF stains relies on commonly available reagents. A crucial step in its design is the removal of antibodies, particularly when using the (increasingly popular) high affinity rabbit monoclonal antibodies developed by some companies. A combination of a strong reducing agent (beta-mercaptoethanol) and a strong detergent (sodium dodecyl sulfate) efficiently strips primary and secondary antibodies [59], provided that the section remains fully hydrated throughout the staining and destaining sequence [47].

11. Several recipes for the preparation of gelatin–glycerol mixtures can be found. These media commonly contain 40–80% water, 5–20% gelatin, glycerol, and preservatives. This mounting medium needs to be warmed up before being applied to the slides. It is important to be aware that adjusting the coverslip while the solution hardens may tear up the section; and that the relative low percentage of water in this medium may also reduce the amount of water in the tissue. Mounting media for fluorescence may be recommended, particularly if multiplexing.
12. These media are water-based, and contain glycerol and other substances. Non-hardening media mounting can be stabilized by applying a clear nail polish around the coverslip border.
13. Warming up the slide prior to this step is not necessary

14. The pH of the buffer used in this step is critical for successful antigen retrieval, and in general terms is largely empirical. Therefore, it is recommended to test different pHs and use proper tissue controls to evaluate which condition better supports the retrieval of the desired epitope. Tris–EDTA pH 9.0 buffer is suitable for a majority of antigens, as is the case for the widely used citrate buffer (pH 6.0).
15. An alternative way to accomplish the heat-induced antigen retrieval is by using a presure cooker or vegetable steamer. We routinely use a microwave cooker (e.g., Nordicware microwave tender cooker) to perform this task. Briefly, we immerse the slides in a slide rack in 1 L of retrieval buffer, seal the lid, and set the microwave (1200 W) for 10 min at high power (position 10). This brings the buffer to boiling point and reach full pressure inside the chamber. We then set the microwave at 30% power for 3–5 more minutes. The 3–5-min time point is somewhat empirical. Shorter times may be insufficient for effective retrieval (no signal), while longer times may be detrimental (increase background) and damage the tissue. After this 3–5-min incubation, we let the cooker stay at room temperature until the pressure is released, then unseal the chamber and let cool down until slides can be retrieved.
16. Keeping the sections hydrated at all times is critical to prevent overstaining and/or high background.
17. Although the levels of endogenous biotin in lymphoid tissues are low, it is recommended to block this molecule prior to staining with biotin-labeled antibodies. We usually use a sequential incubation with avidin and biotin at room temperature to this end (there are several commercial kits for this purpose). Endogenous HRP is abundant in red blood cells, and thus it is recommended when staining spleen sections. We routinely block all sections with 3% H_2O_2 (hydrogen peroxide) in PBS or TBS for 20 min at room temperature, followed by 2–3 5-min washes in TBS-T. For the protein blocking step, we normally use a combination of 3% BSA plus 5% goat serum. The choice of serum species is based on the species in which most (or all) the secondary antibodies have been raised (most often goat or donkey).

    Immunostaining with mouse primary antibodies on mouse lymphoid tissues can be problematic. Anti-mouse immunoglobulins (H + L chain) invariably cause background staining of interstitial immunoglobulins, plasma cells, B cells and other cells with Fc-bound immunoglobulins. However, some alternatives can be used to circumvent this problem: (1) Use monomeric F(ab) conjugated secondaries; (2) Using isotype-specific secondary antibodies can help prevent nonspecific staining (note that most mouse monoclonal antibodies are of the

IgG1 isotype). This approach takes advantage of the relatively lower levels of each IgG isotype in follicular B cells, more obvious in mice raised in pathogen-free conditions. In this case, be aware of possible odd cross-reactivity (e.g., many goat anti-mouse IgG_1 heavy chain do stain rat IgG_{2a} and IgG_{2b}.); (3) Adding an extra step to block the endogenous immunoglobulins and Fc receptors. This is a very useful approach in our hands. It can be easily accomplished by using commercial kits, known as "mouse-on-mouse" currently distributed by different companies, or by incubating the sections with >0.1 mg/mL unconjugated F(ab) fragment goat anti-mouse IgG (H + L) F(ab) for 2 h at room temperature or overnight in cold. (4) Or a combination of some of these approaches (e.g., 2 + 3).

18. When performing double immunostaining with two primary antibodies raised in two different species (i.e., rabbit and mouse), it is common to mix the two antibodies at this step.
19. We routinely use overnight incubation at 4°, which may be required when using higher dilutions of the antibody.
20. Special care should be taken to avoid the introduction of any bubbles while mount the coverslips on the slides. Aqueous glycerol-based mounting media needs to be warmed up prior to use, and it quickly hardens at room temperature. Therefore, one must make sure that the medium is fluid when applying to the slide. Make sure to use excess medium ot avoid any bubbles, and then squeeze out the excess by carefully pressing on the coverslip with a pen or by pressing the coverslipped slide upside down against a paper towel. Excess of medium will alter the focal plane of the slide, which is critical when using high power lenses.
21. AEC and some AP chromogens are light sensitive and fade over time.
22. It is possible to use polyclonal rabbit antibodies in the same staining sequence without much difficulty. Briefly, one of the antibodies will need to be conjugated to either fluorochrome or biotin. Some primary antibodies are already sold in fluorochrome- conjugated or biotin variants, but if not available, it is possible to transiently label these antibodies using fluorochrome or biotin-conjugated anti-rabbit F(ab)s (different companies provide kits for this purpose). In this case, the primary antibody is labeled with conjugated F(ab)s, and then an excess of unlabeled, nonspecific rabbit immunoglobulin is added to block the excess of F(ab) in the solution. This can be then diluted at the desired concentration and applied directly to the slide. Because the F(ab)-antibody interaction is transient, it is important to use the mixture right away and avoid storage. To use two different rabbit primary

antibodies: (1) incubate for the desired time with the first, unlabeled primary; we normally use the one against the least abundant protein. (2) wash sections abundantly as per protocol, and incubate with the desired anti-rabbit secondary antibody (fluorochrome-labeled). (3) Block with a solution of TBS-T and 20% rabbit serum at room temperature (1 h). This will cover all residual free anti-rabbit secondary antibody molecules/variable regions. (4) Wash twice in TBS-T. (5) Apply the second, directly labeled primary rabbit antibody. Incubate for the desired time, normally 1–2 h at RT if labeled with F(ab)s and wash. If planning to add an additional antibody incubation (for example, with a third, mouse or other species primary), and have used a F(ab) labeled primary in the previous step, it is helpul to fix the "signal" by quickly incubating the slides in 10% buffered formalin for 10 min prior to washing and continuing with the next primary incubation.

23. Secondary antibodies should be selected so that: (1) they do not crossreact with any other of the primary or secondary antibodies in the combination; they are conjugated with fluorochromes appropriate for the choice (availability) of filters of the microscope/scanner; (3) they are absorbed against the tissue host and the other immunoglobulins used in the combination.

Acknowledgment

We wish to thank Carla Rossana Scalia, Rossella Gendusa for excellent technical assistance, Maddalena Maria Bolognesi for digital editing of multiplex images, and Franco Ferrario, Hong Yan and Qiong Shen for support and helpful insights.

References

1. Uhlen M, Oksvold P, Fagerberg L, Lundberg E, Jonasson K, Forsberg M, Zwahlen M, Kampf C, Wester K, Hober S, Wernerus H, Bjorling L, Ponten F (2010) Towards a knowledge-based human protein atlas. Nat Biotechnol 28(12):1248–1250. doi:10.1038/nbt1210-1248
2. Uhlen M, Fagerberg L, Hallstrom BM, Lindskog C, Oksvold P, Mardinoglu A, Sivertsson A, Kampf C, Sjostedt E, Asplund A, Olsson I, Edlund K, Lundberg E, Navani S, Szigyarto CA, Odeberg J, Djureinovic D, Takanen JO, Hober S, Alm T, Edqvist PH, Berling H, Tegel H, Mulder J, Rockberg J, Nilsson P, Schwenk JM, Hamsten M, von Feilitzen K, Forsberg M, Persson L, Johansson F, Zwahlen M, von Heijne G, Nielsen J, Ponten F (2015) Proteomics. Tissue-based map of the human proteome. Science 347(6220):1260419. doi:10.1126/science.1260419
3. Andersson S, Nilsson K, Fagerberg L, Hallstrom BM, Sundstrom C, Danielsson A, Edlund K, Uhlen M, Asplund A (2014) The transcriptomic and proteomic landscapes of bone marrow and secondary lymphoid tissues. PLoS One 9(12):e115911. doi:10.1371/journal.pone.0115911
4. Coons AH (1961) The beginnings of immunofluorescence. J Immunol 87:499–503
5. Coons AH, Kaplan MH (1950) Localization of antigen in tissue cells; improvements in a method for the detection of antigen by means of fluorescent antibody. J Exp Med 91(1):1–13

6. Carvajal-Hausdorf DE, Schalper KA, Neumeister VM, Rimm DL (2015) Quantitative measurement of cancer tissue biomarkers in the lab and in the clinic. Lab Investig 95(4):385–396. doi:10.1038/labinvest.2014.157
7. Merlos-Suarez A, Batlle E (2008) Eph-ephrin signalling in adult tissues and cancer. Curr Opin Cell Biol 20(2):194–200. doi:10.1016/j.ceb.2008.01.011
8. Gorgoulis VG, Vassiliou L-VF, Karakaidos P, Zacharatos P, Kotsinas A, Liloglou T, Venere M, DiTullio RA, Kastrinakis NG, Levy B, Kletsas D, Yoneta A, Herlyn M, Kittas C, Halazonetis TD (2005) Activation of the DNA damage checkpoint and genomic instability in human precancerous lesions. Nature 434 (7035):907–913. doi:10.1038/nature03485
9. Bartkova J, Horejsí Z, Koed K, Krämer A, Tort F, Zieger K, Guldberg P, Sehested M, Nesland JM, Lukas C, Ørntoft T, Lukas J, Bartek J (2005) DNA damage response as a candidate anti-cancer barrier in early human tumorigenesis. Nature 434(7035):864–870. doi:10.1038/nature03482
10. Sander S, Chu VT, Yasuda T, Franklin A, Graf R, Calado DP, Li S, Imami K, Selbach M, Di Virgilio M, Bullinger L, Rajewsky K (2015) PI3 kinase and FOXO1 transcription factor activity differentially control B cells in the germinal center light and dark zones. Immunity 43(6):1075–1086. doi:10.1016/j.immuni.2015.10.021
11. Dominguez-Sola D, Kung J, Holmes AB, Wells VA, Mo T, Basso K, Dalla-Favera R (2015) The FOXO1 transcription factor instructs the germinal center dark zone program. Immunity 43 (6):1064–1074. doi:10.1016/j.immuni.2015.10.015
12. Bannard O, Horton RM, Allen CD, An J, Nagasawa T, Cyster JG (2013) Germinal center centroblasts transition to a centrocyte phenotype according to a timed program and depend on the dark zone for effective selection. Immunity 39(5):912–924. doi:10.1016/j.immuni.2013.08.038
13. Victora GD, Nussenzweig MC (2012) Germinal centers. Annu Rev Immunol 30:429–457. doi:10.1146/annurev-immunol-020711-075032
14. Victora GD, Dominguez-Sola D, Holmes AB, Deroubaix S, Dalla-Favera R, Nussenzweig MC (2012) Identification of human germinal center light and dark zone cells and their relationship to human B-cell lymphomas. Blood 120(11):2240–2248. doi:10.1182/blood-2012-03-415380
15. Victora GD, Dominguez-Sola D, Holmes AB, Deroubaix S, Dalla-Favera R, Nussenzweig MC (2012) Identification of human germinal center light and dark zone cells and their relationship to human B-cell lymphomas. Blood 120(11):2240–2248
16. Cattoretti G, Shaknovich R, Smith PM, Jack HM, Murty VV, Alobeid B (2006) Stages of germinal center transit are defined by B cell transcription factor coexpression and relative abundance. J Immunol 177(10):6930–6939
17. Cattoretti G (2006) Nuclear and cytoplasmic AID in extrafollicular and germinal center B cells. Blood 107(10):3967–3975. doi:10.1182/blood-2005-10-4170
18. Cattoretti G, Chang CC, Cechova K, Zhang J, Ye BH, Falini B, Louie DC, Offit K, Chaganti RS, Dalla-Favera R (1995) BCL-6 protein is expressed in germinal-center B cells. Blood 86 (1):45–53
19. Cattoretti G, Pasqualucci L, Ballon G, Tam W, Nandula SV, Shen Q, Mo T, Murty VV, Dalla-Favera R (2005) Deregulated BCL6 expression recapitulates the pathogenesis of human diffuse large B cell lymphomas in mice. Cancer Cell 7 (5):445–455. doi:10.1016/j.ccr.2005.03.037
20. Dominguez-Sola D, Victora GD, Ying CY, Phan RT, Saito M, Nussenzweig MC, Dalla-Favera R (2012) The proto-oncogene MYC is required for selection in the germinal center and cyclic reentry. Nat Immunol 13 (11):1083–1091. doi:10.1038/ni.2428
21. Calado DP, Sasaki Y, Godinho SA, Pellerin A, Kochert K, Sleckman BP, de Alboran IM, Janz M, Rodig S, Rajewsky K (2012) The cell-cycle regulator c-Myc is essential for the formation and maintenance of germinal centers. Nat Immunol 13(11):1092–1100. doi:10.1038/ni.2418
22. Flajnik MF (2002) Comparative analyses of immunoglobulin genes: surprises and portents. Nat Rev Immunol 2(9):688–698. doi:10.1038/nri889
23. Ramos-Vara JA, Kiupel M, Baszler T, Bliven L, Brodersen B, Chelack B, West K, Czub S, Del Piero F, Dial S, Ehrhart EJ, Graham T, Manning L, Paulsen D, Valli VE (2008) Suggested guidelines for Immunohistochemical techniques in veterinary diagnostic laboratories. J Vet Diagn Investig 20(4):393–413. doi:10.1177/104063870802000401
24. Ramos-Vara JA, Miller MA (2013) When tissue antigens and antibodies get along: revisiting the technical aspects of immunohistochemistry—the red, Brown, and blue technique. Vet Pathol 51 (1):42–87. doi:10.1177/0300985813505879
25. Boenisch T (1999) Diluent buffer ions and pH: their influence on the performance of monoclonal antibodies in immunohistochemistry. Appl Immunohistochem Mol Morphol 7 (4):300. doi:10.1097/00129039-199912000-00009

26. Shi SR, Key ME, Kalra KL (1991) Antigen retrieval in formalin-fixed, paraffin-embedded tissues: an enhancement method for immunohistochemical staining based on microwave oven heating of tissue sections. J Histochem Cytochem 39(6):741–748
27. Shi SR, Shi Y, Taylor CR (2011) Antigen retrieval immunohistochemistry: review and future prospects in research and diagnosis over two decades. J Histochem Cytochem 59 (1):13–32. doi:10.1369/jhc.2010.957191
28. Fox CH, Johnson FB, Whiting J, Roller PP (1985) Formaldehyde fixation. J Histochem Cytochem 33(8):845–853. doi:10.1177/33.8.3894502
29. Wolff AC, Hammond MEH, Hicks DG, Dowsett M, McShane LM, Allison KH, Allred DC, Bartlett JMS, Bilous M, Fitzgibbons P, Hanna W, Jenkins RB, Mangu PB, Paik S, Perez EA, Press MF, Spears PA, Vance GH, Viale G, Hayes DF (2014) Recommendations for human epidermal growth factor receptor 2 testing in breast cancer: American Society of Clinical Oncology/College of American Pathologists Clinical Practice Guideline Update. Arch Pathol Lab Med 138(2):241–256. doi:10.5858/arpa.2013-0953-sa
30. Shi S-R, Liu C, Pootrakul L, Tang L, Young A, Chen R, Cote RJ, Taylor CR (2008) Evaluation of the value of frozen tissue section used as "gold standard" for immunohistochemistry. Am J Clin Pathol 129(3):358–366. doi:10.1309/7cxuyxt23e5al8kq
31. Chafin D, Theiss A, Roberts E, Borlee G, Otter M, Baird GS (2013) Rapid two-temperature formalin fixation. PLoS One 8(1):e54138. doi:10.1371/journal.pone.0054138
32. Dotti I, Bonin S, Basili G, Nardon E, Balani A, Siracusano S, Zanconati F, Palmisano S, De Manzini N, Stanta G (2010) Effects of formalin, Methacarn, and FineFIX fixatives on RNA preservation. Diagn Mol Pathol 19 (2):112–122. doi:10.1097/pdm.0b013e3181b520f8
33. Moelans CB, ter Hoeve N, van Ginkel JW, ten Kate FJ, van Diest PJ (2011) Formaldehyde substitute fixatives: analysis of Macroscopy, morphologic analysis, and Immunohistochemical analysis. Am J Clin Pathol 136(4):548–556. doi:10.1309/ajcphh1b0cocb gom
34. Matsuda Y, Fujii T, Suzuki T, Yamahatsu K, Kawahara K, Teduka K, Kawamoto Y, Yamamoto T, Ishiwata T, Naito Z (2011) Comparison of fixation methods for preservation of morphology, RNAs, and proteins from paraffin-embedded human cancer cell-implanted mouse models. J Histochem Cytochem 59(1):68–75. doi:10.1369/jhc.2010.957217
35. Puchtler H, Sweat Waldrop F, Conner HM, Terry MS (1968) Carnoy fixation: practical and theoretical considerations. Histochemie 16(4):361–371. doi:10.1007/bf00306359
36. Puchtler H, Waldrop FS, Meloan SN, Terry MS, Conner HM (1970) Methacarn (methanol-Carnoy) fixation. Histochemie 21 (2):97–116. doi:10.1007/bf00306176
37. Re GG, Hazen-Martin DJ, Bahtimi RE, Brownlee NA, Willingham MC, Garvin AJ (1999) Prognostic significance of Bcl-2 in Wilms' tumor and oncogenic potential of Bcl-XL in rare tumor cases. Int J Cancer 84 (2):192–200. doi:10.1002/(sici)1097-0215(19990420)84:2<192::aid-ijc17>3.0.co;2-1
38. van Dierendonck JH, Keijzer R, van de Velde CJ, Cornelisse CJ (1989) Nuclear distribution of the Ki-67 antigen during the cell cycle: comparison with growth fraction in human breast cancer cells. Cancer Res 49(11):2999–3006
39. Mason DY, Farrell C, Taylor CR (1975) The detection of intracellular antigens in human leucocytes by Immunoperoxidase staining. Br J Haematol 31(3):361–370. doi:10.1111/j.1365-2141.1975.tb00867.x
40. Argentieri MC, Pilla D, Vanzati A, Lonardi S, Facchetti F, Doglioni C, Parravicini C, Cattoretti G (2013) Antibodies are forever: a study using 12-26-year-old expired antibodies. Histopathology 63(6):869–876. doi:10.1111/his.12225
41. D'Amico F, Skarmoutsou E, Stivala F (2009) State of the art in antigen retrieval for immunohistochemistry. J Immunol Methods 341 (1–2):1–18. doi:10.1016/j.jim.2008.11.007
42. Kajiya H, Takekoshi S, Takei M, Egashira N, Miyakoshi T, Serizawa A, Teramoto A, Osamura RY (2009) Selection of buffer pH by the isoelectric point of the antigen for the efficient heat-induced epitope retrieval: re-appraisal for nuclear protein pathobiology. Histochem Cell Biol 132(6):659–667. doi:10.1007/s00418-009-0635-8
43. Shi SR, Imam SA, Young L, Cote RJ, Taylor CR (1995) Antigen retrieval immunohistochemistry under the influence of pH using monoclonal antibodies. J Histochem Cytochem 43(2):193–201
44. Freeland JH, Harvey MA (2013) Dewaxing buffer containing a water-soluble organic solvent and methods of use thereof. US Patent app. 13/356,491
45. Zhang G, Yu CZ, Su SH, Kalra K, Zhou D (2006) Deparaffinization compositions for

dewaxing tissue specimens. US Patent app. 11/186,392

46. Scalia CR, Gendusa R, Cattoretti G (2015) A 2-step Laemmli and antigen retrieval method improves Immunodetection. Appl Immunohistochem Mol Morphol. doi:10.1097/PAI.0000000000000203

47. Boi G, Scalia CR, Gendusa R, Ronchi S, Cattoretti G (2015) Disaccharides protect antigens from drying-induced damage in routinely processed tissue sections. J Histochem Cytochem 64 (1):18–31. doi:10.1369/0022155415616162

48. Li CY, Ziesmer SC, Lazcano-Villareal O (1987) Use of azide and hydrogen peroxide as an inhibitor for endogenous peroxidase in the immunoperoxidase method. J Histochem Cytochem 35(12):1457–1460. doi:10.1177/35.12.2824601

49. Mason DY, Micklem K, Jones M (2000) Double immunofluorescence labelling of routinely processed paraffin sections. J Pathol 191 (4):452–461. doi:10.1002/1096-9896(2000)9999:9999<::aid-path665>3.0.co;2-o

50. Fountaine TJ, Wincovitch SM, Geho DH, Garfield SH, Pittaluga S (2006) Multispectral imaging of clinically relevant cellular targets in tonsil and lymphoid tissue using semiconductor quantum dots. Mod Pathol 19 (9):1181–1191. doi:10.1038/modpathol.3800628

51. Buscone S, Argentieri MC, Pilla D, Cattoretti G (2014) Whole-slide, quadruple immunofluorescence labeling of routinely processed paraffin sections. Appl Immunohistochem Mol Morphol 22(4):e1–e7. doi:10.1097/pai.0b013e31829928e7

52. Pang Z, Laplante NE, Filkins RJ (2012) Dark pixel intensity determination and its applications in normalizing different exposure time and autofluorescence removal. J Microsc 246(1):1–10. doi:10.1111/j.1365-2818.2011.03581.x

53. Pang Z, Barash E, Santamaria-Pang A, Sevinsky C, Li Q, Ginty F (2013) Autofluorescence removal using a customized filter set. Microsc Res Tech 76(10):1007–1015. doi:10.1002/jemt.22261

54. Constantinou P, Dacosta RS, Wilson BC (2009) Extending immunofluorescence detection limits in whole paraffin-embedded formalin fixed tissues using hyperspectral confocal fluorescence imaging. J Microsc 234 (2):137–146. doi:10.1111/j.1365-2818.2009.03155.x

55. Van de Lest CH, Versteeg EM, Veerkamp JH, Van Kuppevelt TH (1995) Elimination of autofluorescence in immunofluorescence microscopy with digital image processing. J Histochem Cytochem 43(7):727–730. doi:10.1177/43.7.7608528

56. Davis AS, Richter A, Becker S, Moyer JE, Sandouk A, Skinner J, Taubenberger JK (2014) Characterizing and diminishing autofluorescence in formalin-fixed paraffin-embedded human respiratory tissue. J Histochem Cytochem 62(6):405–423. doi:10.1369/0022155414531549

57. Gerdes MJ, Sevinsky CJ, Sood A, Adak S, Bello MO, Bordwell A, Can A, Corwin A, Dinn S, Filkins RJ, Hollman D, Kamath V, Kaanumalle S, Kenny K, Larsen M, Lazare M, Li Q, Lowes C, McCulloch CC, McDonough E, Montalto MC, Pang Z, Rittscher J, Santamaria-Pang A, Sarachan BD, Seel ML, Seppo A, Shaikh K, Sui Y, Zhang J, Ginty F (2013) Highly multiplexed single-cell analysis of formalin-fixed, paraffin-embedded cancer tissue. Proc Natl Acad Sci 110(29):11982–11987. doi:10.1073/pnas.1300136110

58. Mansfield JR (2014) Multispectral imaging: a review of its technical aspects and applications in anatomic pathology. Vet Pathol 51(1):185–210. doi:10.1177/0300985813506918

59. Gendusa R, Scalia CR, Buscone S, Cattoretti G (2014) Elution of high-affinity (>10-9 KD) antibodies from tissue sections: clues to the molecular mechanism and use in sequential Immunostaining. J Histochem Cytochem 62 (7):519–531. doi:10.1369/0022155414536732

60. Glass G, Papin JA, Mandell JW (2009) SIMPLE: a sequential Immunoperoxidase labeling and erasing method. J Histochem Cytochem 57(10):899–905. doi:10.1369/jhc.2009.953612

61. Pirici D, Mogoanta L, Kumar-Singh S, Pirici I, Margaritescu C, Simionescu C, Stanescu R (2009) Antibody elution method for multiple immunohistochemistry on primary antibodies raised in the same species and of the same subtype. J Histochem Cytochem 57 (6):567–575. doi:10.1369/jhc.2009.953240

62. Toth ZE, Mezey E (2007) Simultaneous visualization of multiple antigens with Tyramide signal amplification using antibodies from the same species. J Histochem Cytochem 55 (6):545–554. doi:10.1369/jhc.6a7134.2007

Chapter 2

Detecting Gene Expression in Lymphoid Microenvironments by Laser Microdissection and Quantitative RT-PCR

Yang Zhang, Laura Garcia-Ibanez, Geoffrey Brown, and Kai-Michael Toellner

Abstract

Quantitative reverse transcription-polymerase chain reaction (qRT-PCR) is a valuable tool for measuring gene expression in cells and tissues. Unique challenges are encountered when studies are performed on cells microdissected from small specific areas of frozen animal or human tissue. This chapter describes the analysis of gene expression of chemokines and cytokines that are important for the differentiation and migration of germinal center (GC) derived plasmablasts/plasma cells and memory B cells by using laser capture microdissection (LCM) and qRT-PCR to examine tissue sections.

Key words Quantitative RT-PCR, Laser capture microdissection, Germinal center, B cells

1 Introduction

Exposure to vaccines or pathogens results in the activation of B lymphocytes. This ultimately leads to the production of high affinity antibodies that are protective against invading antigens. After antigen exposure, activated B lymphocytes migrate through different microenvironments in lymphoid tissues, where they mature into high affinity antibody forming cells—the plasma cell [1]. Maturation to high affinity happens in a specialized microenvironment: the germinal center (GC). This maturation process represents a very rapid Darwinian evolution at a cellular level: involving B cell proliferation, hypermutation of the genes that encode antibody, and the antigen-dependent selection of B cells with genetic information coding for the highest affinity antibodies [2].

How activated B lymphocytes travel through the different microenvironments, which signals (including molecular and cellular signals) regulate the GC reaction, and which signals regulate the production and migration of plasma cells and memory cells from GC remains unanswered. Techniques such as flow cytometry,

Dinis Pedro Calado (ed.), *Germinal Centers: Methods and Protocols*, Methods in Molecular Biology, vol. 1623,
DOI 10.1007/978-1-4939-7095-7_2, © Springer Science+Business Media LLC 2017

quantitative gene expression analysis, and others are able to provide precise information to answer these questions. However, some types of cells in tissue are still not so easy to isolate by traditional techniques, for example, the stroma that surrounds differentiating lymphocytes in lymphoid tissues. Some chemokines and cytokines are produced at a specific time and location. The location of signals involved in the differentiation of GC B cell, plasma cells, and memory B cells can be identified by defining the local environment by histology and isolating small areas of tissues.

Laser microdissection is a powerful tool for the isolation of targeted cell populations (or even single cells) from stained sections of both formalin-fixed, paraffin-embedded, and frozen tissues. The material is suitable for a wide range of downstream assays, particularly gene expression analysis as to mRNA [3, 4]. This chapter describes the use of laser capture microdissection from frozen sections, and the protocol of qRT-PCR analysis using small amounts of mRNA from laser microdissected lymphoid tissue microenvironments.

2 Materials

2.1 Immunization (See Note 1)

1. Chicken gamma globulin (CGG; e.g., Jackson ImmunoResearch).
2. Alum $AlK(SO_4)_2$ (e.g., Sigma).
3. 10 M NaOH.
4. Heat inactivated *Bordetella pertussis* (b.p.) bacteria (e.g., BD Lee Laboratories).
5. NP-O-succinimide esther (NP-OSu) (e.g., LGC BioSearch Technologies)
6. Dimethylformamide (DMF).
7. 0.1 M sterile sodium bicarbonate ($NaHCO_3$).
8. 10,000 MW dialysis cassettes (e.g., Slide-A-Lyzer™, Thermo Fisher)

2.2 Immunohistology of Adjacent Tissues

1. Tissue-Tek OCT compound (e.g., Sakura).
2. Cryospray (e.g., Bright).
3. Four-spot polytetrafluoroethylene (PTFE)-coated glass slides (e.g., Hendley-Essex).
4. Acetone.
5. Glass wash trough.
6. Humid chamber to incubate slides.
7. Wash buffer: mix 1 Vol 0.2 M Tris (Trizma base), 1.4 Vol 0.1 N HCl, 1.6 Vol 0.855 NaCl. Wash buffer should have pH 7.6.

8. Normal mouse serum (e.g., Sigma).
9. 3,3′-Diaminobenzidine hydrochloride (DAB) tablets (e.g., Sigma), H_2O_2. 1 tablet dissolved in 15 ml Tris–HCl (pH 7.6).
10. Alkaline phosphatase (AP) substrate Fast Blue: dissolve 8 mg levamisole hydrochloride (e.g., Sigma) in 10 ml Tris–HCl (pH 9.2). Mix with 340 μl of DMF containing 4 mg naphthol AS-MX phosphate. Finally, add 10 mg Fast Blue BB salt and shake until dissolved. Filter and use immediately.
11. Glycerol gelatin as mounting medium.
12. 22-mm × 65-μm glass coverslips.
13. Cryostat (e.g., Bright).
14. Goat anti-mouse IRF4 (e.g., Santa Cruz Biotechnology).
15. Rat anti-mouse IgD (e.g., BD Bioscience).
16. Biotinylated donkey anti-sheep goat (e.g., Binding Site).
17. Peroxidase (PX) labeled rabbit anti-rat (e.g., DakoCytomation).

2.3 Cresyl Violet Staining of Microdissected Tissues

1. PALM membrane slides NF (polyethylene naphthalate (PEN) membrane coated slides) from Zeiss.
2. PALM Liquid Cover Glass N from Zeiss.
3. Cresyl violet acetate (e.g., Sigma).
4. Ethanol (RNA free) for molecular biology.
5. Injection water (e.g., B. Braun).
6. RNase-free water.
7. Slide scanner (e.g., AxioScan Z1, Zeiss).

2.4 Microdissection

1. Eight-well strips of 200 μl PCR tubes (e.g., Alpha laboratories).
2. Eight-well cap stripes (e.g., ABgene™, Thermo Fisher).
3. Lysis buffer: RLT buffer from Rneasy@ Micro Kit from Qiagen.
4. Low Biding barrier pipette tips (20 μl) (e.g., Multiguard).
5. Dry ice.
6. PALM Microbeam HT microscope from Zeiss.

2.5 RT-PCR

1. RNase Free DNase.
2. RNeasy Micro Kit from Qiagen.
3. Shredder spin column from Qiagen.
4. Microcentrifuge allowing max speed 15871 × *g*.
5. Random hexanucleotide primers.
6. 100 mM dNTP.

7. RNase Inhibitor (e.g., RNasin Plus, Promega).
8. 200 U/μl Moloney murine leukemia virus (M-MLV) reverse transcriptase (e.g., Invitrogen) containing 5 × first strand buffer, 0.1 M DTT.
9. 96-well PCR machine.
10. Real-Time PCR master Mix (e.g., TaqMan Universal PCR master Mix, Applied Biosystems).
11. Clear adhesive foil (e.g., Applied Biosystems).
12. Real-time PCR probes for genes of interest and housekeeping genes, such as β2-microglobulin (e.g., Applied Biosystems).
13. Real-Time PCR System with 384-Well Block Module (e.g., 7900HT, Applied Biosystems).
14. Real-Time PCR system software (e.g., SDS 2.4, Applied Biosystems).
15. Real-Time PCR analysis software (e.g., RQ manager 1.2.1, Applied Biosystems).

3 Methods

3.1 Immunization and Tissue Preparation

1. C57BL/6 mice are primed with 50 μg CGG in alum precipitated with 2×10^7 heat inactivated *B. pertussis* bacteria as an adjuvant.
2. After 5 weeks, the mice are boosted with 50 μg soluble NP_{18}-CGG in PBS, via intraperitoneal (*i.p.*) injection.
3. Termination of the response is from 5 days after immunization with soluble NP-CGG.
4. Spleens are removed, placed on a piece of aluminium foil, and snap frozen in liquid N_2.
5. After wrapping the frozen tissue in aluminium foil, spleens are stored at −80 °C in grip-seal polythene bags until sections are cut (*see* **Note 2**).
6. The spleen is mounted on a cryostat holder using a small amount of OCT compound (*see* **Note 3**).
7. Trim off any excess of OCT compound (*see* **Note 4**).
8. Eight micrometers thick spleen sections are distributed equally onto a NF membrane slide (*see* **Note 5**). Approximately every 10 cuts, 1 or 2 sections (6 μm) are mounted onto 4-spot glass slides in order to perform immunohistochemistry of these tissues. These will allow the identification of the different regions in regard to the adjacent tissue sections on the membrane slides.

9. Membrane and 4-spot glass slides are air-dried under a fan for 30 min then fixed in fresh analytical-grade acetone at 4 °C for 20 min, air-dried for 10 min (*see* **Note 6**).
10. Glass slides are stored in sealed polyethylene bags at −20 °C until use. Membrane slides are stored in −80 °C in the original manufacturer's storage box.

3.2 Immunohistological Staining of Glass Slides

The 4-spot glass slides are stained by immunohistochemistry to identify different splenic regions (Fig. 1a, b). The example described here works well to differentiate white pulp into T zones, follicles, germinal centers, and plasma cell areas. Particularly GCs cannot be differentiated from follicles by cresyl violet staining

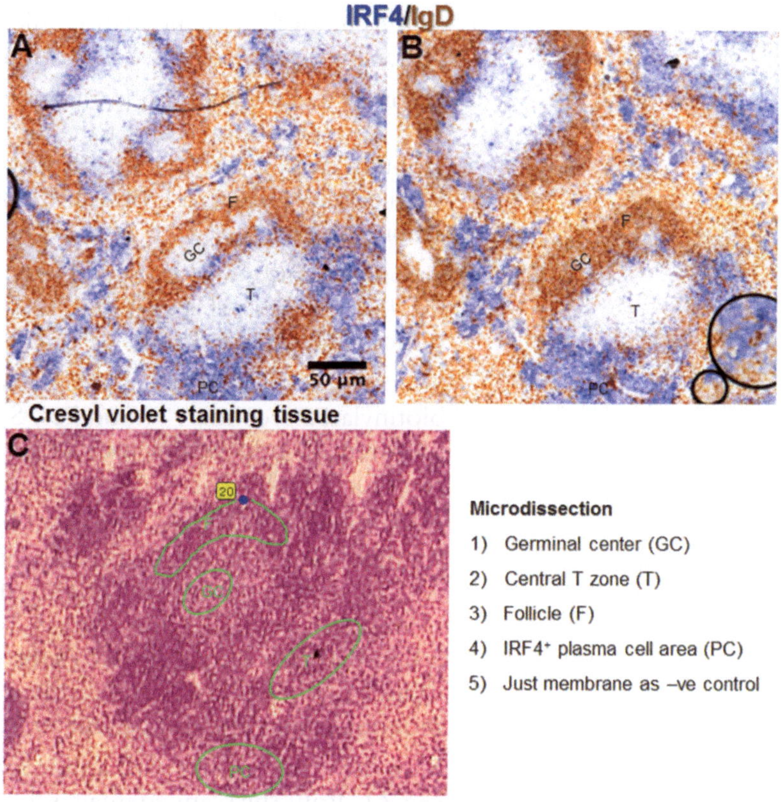

Fig. 1 Appearance of the sections from a consecutive series of tissue sections, stained by immunohistochemistry for IgD and IRF4, or with cresyl violet. Photomicrographs are from a spleen 5 days after NP-CGG immunization of a carrier-primed mouse. (**a**, **b**) For orientation, every tenth serial section is stained by immunohistochemistry for IgD (*brown*, B cell follicles) and IRF4 (*blue*, plasma cells). Scale bar: 50 μm. (**c**) A section from the same series stained with cresyl violet, showing the white pulp area in the bottom right of **a**, and **b**. This shows intense purple staining of the B cell follicles and plasma cell areas, slightly weaker staining of the GC and T zone, and light staining of the red pulp

alone. Optimum dilutions of all reagents are predetermined to give clear positive staining with minimum background. Spleen sections are stained with IgD and IRF4. Germinal centers are IgD negative areas within follicles, central areas containing central arterioles are T zone. IRF4 stains plasmablasts and plasma cells. A detailed staining protocol has been described before [5].

1. Defrost microscope slides in unopened polyethylene bags for 10 min under a fan.
2. Rehydrate slides by immersing in Tris buffer (pH 7.6) for 5 min in a wash trough at room temperature.
3. Rehydrated sections are covered with diluted primary antibodies: Goat anti-mouse IRF4, rat anti-mouse IgD. Incubate slides for 1 h at room temperature in a humid chamber.
4. At the end of the first incubation, slides are washed twice in Tris buffer for 5 min, using a wash trough and a magnetic stirrer.
5. Secondary antibodies are pre-absorbed 30 min before use in 10% normal mouse serum to eliminate anti-mouse immunoglobulin cross-reactivity. The secondary antibodies are biotinylated donkey anti-goat biotin, and horseradish peroxidase (PX) rabbit anti-rat.
6. Secondary antibodies are added to the sections and left to react for 45 min. After this excess antibodies are washed off in Tris buffer (pH 7.6) for 5 min.
7. Streptavidin alkaline phosphatase (AP) complex is added for biotinylated secondary antibodies. Streptavidin AP complex is prepared 30 min before use, added onto section and left to react for 30 min.
8. After washing, stains are developed using first the PX substrate solution for PX conjugated to secondary antibodies, and then AP substrate solution for biotinylated antibodies. Substrate solutions have to be made fresh before use, and wash for 5 min between adding the two substrate solutions. Developed slides are washed twice in distilled H_2O (dH_2O) and then mounted using glycerol gelatin and coverslips.
9. Scan the stained slides with a slide scanner, e.g., a Zeiss AxioScan Z1. Print the scanned images for reference to help identifying different areas on membrane slide when performing microdissection.

3.3 Cresyl Violet Staining of Membrane Slides

An ethanol-based cresyl violet staining is used for membrane slides. The use of an ethanol based dye provides maximal protection of RNA from degradation (Figs. 1c and 3) [6].

1. Dissolve 0.1 g of cresyl violet acetate in 10 ml of 75% ethanol (molecular biology grade) (1% W/V).

2. Leave in the dark at room temperature, rotate overnight to dissolve completely and then filter the next day.
3. Make 100% ethanol, 70% ethanol in water (aqua ad iniectabilia, pH balanced around 7.0) (*see* **Note 7**).
4. Defrost the membrane slides in the unopened container for 10 min under a fan.
5. Rehydrate the membrane slide in 70% ethanol; just leave the slide in the tube for seconds and not for longer than 30 s.
6. Leave the slide in a chamber, incubate for 1–2 min with cresyl violet solution (*see* **Note 8**) and then tip off the remaining liquid from the slide.
7. Dehydrate the slide via subsequent washes in 70%, and 100% ethanol.
8. Leave the membrane slide to air dry. The slide is ready for microdissection immediately after staining and can also be stored at −20 °C until use.
9. PALM Liquid Cover Glass N (Zeiss) is used to improve the visible details of tissue sections before microdissection. The working solution is prepared according to manufacturer's instructions (*see* **Note 9**).

3.4 Microdissection of Different Areas on the Spleen Section

White pulp (dark areas) and red pulp (light areas) can be identified on sections stained with cresyl violet (Fig. 1). The location of B cell follicles, germinal centers, T zones and plasma cell areas can be confirmed by checking the printed images of the sections immunohistochemically stained with IRF4 and IgD. A microdissection experiment may include the identification of GC (IgD negative area), follicle (IgD positive), T zone, and plasmablast/plasma cell rich extrafollicular foci (IRF4 positive) (*see* **Note 10**).

1. A Microbeam HT microscope with software Palm@robo V3.0 (Zeiss) is used (Fig. 2). The slides are mounted in the slide holder with the tissue section facing upwards. Initially an accurate overlap between the pointer in the Palm Robo software and the actual cutting position needs to be made (Fig. 3a, b). Then, the laser power for cutting and laser pressure catapulting (LPC) are adjusted (Fig. 3) (*see* **Note 11**). The settings are saved for future use. Cutting efficiency may be inhibited by OCT compound (*see* **Note 4**).
2. Identify the same white pulp area in the adjacent sections on the membrane slides (Fig. 1).
3. Prepare 8-well cap strips containing 20 μl of RNeasy RLT buffer. Make sure the buffer covers the cap completely.

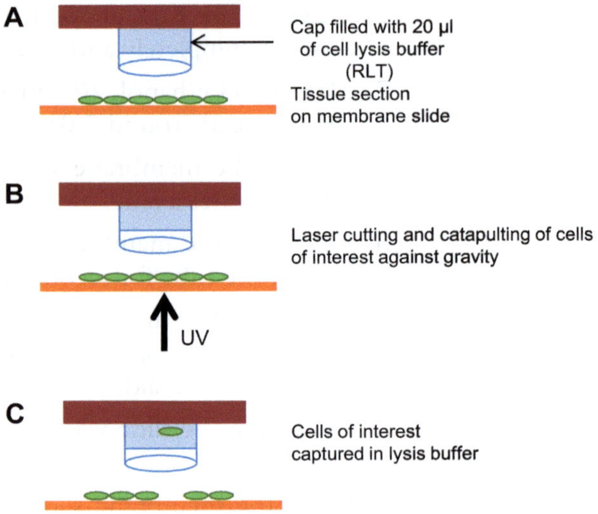

Fig. 2 Mechanism of tissue laser capture using the PALM microscope. (**a**) When isolating cells with the PALM–LCM system, tissue sections are firstly prepared on membrane-coated microscope slides and collecting caps are filled with 20 µl RLT lysis buffer. (**b**) The UV laser is focused onto the section and used to cut around the area of interest. The laser then automatically focuses below the area of interest and a single stronger laser pulse is fired. (**c**) Catapulting of tissue into the overlying microcentrifuge tube cap

4. In order to get a good RT-PCR signal, normally 10–20 similar areas are selected for adjacent sections for each of the areas identified and catapulted into the same cap (Fig. 1a, b). Catapult different areas into separate caps (*see* **Note 12**).
5. After collecting the tissue clip the 8-well strips of caps onto 200 µl RNAse-free 96-well PCR plates.
6. Cut membrane only (tissue free area) into separate caps as a negative control for RNA contamination after finishing each type of area (*see* **Notes 13** and **14**) (Fig. 4).
7. Spin down by high speed, leave on dry ice, and then store in −20 °C freezer until use.

3.5 RNA Isolation

It is important to extract and isolate RNA from the microdissected samples prior to qRT-PCR analysis to reduce the failure rate of qRT-PCR.

1. RNA is isolated by using the RNeasy Micro kit (*see* **Note 15**). RNA carrier in the kit is used to enhance RNA extraction from the microdissected tissue. For tubes with less than 5000 cells, 20 ng of carrier RNA (5 µl of a 4 ng/µl solution) are added to the lysate before homogenization (*see* **Note 16**).

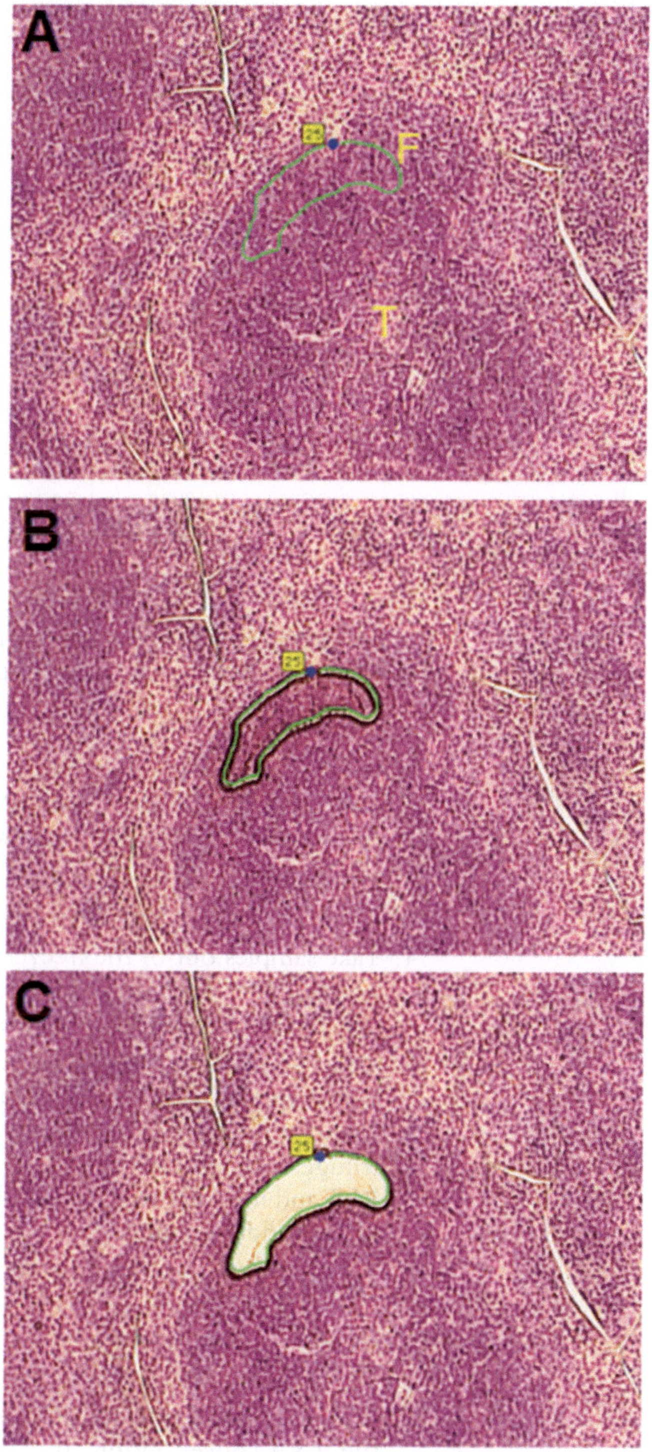

Fig. 3 Identification and microdissection of an identified area from a spleen section. (**a**) Follicular area identified and marked. (**b**) After laser cut, a small bridge at the top still connects the area to the rest of the section. (**c**) After laser pressure catapulting

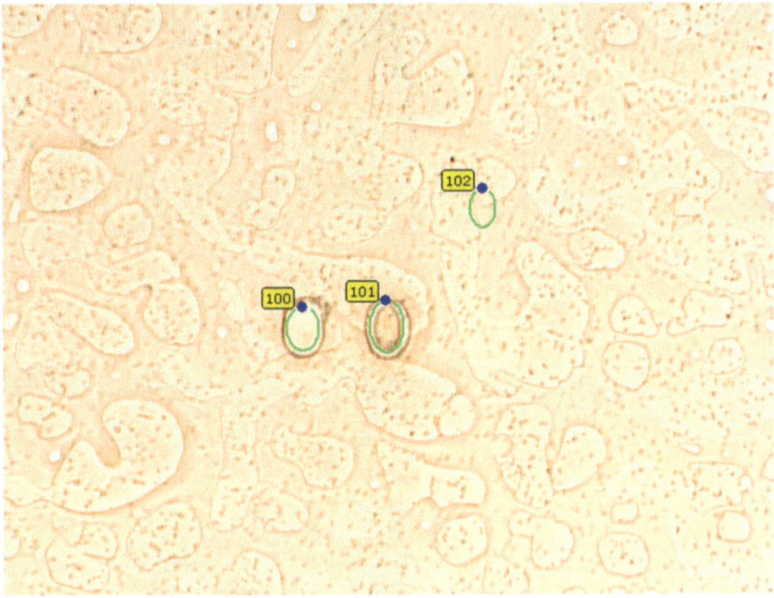

Fig. 4 Microdissection of membrane only Identification and microdissection of tissue-free area (membrane only), as a negative control

2. Prepare an appropriate volume of buffer containing 10% β2-mercaptoethanol (β-ME) in RLT buffer plus RNA carrier. Always make this buffer fresh (*see* **Note 17**).
3. Prepare an appropriate volume of 70% and 80% ethanol in RNase-free water (*see* **Note 18**).
4. Take samples out of the freezer and leave on dry ice until use. To minimize RNA degradation, avoid the prolonged exposure of samples to room temperature.
5. Add 150 μl of RTL buffer to the sample in a 200 μl PCR tube before the tissue is completely thawed. Mix well by pipetting. Transfer the sample and RLT buffer into a 1.5 ml microcentrifuge tube, adjust the sample volume to 350 μl with buffer RLT and mix very well. Finally, transfer the total volume into a QIAshredder column.
6. Follow the kit instructions for the subsequent steps. During the RNA extraction, DNase treatment for 15 min is recommended.
7. The protocol ends with the RNA elution. Add just 10–15 μl of RNA free water into the RNeasy column in order to increase the RNA concentration. If cDNA can't be prepared immediately then store the RNA at −80 °C for the required period (*see* **Note 19**).

3.6 cDNA Preparation

1. Prepare a sufficient number of 0.2 ml 8-well strips of PCR tubes.
2. 10 μl of RNA solution is mixed with 1 μl of random primers (0.5 μg/μl).
3. Denature the samples for 10 min at 70 °C in the thermocycler. Place on ice immediately after the 10 min.
4. 9 μl of reverse transcription mix are added and mixed well. The following reagents are added to each sample: 0.5 μl dH_2O (RNA free water); 4 μl 5× first strand buffer; 2 μl 0.1 M DTT; 1 μl dNTP (10 Mm) (diluted from 100 mM set); 0.5 μl RNAse Inhibitor (40 μg/μl); 1 μl Moloney murine leukemia virus (M-MLV) reverse transcriptase.
5. Take the samples back to thermocycler for the reverse transcription of RNA into cDNA. Heat for 1 h at 41 °C, and then for 10 min at 90 °C to inactivate reverse transcriptase.
6. The cDNA is stored at −20 °C.

3.7 Semiquantitative Real Time PCR (qRT-PCR)

qRT-PCR is used to quantify the expression of genes of interest such as immunoglobulin heavy chain germ line transcripts, chemokines, or cytokines in cDNA prepared from microdissected tissue [7, 8].

1. PCR is performed by adding 1 μl of cDNA to each well of a 384 well plate followed by the required primers and FAM-labeled probe mixtures and 3 μl Universal qRT-PCR master Mix (final volume is 6 μl) (*see* **Notes 20–22**).
2. Cover the plate with clear adhesive foil, and centrifuged to remove any air bubbles (689 × *g*, a few seconds). Along with the amplification of the targeted gene, the internal housekeeping gene β2-microglobulin is amplified so that the relative quantities of gene amplification can be assessed regardless of the number of cells harvested to produce each cDNA sample (*see* **Note 23**).
3. Run plate in an Real-Time PCR System with 384-Well Block Module with a cycling program as follows: 2 min at 50 °C; 10 min at 95 °C; Then 40 cycles of 15 s at 95 °C and final step of 1 min 60 °C.
4. Analyze data using an appropriate software, and set a threshold within the logarithmic phase of the PCR. The cycle number (Ct) at which the signals for target gene and housekeeping gene reach the threshold are recorded for each sample. Assuming gene of interest and the housekeeping gene amplify with similar efficiency, the relative quantity of expressed target gene mRNA is estimated by subtracting the Ct for the housekeeping gene from the Ct of the target gene (ΔCt) and then calculating $2^{-\Delta Ct}$ (Fig. 5).

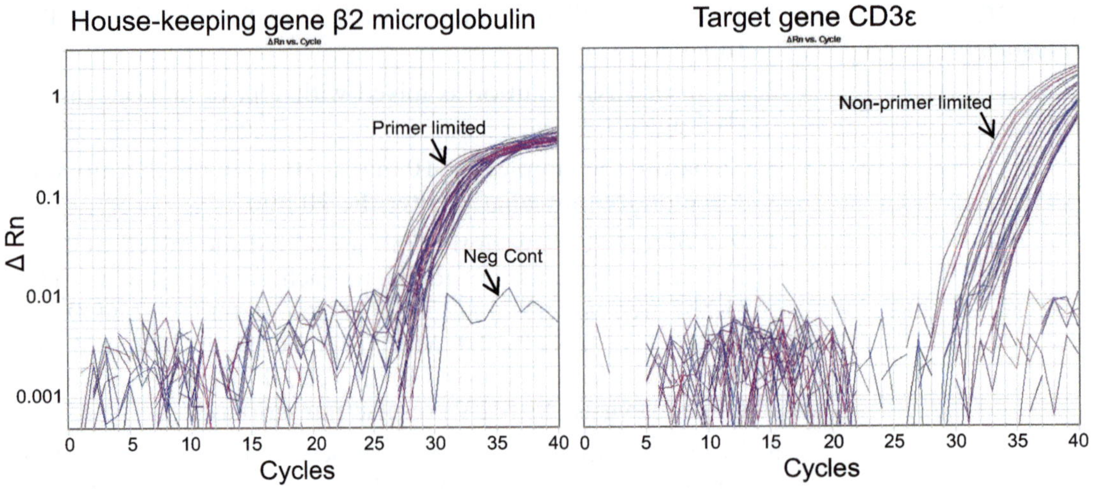

Fig. 5 Typical TaqMan qPCR amplification plots from 384-well setup from microdissected tissue. Background subtracted fluorescence signal (ΔRn) versus cycle number. Primer/probe concentrations for TaqMan PCR are usually optimized for maximal ΔRn and lowest Ct values. *Left*: The housekeeping gene is amplified using primer limiting conditions and a VIC-labeled probe. Limiting primers specific for the stronger expressed housekeeping gene prevents the PCR being depleted of nucleotides before the target gene specific fluorescence becomes detectable. *Right*: Genes of interest are measured using FAM-labeled non-primer limited primer/probe sets. Triplicate technical replicates result in good reproducibility. Neg Cont: Negative control (microdissected membrane only)

4 Notes

1. Chicken gamma globulin (CGG) is precipitated in alum (detailed protocol in Toellner et al. 2004). Briefly, CGG is made up as a 5 mg/ml solution in sterile water and added to an equal volume of sterile 9% $AlK(SO_4)_2$ in water. The protein/alum mix is precipitated with 10 M NaOH to pH 6.5, and left in dark for 1 h at room temperature, rotating to allow for maximum precipitation. Important: NP-CGG alum is precipitated with 10 M NaOH to pH 6.5 and 0.1 N HCl. If the pH is too high (e.g., pH 10.0), just discard the mixture and restart again. The precipitate is washed twice in sterile PBS. The pellet is suspended to a final concentration of 50 μg/200 μl in sterile PBS with 5×10^7 heat inactivated *Bordetella pertussis* (b.p.) bacteria as an adjuvant for intraperitoneal (i.p.) injection. The preparation of 4-hydroxy-3-nitrophenylacetyl (NP)-CGG has been described before [5, 9]. NP-O-succinimide esther (NP-OSu) is dissolved at 10 mg/ml in dimethylformamide (DMF). CGG is dissolved at 10 mg/ml in 0.1 M sterile $NaHCO_3$ on ice. NP-OSu is added drop by drop, while stirring, into the protein solution at a ratio of 1 mg NP: 20 mg protein and then incubated at room temperature in the dark for 2 h on a rotating mixer. The NP-Protein solution is dialyzed against 0.1 M $NaHCO_3$ overnight and then against sterile PBS (pH 7.4)

overnight using dialysis cassettes (10,000 MW). This step removes any non-conjugated NP from the mixture. This procedure gives an approximate substitution rate of 18 NP per molecule of protein.

2. Termination of response is at day 5, or day 8, depending on the experiment. Freeze tissue samples slowly by repeated dipping in liquid N_2. Snap-frozen samples provide the highest quality of DNA, mRNA, and protein for analysis [10]. Do not store the tissue in the freezer for too long, even at −80 °C.
3. Wear gloves and a laboratory coat. Put a small strip of OCT compound on the cold cutting block and place the spleen on top, freeze immediately using freeze spray. Trim the spleen a little bit until the white pulp is visible, before starting to mount sections onto microscope slides.
4. Make sure any OCT compound that may come into contact with the microtome knife is carefully trimmed off before cutting. OCT can seriously inhibit microdissection efficiency by interfering with the microdissection laser light.
5. If the tissue does not stick well to membrane slides, warm the membrane slides by putting a finger on the back of the slide when mounting the section from the cryostat knife. Then dry the slide at room temperature under a fan.
6. Fix membrane slides and normal glass slides in separate glass troughs. Especially membrane slides should be fixed in a RNAse/DNAse-free glass trough.
7. For cresyl violet staining, make up 100% ethanol and 70% ethanol in water solutions in separate 50 ml FALCON tubes. Then just soak the slides in these tubes for seconds. Here *aqua ad iniectabilia* is used (pH 7.0). The pH value may affect the quality of cresyl violet staining and RNA extraction [6].
8. The timing for cresyl violet staining should be optimized using spare tissue sections. Normally for murine lymph node and spleen, 1–2 min are sufficient, while murine kidney stains well within a few seconds. Dependent on the tissue cresyl violet in 50% ethanol in water may give better results.
9. Make a working solution in a spray bottle as a stock solution to thinning solution in a ratio of 1:6. Spray once at a distance of 10–20 cm onto the membrane slides and allow to dry for approximately 5 min. For spleen and lymph node sections, the manufacturer's recommended concentration works well. Other types of tissues may need optimization with a different dilution of stock solution or repeated spaying and drying cycles.
10. Have separate pipettes for performing microdissection and cDNA preparation. Clean the pipettes, working area, slide holder, collection holder of microdissection microscopy with RNAseZap@ RNase Decontamination Solution (Ambion, Thermo Fisher).

11. Microscope is set to best focus with minimal cutting power, which should result in minimal tissue damage. Adjust the collection tube holder to make sure the microdissected tissue can be laser pressure catapulted into the cap or PCR tube. Use the Robo LPC mode for cutting and shooting tissue. Robo LPC mode cuts around the selected tissue area, leaving a small bridge which is destroyed during the subsequent laser pressure catapulting step. Set the right bridge width for optimal catapulting of the microdissected area. If the bridge is too big, the selected area may not catapult well and stick to the tissue. If the bridge is too narrow, the dissected section may drop off during cutting. *See also* **Note 4**.
12. Marking and microdissection of all corresponding areas in all adjacent sections will create easily identifiable landmarks to the next series of areas of interest.
13. If the tissue does not catapult into the cap, cut once more with same laser power. *See also* **Note 4**. Clean the membrane slides from residual contaminating tissue particles by placing under a fan for seconds before microdissecting other areas.
14. Cut one tissue section, do the cDNA prep, and use qRT-PCR to check whether the cut area is correct and one can get an appropriate amplification signal.
15. Normally, keep the higher concentration stock (400 μg/ml) in freezer (−20 °C). Dissolve RNA carrier at higher concentration solution (400 μg/ml) by using RNase-free water, and then save at −20 °C until use.
16. β-ME is toxic, so dispense in a fume hood and wear appropriate protective clothing. It is advisable to work in a fume hood when using buffer containing β-ME.
17. The RNase-free technique needs to be followed stringently to prevent contamination. Wear gloves, a clean laboratory coat and change gloves often. Clean the bench tops and pipettes with RNAseZap@ RNase Decontamination Solution. Ideally a dedicated working area is required.
18. Here β2-microglobulin is chosen as the only housekeeping gene. If performing multiplex reactions, it is recommended to use primer-limited primer/probe sets for the amplification of the endogenous control gene (Fig. 5).
19. TaqMan® chemistry based primer/probe sets are chosen in our experiment. Primer/probe combinations are strongly recommended for use with RNA from laser microdissected samples because of the improved specificity over primer only based detection chemistries (e.g., SYBR green detection) [10]. Primer and probe design has been discussed extensively in other papers, for example [11].

20. In our experience the lowest volume that reliably produces efficient qRT-PCR amplification in 384-well plates is 1 μl of cDNA with 5 μl TaqMan mix. These volumes are also easily pipetted with common electronic multichannel pipettes.
21. After covering the plate with adhesive sealing film, shake the plate on a vortex mixer with a plate adapter to mix the cDNA with the reaction buffer and then centrifuge at a high speed for seconds. Centrifugation after shaking is essential for successful RT-PCR.
22. Always use cDNA extracted from whole spleen or lymph node sections as a positive control and RNase-free water as negative control for qRT-PCR. CD3ε and CD19 are chosen as positive controls for successful dissection of different areas prior to testing the expression of other genes (Fig. 6).
23. We use a probe for β2-microglobulin either VIC or NED labeled (e.g., Applied Biosystems) so as not to interfere with the FAM-labeled.

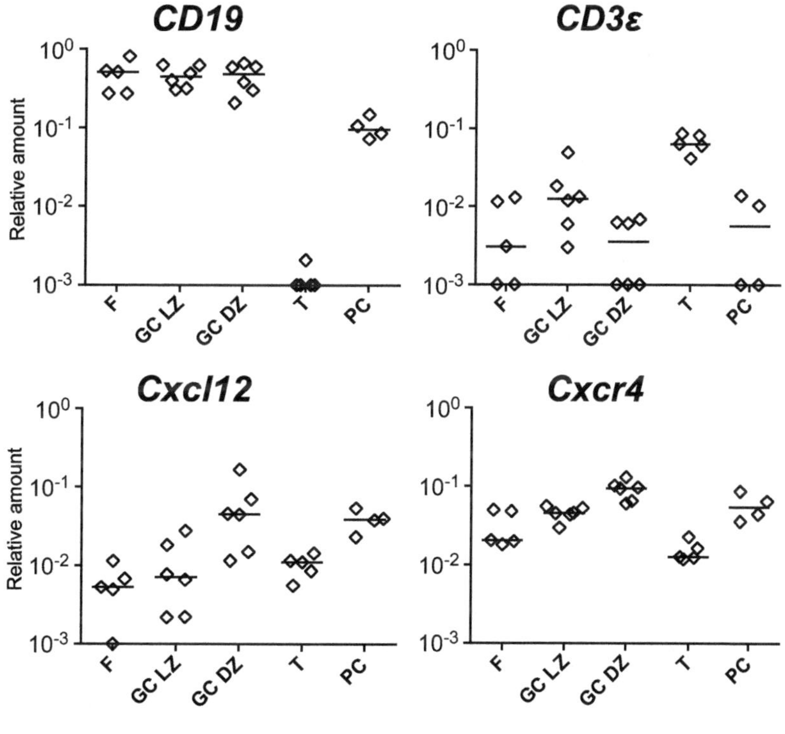

Fig. 6 Validation of qRT-PCR method from different areas of a laser microdissected spleen section. Relative expression levels for *Cd3ε*, *Cd19*, *Cxcl12*, and *Cxcr4* from different microdissected areas. Expression levels are given as the amount of mRNA relative to the level of *β2m* mRNA. *F* Follicle, *GC LZ* Germinal center light zone. *GC DZ* Germinal center dark zone, *T* T zone, *PC* IRF4⁺ Plasma cell area. Each *diamond* represents the pooled microdissected material from ten consecutive sections from one identified area

Acknowledgment

This work was supported by grants from the MRC (Topjabs, G1001390) and BBSRC (BB/M025292/1), and received funding from the People Programme (Marie Curie Actions) of the European Union's Seventh Framework Programme FP7/2007-2013 under Research Executive Agency grant agreement № 315902. LGI gratefully acknowledges receipt of a Marie Curie Research Associate post. GB and K-MT are partners within the Marie Curie Initial Training Network DECIDE (Decision-making within cells and differentiation entity therapies).

Competing Financial Interests
The authors declare no competing financial interests.

References

1. Toellner KM (2014) Cognate interactions: extrafollicular IL-4 drives germinal-center reactions, a new role for an old cytokine. Eur J Immunol 44(7):1917–1920. doi:10.1002/eji.201444825
2. MacLennan IC (1994) Germinal centers. Annu Rev Immunol 12:117–139. doi:10.1146/annurev.iy.12.040194.001001
3. Fink L, Bohle RM (2005) Laser microdissection and RNA analysis. Methods Mol Biol 293:167–185
4. Datta S, Malhotra L, Dickerson R, Chaffee S, Sen CK, Roy S (2015) Laser capture microdissection: big data from small samples. Histol Histopathol 30(11):1255–1269. doi:10.14670/HH-11-622
5. Toellner KM, Khan M, Sze DM (2004) Analysis of the germinal center reaction and in vivo long-lived plasma cells. Methods Mol Biol 271:111–125. doi:10.1385/1-59259-796-3:111
6. Cummings M, McGinley CV, Wilkinson N, Field SL, Duffy SR, Orsi NM (2011) A robust RNA integrity-preserving staining protocol for laser capture microdissection of endometrial cancer tissue. Anal Biochem 416(1):123–125. doi:10.1016/j.ab.2011.05.009
7. Marshall JL, Zhang Y, Pallan L, Hsu MC, Khan M, Cunningham AF, MacLennan IC, Toellner KM (2011) Early B blasts acquire a capacity for Ig class switch recombination that is lost as they become plasmablasts. Eur J Immunol 41(12):3506–3512. doi:10.1002/eji.201141762
8. Zhang Y, Meyer-Hermann M, George LA, Figge MT, Khan M, Goodall M, Young SP, Reynolds A, Falciani F, Waisman A, Notley CA, Ehrenstein MR, Kosco-Vilbois M, Toellner KM (2013) Germinal center B cells govern their own fate via antibody feedback. J Exp Med 210(3):457–464. doi:10.1084/jem.20120150
9. Nossal GJ, Karvelas M (1990) Soluble antigen abrogates the appearance of anti-protein IgG1-forming cell precursors during primary immunization. Proc Natl Acad Sci U S A 87(4):1615–1619
10. Erickson HS, Albert PS, Gillespie JW, Rodriguez-Canales J, Marston Linehan W, Pinto PA, Chuaqui RF, Emmert-Buck MR (2009) Quantitative RT-PCR gene expression analysis of laser microdissected tissue samples. Nat Protoc 4(6):902–922. doi:10.1038/nprot.2009.61
11. Nolan T, Hands RE, Bustin SA (2006) Quantification of mRNA using real-time RT-PCR. Nat Protoc 1(3):1559–1582. doi:10.1038/nprot.2006.236

Chapter 3

Tracking Plasma Cell Differentiation in Living Mice with Two-Photon Microscopy

Carolin Ulbricht, Randall L. Lindquist, Laura Tech, and Anja E. Hauser

Abstract

Due to the multitude of cell types involved in the differentiation of plasma cells during the germinal center reaction, and due to a lack of in vitro systems, which recapitulate germinal centers, the most suitable way to study plasma cell generation in germinal centers is in vivo. In this chapter we describe how to induce humoral immune responses to defined model antigens and how to visualize and track plasma cells and their interactions with other cells in the lymph nodes of living mice.

Key words Fluorescent reporter mice, Two-photon microscopy, B cells, Plasma cells, Lymph node, Germinal center, Intravital imaging

1 Introduction

Intravital imaging allows cells and their interactions to be visualized in their native context and followed over time [1]. This is critical to understand complex processes like the affinity maturation of humoral immune responses, which occurs only after multiple specific interactions between rare antigen-specific B and T lymphocytes have induced the formation of a germinal center. Inside the germinal center, B cells interact closely with follicular dendritic cells (FDCs), which bind antigen in the form of immune complexes on their surface. Continuous selection for higher-affinity B cell receptor (BCR) variants in germinal centers ultimately leads to the generation of high-affinity memory B cells and antibody-secreting plasma cells. Motile plasma blasts leave germinal centers; interestingly, plasma cells that have taken up residence elsewhere are largely nonmotile [2, 3]. Sessile plasma cells are mainly found in the medullary cords of lymph nodes and in splenic extrafollicular foci;

Authors Carolin Ulbricht and Randall L. Lindquist contributed equally to this work.

Dinis Pedro Calado (ed.), *Germinal Centers: Methods and Protocols*, Methods in Molecular Biology, vol. 1623, DOI 10.1007/978-1-4939-7095-7_3, © Springer Science+Business Media LLC 2017

however, a fraction of plasma blasts migrate from secondary lymphoid organs via the bloodstream to the bone marrow, where they become sessile and can survive for extended time periods. These long-lived plasma cells secrete the antibodies that can provide lifelong protection, and are critical for protective immunity; however, they can also contribute to devastating autoimmune conditions. The precise mechanism by which a subset of GC B cells develop into long-lived plasma cells remains unclear, but given the exquisite complexity of the humoral immune response, this is best studied in situ in the native context.

2 Materials

2.1 Animals

1. The appropriate regulatory committee for the jurisdiction in which the experiments are performed, e.g., the Institutional Review Board or state committee for animal welfare, must approve all animal experiments. An approved animal protocol is therefore necessary before beginning any animal experiments.
2. Various strains of reporter mice have been developed that bear fluorescent proteins knocked into a locus of interest or under the control of a specific promoter. For studies of plasma cells, Blimp-GFP [4] Blimp1-YFP [3], and Cγ1-Cre crossed onto strains containing fluorescent protein genes inserted into the *Rosa26* locus with an upstream *loxp*-flanked STOP sequence (Rosa26: YFP [5] or RFP [6]) are all of interest. The suitability of various fluorescent proteins is in turn determined by the excitation wavelengths available from the pulsed laser with which your two-photon microscope is equipped. Red and far-red fluorescent proteins are optimally excited by the longer wavelengths only an optical parametric oscillator (OPO) can provide [7]. It is recommended to test for a possible rejection of fluorescent cells in the recipient, as for example a rejection of GFP can occur in Balb/c mice [8] This can be avoided by using host mice in which a GFP variant is expressed in an extralymphoid tissue, so the animal is centrally tolerized to GFP and its variants.
3. Antigen-specific immune responses can be studied by breeding the above strains of mice to another strain expressing a specific antigen receptor. For example, breeding Blimp1-GFP mice with B1–8^{hi} mice, in which all lambda^{+} B cells express the nitrophenyl (NP)-specific B1–8 BCR, yields mice with NP-specific B cells in which plasma cells express EGFP and produce NP-specific antibodies [9, 10].
4. GC cells can be visualized either by specific labeling with fluorescent reporters such as Fucci-red, or by labeling the rest of the B cell follicle and FDCs in separate channels (*see* **Notes 1** and **2**).

2.2 Buffers and Solutions

1. 1× PBS.
2. PBS with 0.5% BSA.
3. Tris-buffered ammonium chloride (ACT): 0.16 M NH_4Cl in H_2O, 0.17 M Tris in H_2O pH 7.65, dilute Tris 1:10 in NH_4Cl, adjust pH 7.2.
4. 0.9% saline.
5. Transfer buffer: 100 ml sterile PBS, 1 ml 1 M HEPES, 50 U/ml penicillin/streptomycin (0.5 ml of 10,000 U/ml).
6. Acid-Citrate-Dextrose formula A (ACD-a, Merck).

2.3 Single Cell Suspensions

1. Ice bath.
2. Spleens of donor animals.
3. Cell strainer 100 μm (Falcon).
4. 50-ml tubes.
5. Plunger of a 3-ml syringe.
6. Cooling centrifuge (e.g., Heraeus).
7. Cell counting device (e.g., Neubauer chamber, flow cytometer).
8. Mouse B cell isolation kit (e.g., Miltenyi).
9. MACS MS Columns (Miltenyi, other magnetic cell sorting protocols might apply).
10. MACS Magnets.
11. Magnetic stand.

2.4 Cell Transfer, Immunization and In Vivo Labeling

1. Isolated B cells in transfer buffer w/ 2,5% ACD-a.
2. 1-ml syringes.
3. 27G × ¾″ cannulas.
4. Restrainer for injections.
5. Warming lamp.
6. Chicken gamma globulin (CGG), conjugated to NP, Ratio 20–29 (Biosearch technologies).
7. Complete Freund's adjuvant (CFA).
8. Anti CD21/35 F(ab)-Fragments labeled with fluorescent dye of choice that does not interfere with GFP: e.g., Alexa 568, Atto 590, Alexa 647, (*see* **Note 2**).

2.5 Surgical Procedures

1. Anesthesia (ketamine 50 mg/ml, xylazine 20 mg/ml, sterile 0.9% saline for injection). Mix working solution of anesthetic: 10 mg/ml ketamine, 1 mg/ml xylazine, in 0.9% saline for injection. Weigh the mouse, and inject 10 μl per gram body weight. This yields a dose of 100 mg/kg ketamine, 10 mg/kg xylazine, and should provide ca. 45 min of anesthesia for a WT B6 mouse (*see* **Note 3**).

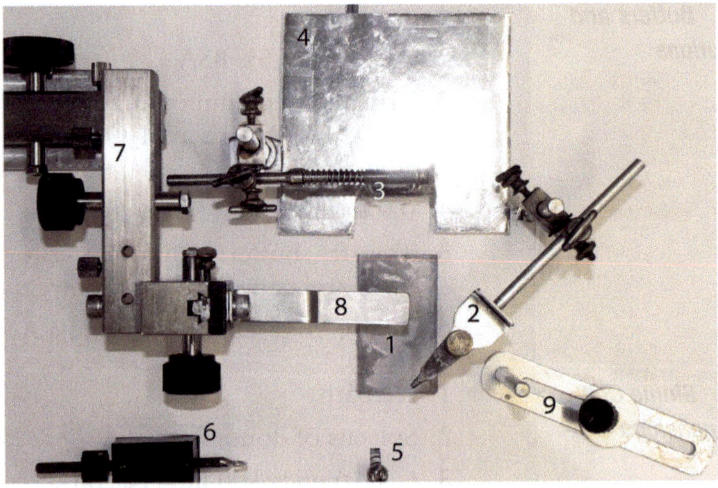

Fig. 1 Example of a custom-built imaging stage for intravital microscopy. (*1*) platform supporting right positioning of the hind leg of the imaged mouse, (*2*) femoral fixator, (*3*) spinal fixator, (*4*) heating device that is to be attached to a power supply, (*5*) screw for winding the thread that is attached to a (*6*) movable hook for spanning the leg, (*7*) 3D–adjustment for (*8*) holder of the coverslip, (*9*) holder for heating coil

2. Eye ointment.
3. Electric razor.
4. Depilating cream and applicator.
5. Scissors (e.g., fine iris scissors, 14 mm (blunt) and Vannas spring scissors (straight) 2.5 mm, all by FST).
6. Thread.
7. 2× forceps INOX size 5 titanium med biology (e.g., Dumont by FST).
8. Custom-built stage (Fig. 1).
9. Tape.
10. Modeling clay.
11. Vacuum grease or Vaseline.
12. Round coverslips (ca. $d = 19$ mm).

2.6 Imaging

1. Dissecting microscope.
2. Stage heater, objective heater, heating coil.
3. Two-photon-microscope with pulsed lasers. For optimal excitation of red-to near-IR FPs, an OPO is recommended (*see* **Note 4**).
4. Analysis software: Microscopes may or may not save image data in a proprietary image format, but there are many commercial and open-source programs that can successfully import images

and the associated metadata. Imaris (Bitplane) and Volocity (PerkinElmer) are commercial programs that are often used to analyze intravital imaging data, and ImageJ is an extensively documented open-source program.

3 Methods

An overview of the experimental procedures and timeline summing up the methods described here is shown in Fig. 2.

3.1 Purification and Transfer of Fluorescently Labeled B Cells (Day −1)

1. Sacrifice donor mice and harvest their spleens and/or lymph nodes. To visualize and track NP-specific plasma cells, use B1–8$^{+/+}$ J$\kappa^{-/-}$ mice crossed onto Blimp1:GFP reporter mice, which indicates plasma cell differentiation [3, 4]. A second fluorescent reporter ("double reporter mice") for labeling germinal center B cells (such as Fucci-red mice [11]) is used here,

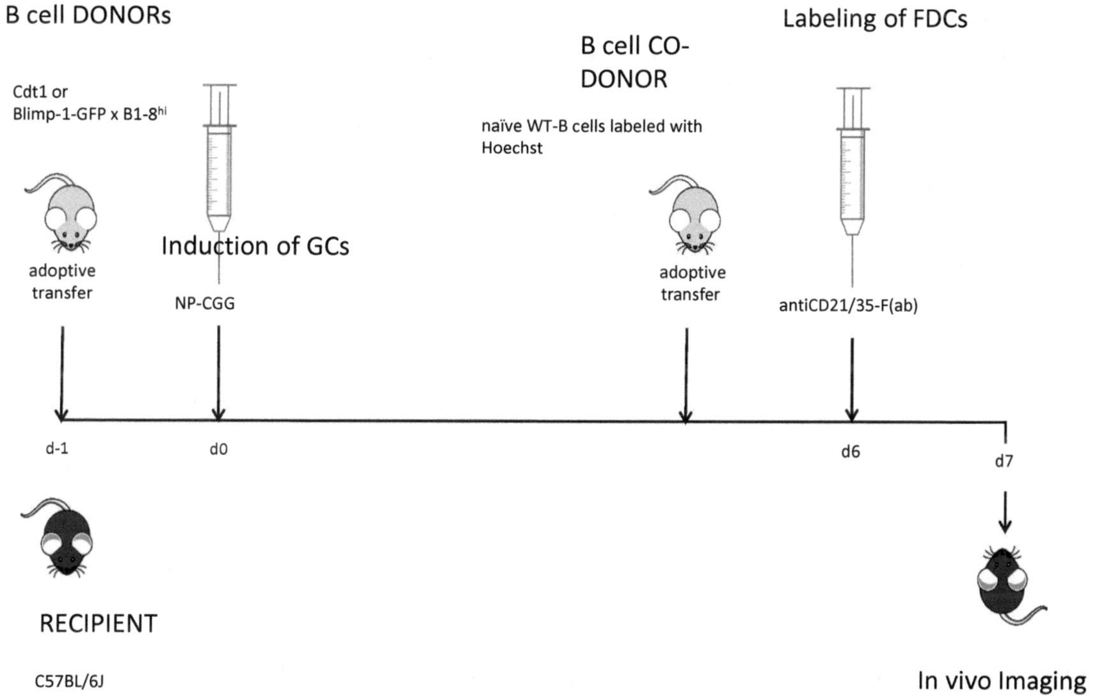

Fig. 2 Outline of the experimental procedure for intravital imaging of germinal centers and plasma blasts in the popliteal lymph node. A mixture of purified fluorescent B cells is injected via the tail vein of wild type animals 1 day prior to immunization with NP-CGG in the footpad. Wait for 5–10 days for the establishment of germinal centers. Two days prior to imaging, transfer labeled naive wild type B cells to demarcate the B cell follicle. Label FDCs 12–24 h before imaging via injection of fluorophore-coupled anti-CD21/35-F(ab)-fragments in the footpad

but optional (*see* **Note 1**). From this step on, sterile conditions should be maintained and cells should be kept on ice for all steps unless noted otherwise.

2. Lyse erythrocytes by resuspending the cell pellet in ACT buffer and incubate for 2 min at room temperature. Add serum-containing buffer (PBS/BSA) and wash by centrifugation (300 × *g*, 8 min).
3. Purify B cells immunomagnetically by negative selection from the organs harvested above: mash the spleens through a 100 μm filter using the plunger of a 3-ml syringe, wash, and centrifuge at 300 × *g* for 8 min. Perform an immunomagnetic separation using the desired separation kit according to the manufacturer's instructions. The purity of the B cells should be >95%, as checked by flow cytometry before proceeding with adoptive transfers. Inject intravenously (using restrainer and warming lamp) 3 × 10^6 NP-specific B cells in transfer buffer and add 2.5% anticoagulant ACD-a freshly before each injection) into sex-matched C57Bl/6J recipients.

3.2 Immunization to Induce a Germinal Center Response LN (Day 0)

Inject 10 μg NP-CGG emulsified in Complete Freund's adjuvant (CFA) in the right hind footpad (*see* **Note 3**). As CFA is very sticky, leading to significant retention in the syringe dead space, it is advisable to prepare twice the quantity required.

3.3 Fluorescent Labeling of Non-Transgenic B Cells (Day 4–7)

In order to demarcate B cell follicles, non-transgenic B cells are fluorescently labeled and transferred 24–48 h prior to imaging (day 4–7). Note that you might need more than one donor spleen for the respective cell numbers. Naive B cells are purified from non-transgenic mice immunomagnetically as described under Subheading 3.1.

1. To label cells with the blue nuclear dye Hoechst 33342, resuspend cells at 2 × 10^7 cells per ml in PBS/0.5% BSA with 2.5 μM Hoechst 33342 in a 14 ml Polystyrene tube. Stain for 30 min. At room temperature, wash twice with PBS and centrifuge (300 × *g*, 8 min).
2. Resuspend in transfer buffer and inject 6 × 10^7 labeled cells into the tail vein.

3.4 In Vivo Labeling of FDCs (Day 4–7)

To highlight the network of FDCs within the germinal center, label FDCs with fluorescent F(ab) fragments by s.c. injection 12–24 h prior to imaging (day 4–7).

1. Generate F(ab) fragment of anti-CD21/35 clone 7G6 or comparable antibody.
2. Label F(ab) fragment with Atto590-succimidyl ester or other desired flurophore according to the manufacturer's protocol; Alexa dyes may be used in place of Atto dyes (*see* **Note 2**).

3. Inject 10 µg of fluorochrome-labeled F(ab) fragments into the right hind footpad of the mice 12–24 h before intravital imaging is performed.

3.5 Preparation of Popliteal LN for Imaging (Day 5–8)

1. Anesthetize mouse by i.p. injection of 100 mg/kg ketamine/10 mg/kg xylazine, with the volume injected depending on its weight. For C57B/6 mice, 10 µl per gram body weight of a 10 mg/ml ketamine, 1 mg/ml xylazine solution should provide ca. 45 min of anesthesia (*see* **Note 4**). Apply eye ointment to reduce any pain from drying out eye bulbs.
2. Test reflexes to monitor the depth of anesthesia over the whole imaging period. Subsequent doses of injected anesthetic should be half of the induction dose, and will provide a further 30–45 min of anesthesia.
3. Shave leg, back, and right flank of mouse rigorously and use depilating cream to remove residual hair completely (*see* **Notes 5** and **6**) (Figs. 3a, 3b).
4. Fix right trochanter major (femoral bone): locate bony chip with fingers, set a small skin incision with scissors until a small

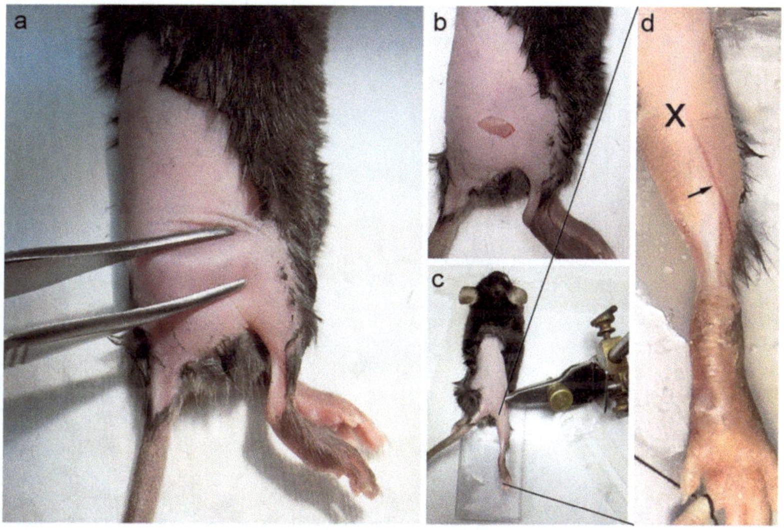

Fig. 3 Identifying the location for skin incisions in order to expose the popliteal lymph node. Pinching the right thigh with a forceps helps locating the femoral bone (**a**). The bone should point out as a small ridge. Hold up the skin right there and make a small incision (**b**). You should see a white bony structure emerging. Attach the femoral fixator around it like described in the protocol (**c**) and adjust the angle of the femoral fixator till the leg is stretched out and the back of the knee is planar to the stage surface. You will now be able to see the superficial popliteal vein covering the lower leg (**d**, *arrow*). The popliteal lymph node is located just medial to where the vein disappears into the tissue (d, *cross*). Make an incision here

white triangle-shaped fascia appears. Loosen the muscular tissue above the fascia (Fig. 3c).

5. Clamp bony chip beneath this fascia with pointed tweezers of the restrainer.
6. Fix the spine in lumbar area using toothed tweezers.
7. Fix right hind leg with a thread around a toe on the restrainer and tighten the screw.
8. Tape tail to keep the leg in the correct position.
9. At the caudal side of the thigh, create an incision in the skin, in the area where the prominent popliteal vein enters the tissue, separate the skin from the underlying tissue with blunt tweezers and follow the popliteal vein from proximal to distal (Fig. 3d).
10. Expose the lymph node using blunt tweezers, try to avoid cutting in order to protect the fine lymphatic vessels (keep the lymph node submerged in PBS at all times). Push up the lymph node by building a pile out of modeling clay underneath the leg. The lymph node should now visibly emerge from the tissue (Fig. 4).
11. Create a circle of vacuum grease around the lymph node, fill with a puddle of PBS (*see* **Note** 7).

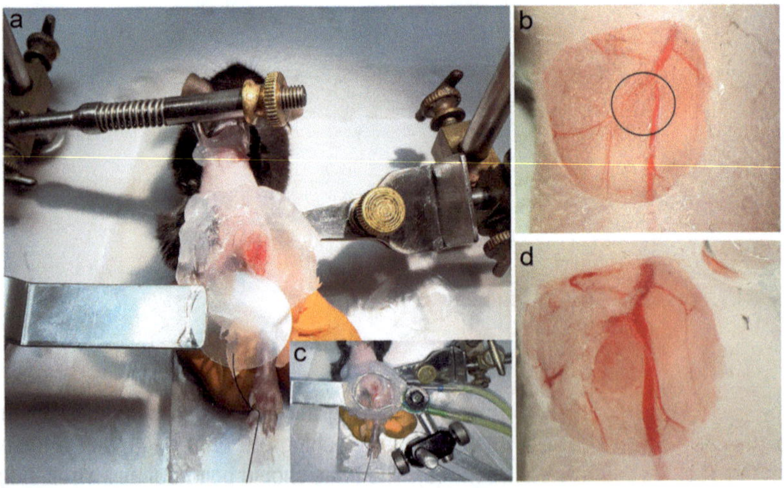

Fig. 4 Mouse with lymph node exposed fixed to stage with coverslip attached. A tub is formed out of vacuum grease around the lymph node and filled up with NaCl. The coverslip holder is put in place that it centers the lymph node (**a**). The lymphnode is expected in the *circled area* (**b**) and exposed (**d**). Above, a heating coil is attached to the respective holder (**c**). This again is sealed with vacuum grease and filled up with water for immersion. The objective of the microscope is dipped in

12. Fix coverslip with superglue to holder and adjust its position. Place the thermometer-probe underneath, in proximity to the lymph node, but not touching it (keep temperature at 37 °C, otherwise the cell velocity will dramatically vary).
13. Along the upper edge of the coverslip, add a circle of vacuum grease.
14. Put the heating coil in its position on the vacuum grease, create a seal in a similar fashion as before and add water/PBS on the glass coverslip to dip the objective into (*see* **Note 8**).
15. After imaging, the mouse is kept under anesthesia, given a lethal dose of ketamine–xylazine, and sacrificed by cervical dislocation (*see* **Note 9**).
16. After sacrificing the animal, the imaged lymph node can be harvested and subsequently analyzed by flow cytometry or fixed and analyzed by histology, if desired (Fig. 5) (*see* **Note 10**).

3.6 Intravital Imaging

1. Ensure that the lasers and detectors are functioning optimally by quickly imaging a known pre-stained test or calibration slide. Check the basic imaging conditions (*see* **Note 11**).
2. Fluorescent proteins that emit in red or longer wavelengths typically are optimally excited at wavelengths >1050 nm; this is beyond the range of most TiSa lasers and thus requires an

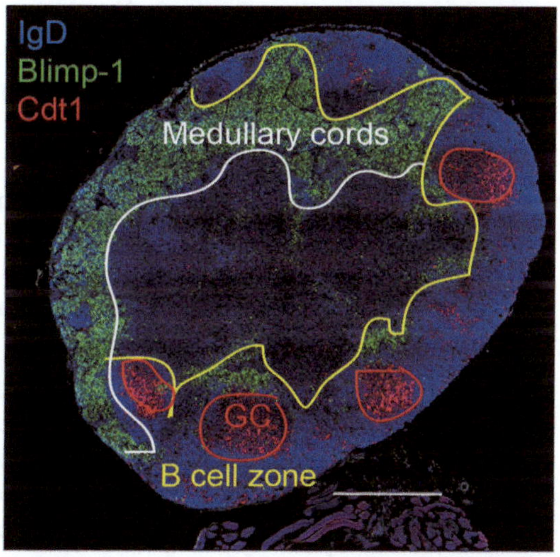

Fig. 5 Overview of the position of B cell subsets in the popliteal lymph node. Naive B cells (*blue*) surround germinal centers (*red*). Plasma blasts/cells (*green*) can be found at the T-B border and in medullary cord areas. Histological section from a mouse that received an adoptive transfer of Cdt1:KusOrange (Fucci-red)/ Blimp-1:GFP double reporter B cells, was immunized as described in Fig. 2 and analyzed by confocal microscopy at day 6 after immunization. Naive B cells were stained with IgD (*blue*). *Scale bar* 500 μm

OPO. This provides better excitation at longer wavelengths, with the benefit of increased tissue penetration and decreased phototoxicity.

3. Find the area of interest within the LN by eye using the injected labeled FDC marker, or by the presence of large, round plasma cells in the medullary cords. Minimize LN exposure to short-wavelength illumination from the arc lamp.
4. Determine the size of the GC, and acquire a stack of images encompassing as much of the GC as possible, typically 500 × 500 μm, 3–5 μm spacing in Z, for a total depth of 50–70 μm. The excitation power can be adjusted as a function of depth to maintain approximately equal illumination intensities in all image planes.
5. Acquire a time series of image stacks, from 20 to 60 min, with acquisition every 15–30 s. Longer durations between successive image stacks will compromise the tracking of rapidly moving cells. Perfusion of the lymph node can be checked by intravenously injecting fluorescent dextran or quantum dots, which are retained in the vasculature (*see* **Note 10**).

3.7 Analysis of Imaging Data

Two recent reviews [12, 13] give an excellent, detailed explanation of cell tracking and data analysis, especially from intravital imaging experiments. Our workflow is very similar, and further details can be found therein. Open an image series, and verify the integrity of the preparation: limited or no gross image drift; no visible photo-damage; control cells maintain their morphology throughout the entire image sequence (Fig. 6).

3.7.1 Define the Cells of Interest and Track Them over Time

1. Define a combination of cell size and signal intensity that distinguishes the cells from background over the entire imaging volume and time. It is better to err on the side of too low a threshold, as the spots will be manually reviewed: it is easier to remove incorrect spots than to add spots manually.
2. Track the cells over time using the cell tracking function in the data analysis program of choice (Imaris, Volocity, CellProfiler, etc.). We strongly prefer automated tracking, as it is less subject to inadvertent user-introduced bias. However, in cases with few or rare cells, it may be sufficient to track cells manually. This can be done using the Manual Tracking or MtrackJ plugins in ImageJ.

3.7.2 Verify Manually that the Tracks are Correct and Representative

Visually inspect all the tracks to ensure they are correct. Common errors made by cell tracking software include: classifying cells close together as one large cell, leading to switched tracks, and not connecting all data points from one cell, leading to the fragmentation of one track into many. Incorrectly classified cells and tracks should be manually split and connected by hand.

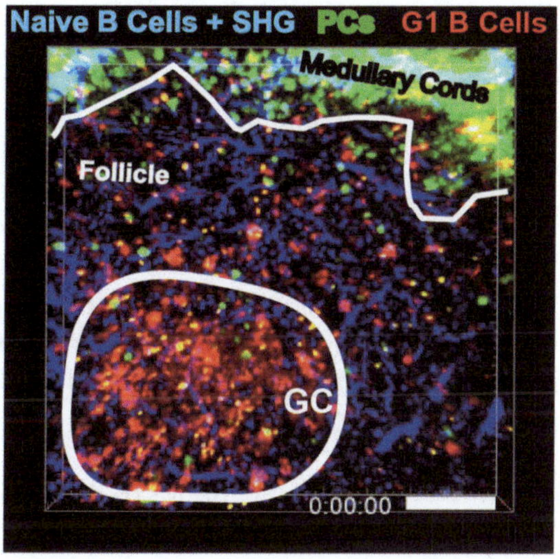

Fig. 6 Tracking plasma blasts/plasma cells and germinal center B cells by intravital microscopy. B cells in the G1 phase of the cell cycle exhibit orange fluorescence, whereas antibody secreting cells characterized by the expression of Blimp-1, are shown by green fluorescence. Nuclei from naive B cells are labeled in *blue*, and collagen fibers also appear blue fluorescent due to second harmonic generation (SHG). *Scale bar* 100 μm

3.7.3 Export the Parameters of Interest

1. From the spatial coordinates measured in the process of cell tracking, it is possible to define various motility parameters, such as mean velocity, displacement, and the directionality of movement. This can be done in the analysis software, or can be done in Microsoft Excel or any other statistical analysis program.
2. The duration of cell-cell contact is a critical determinant of much intracellular communication. Contacts between adjacent cells can be tracked by hand, to record their duration. Alternatively, if the cells are close and bright enough that they give signal in the same pixels, a co-localization channel can be defined and measured to quantify the area, duration, and frequency of cell-cell contacts. This can be done by defining the cells as "Surfaces" in Imaris.

4 Notes

1. Instead of using reporters for labeling GC B cells, GCs can also be visualized as non-fluorescent, round areas that are surrounded by naive B cells in follicular regions [14], and that contain FDCs (as visualized by FDC labeling).

2. For labeling of FDCs, CD35/clone 8C12 or FCD-M1/clone 4C11 [15] can be used instead of CD21/35. As an alternative to ATTO dyes, Alexa dyes can be used; it is critical to choose combinations of fluorescent markers with compatible excitation wavelengths and minimal spectral overlap. If you want to locate the imaging region of interest by eye depending on your FDC staining, note that not near-infrared fluorescent dyes are not visible through the ocular.
3. Footpad injection is appropriate when imaging of the popliteal lymph node is performed. An alternative is imaging of the inguinal lymph nodes [16], where flank immunization is performed.
4. The precise dosing for injectable and inhaled anesthetics will vary by mouse strain and substrain; for instance, Balb/c mice typically have a longer duration of anesthesia, and B10/BR mice are markedly more sensitive to ketamine/xylazine anesthesia. The depth of anesthesia must therefore be frequently assessed. Alternatively, inhaled anesthetics such as isoflurane may be used, with continuous monitoring of respiratory or cardiac function; these provide the ability to more rapidly titrate the anesthesia dosage to the animal's degree of sedation.
5. Before starting the experiment, you should warm up the surrounding technical instruments (objective, stage heater) to a value above 37 °C (40–42 °C), depending on the temperature in the microscope room. Check periodically by means of the temperature probe placed on the tissue near the lymph node. As you proceed with the experiment, you might have to decrease the temperature again after a while, as the operating laser will heat up your experimental set up. It is also possible to connect the thermocouple directly to the heating controller to maintain a constant temperature.
6. When shaving the back and the leg of the animal, take care not to apply too much pressure, especially on the area of the skin, under which you expect the lymph node. Particularly, you should pay attention when removing the depilatory cream with the applicator, to avoid causing a hematoma that will negatively influence your experiment.
7. Prefilling a certain amount of vacuum grease in a 50 ml syringe (without needle) will drastically improve handling as you might wish to operate one-handed, so you can use your other hand for modeling the immersion tub. Furthermore, you will be able to use it more precisely, as the narrow opening of the syringe doses the grease accurately.
8. Many adjustments will be necessary when determining how to best place the mouse in the stage in respect to the various fixators; you might have to change angles and heights, or the

position in xy direction. This should be done before fixing the toe of the mouse and before building the supporting pile underneath the knee. Ensure that the lymph node is centered under the coverslip. Furthermore, the leg should be absolutely planar to the imaging surface. This is sometimes hampered by the position of the tail. You might fix it to the stage with tape.

9. Always take a considerable amount of acetone (e.g., 50 ml tube) with you. After ending the experiment you should put any instruments that got in contact with superglue into the solvent, so remaining fragments of glass from the coverslip will be removed easily.
10. In order to verify proper perfusion of the lymph node, it is recommended to inject either fluorochrome coupled high molecular weight dextran or Qdots, which remain intravascular and can be used to demarcate blood vessels.
11. You should verify beforehand that the imaging area fits your experimental conditions. That means: Do you have enough working distance under your objective to fit the stage with the animal on top? Do the heating elements or fixators somehow prevent the objective from moving freely? Are there cables or tubing providing excessive tension, leading to drift of the imaging area?

Acknowledgement

We would like to thank Raluca Niesner for support with intravital microscopy and Robert Günther and Ralf Uecker for excellent technical assistance. A.E.H is supported by grants from the DFG: TRR130 TP17, HA5354/6-1 (FOR2165) HA5354/8-1 and EXC 257 NeuroCure.

References

1. Mempel TR, Scimone ML, Mora JR, von Andrian UH (2004) In vivo imaging of leukocyte trafficking in blood vessels and tissues. Curr Opin Immunol 16(4):406–417. doi:10.1016/j.coi.2004.05.018
2. Allen CD, Okada T, Tang HL, Cyster JG (2007) Imaging of germinal center selection events during affinity maturation. Science 315 (5811):528–531. doi:10.1126/science.1136736
3. Fooksman DR, Schwickert TA, Victora GD, Dustin ML, Nussenzweig MC, Skokos D (2010) Development and migration of plasma cells in the mouse lymph node. Immunity 33 (1):118–127. doi:10.1016/j.immuni.2010.06.015
4. Kallies A, Hasbold J, Tarlinton DM, Dietrich W, Corcoran LM, Hodgkin PD, Nutt SL (2004) Plasma cell ontogeny defined by quantitative changes in blimp-1 expression. J Exp Med 200(8):967–977. doi:10.1084/jem.20040973
5. Srinivas S, Watanabe T, Lin CS, William CM, Tanabe Y, Jessell TM, Costantini F (2001) Cre reporter strains produced by targeted insertion of EYFP and ECFP into the ROSA26 locus. BMC Dev Biol 1:4
6. Luche H, Weber O, Nageswara Rao T, Blum C, Fehling HJ (2007) Faithful activation of an extra-bright red fluorescent protein in "knock-in" Cre-reporter mice ideally suited for lineage tracing studies. Eur J Immunol 37(1):43–53. doi:10.1002/eji.200636745

7. Herz J, Siffrin V, Hauser AE, Brandt AU, Leuenberger T, Radbruch H, Zipp F, Niesner RA (2010) Expanding two-photon intravital microscopy to the infrared by means of optical parametric oscillator. Biophys J 98 (4):715–723. doi:10.1016/j.bpj.2009.10.035
8. Bubnic SJ, Nagy A, Keating A (2005) Donor hematopoietic cells from transgenic mice that express GFP are immunogenic in immunocompetent recipients. Hematology 10(4):289–295. doi:10.1080/10245330500093468
9. Hauser AE, Junt T, Mempel TR, Sneddon MW, Kleinstein SH, Henrickson SE, von Andrian UH, Shlomchik MJ, Haberman AM (2007) Definition of germinal-center B cell migration in vivo reveals predominant intrazonal circulation patterns. Immunity 26(5):655–667. doi:10.1016/j.immuni.2007.04.008
10. Roth K, Oehme L, Zehentmeier S, Zhang Y, Niesner R, Hauser AE (2014) Tracking plasma cell differentiation and survival. Cytometry A 85(1):15–24. doi:10.1002/cyto.a.22355
11. Aiba Y, Kometani K, Hamadate M, Moriyama S, Sakaue-Sawano A, Tomura M, Luche H, Fehling HJ, Casellas R, Kanagawa O, Miyawaki A, Kurosaki T (2010) Preferential localization of IgG memory B cells adjacent to contracted germinal centers. Proc Natl Acad Sci U S A 107 (27):12192–12197. doi:10.1073/pnas.1005443107
12. Kitano M, Okada T (2012) Four-dimensional tracking of lymphocyte migration and interactions in lymph nodes by two-photon microscopy. Methods Enzymol 506:437–454. doi:10.1016/B978-0-12-391856-7.00047-0
13. Meijering E, Dzyubachyk O, Smal I (2012) Methods for cell and particle tracking. Methods Enzymol 504:183–200. doi:10.1016/B978-0-12-391857-4.00009-4
14. Qi H, Cannons JL, Klauschen F, Schwartzberg PL, Germain RN (2008) SAP-controlled T-B cell interactions underlie germinal centre formation. Nature 455(7214):764–769. doi:10.1038/nature07345
15. Schwickert TA, Lindquist RL, Shakhar G, Livshits G, Skokos D, Kosco-Vilbois MH, Dustin ML, Nussenzweig MC (2007) In vivo imaging of germinal centres reveals a dynamic open structure. Nature 446(7131):83–87. doi:10.1038/nature05573
16. Lindquist RL, Shakhar G, Dudziak D, Wardemann H, Eisenreich T, Dustin ML, Nussenzweig MC (2004) Visualizing dendritic cell networks in vivo. Nat Immunol 5 (12):1243–1250. doi:10.1038/ni1139

Chapter 4

Microanatomical Labeling of Germinal Center Structures for Flow Cytometry Using Photoactivation

Johanne T. Jacobsen and Gabriel D. Victora

Abstract

Germinal centers (GCs) are inducible microanatomical structures required for the generation of high-affinity antibodies. Migration of B and T cells within and into/out of GCs plays a key role in the evolutionary process that underlies affinity maturation, and thus microanatomical location of cells is an important variable when studying GC processes. We describe a protocol in which in situ photoactivation by multiphoton microscopy can be used to add microanatomical information to flow cytometry, allowing for identification and isolation of GC cells based on their location. Cells in different microanatomical compartments can then be sorted and analyzed for surface marker and mRNA expression.

Key words B cell, Flow Cytometry, Germinal Center, Multiphoton Microscopy, Photoactivation

1 Introduction

Antibody affinity maturation takes place in germinal centers (GCs) through a competitive process in which B cells which have acquired higher affinity through somatic hypermutation are selected to proliferate in the GC as well as to differentiate into plasma cells and memory B cells [1, 2]. Selection of high affinity B cells in the GC is intimately linked to microanatomy. Within individual GCs, B cells migrate from light zone to dark zone as affinity maturation progresses, and successful clones expand as others disappear. At the level of the secondary lymphoid organ, ensembles of several GCs evolve in parallel, which has important consequences to the selective process [3]. Given the key role of these anatomical features, distinguishing cells based on their location has become an important goal in the GC field.

We have developed a method in which this microanatomical information can be read out by flow cytometry by "optical highlighting" of areas of interest using in situ photoactivation of PA-GFP (for photoactivatable green fluorescent protein) [4]. PA-GFP is a photoactivatable variant of the *Aquorea victoria* green

Dinis Pedro Calado (ed.), *Germinal Centers: Methods and Protocols*, Methods in Molecular Biology, vol. 1623,
DOI 10.1007/978-1-4939-7095-7_4, © Springer Science+Business Media LLC 2017

fluorescent protein that changes spectral properties in response to irradiation with a specific wavelength of light, converting from a low to a bright fluorescent state with a ~ 100 fold increase in fluorescence intensity [5]. Peak excitation wavelength shifts permanently from ~415 nm to ~495 nm upon one-photon irradiation at ~415 nm or two-photon irradiation at ~720–850 nm [5, 6]. We generated a transgenic mouse which expresses high levels of PA-GFP in all hematopoietic cells (and potentially in other cell types as well, although this has not been verified systematically) [4]. Cells within lymph nodes of these mice can be readily photoactivated with microanatomical precision using a multiphoton laser, both in living mice and in explanted organs. Photoactivated nodes are then processed into single-cell suspensions for flow cytometry, and positivity for the photoactivation signal (observable using an EGFP fluorescence channel) is indicative that cells originated from the photoactivated volume of interest. In recent years, we have used this method to (1) isolate B cells from GC light and dark zones [4]; (2) label T cells based on residency within GCs [7]; and (3) isolate B cells from individual GCs (rather than from the multiple GCs normally found in a LN) [3]. These studies have yielded information on the phenotypes and clonal identities of cells based on their anatomical location rather than simply on expression of cell surface markers, providing a further layer of resolution for the study of GC cellular dynamics.

2 Materials

2.1 Mice Immunization and Dissection

1. UBC PA-GFP mouse (Jackson 022486) [4].
2. 4-hydroxy-3-nitro-phenylacetyl-ovalbumin (NP-OVA) or antigen of choice.
3. Imject Alum (Thermo Scientific) or adjuvant of choice.
4. Phosphate-buffered saline (PBS).
5. Anti-CD35 (clone 8C12) conjugated to Texas Red X for labeling of GC follicular dendritic cell (FDC) networks [8].

2.2 Preparation of Lymph Nodes for Photoactivation

1. Dissecting microscope.
2. Lens cleaning paper.
3. High vacuum grease (Dow Corning)
4. Coverslips (22 × 22 mm), 0.16 mm thickness.
5. Mouse dissection tools: forceps, scissors.
6. Fine-tipped Dumont forceps.
7. Aluminum block (e.g., VWR).
8. Syringes: 5-ml syringe fitted with 16G cut and blunted tip for dispensing vacuum grease and a 10-ml syringe with cut and blunted 19G tip for dispensing PBS.

2.3 Flow Cytometry and Microscopy

1. PBS supplemented with 0.5% BSA and 1 mM EDTA (PBE).
2. 1.5 ml microfuge tubes.
3. Plastic pestle (e.g., Axygen).
4. Nylon mesh (100 μm) for filtering samples (e.g., Sefar Lab Pak), cut into 1× 1 cm squares.
5. Anti-mouse CD16/32 (Fc Block).
6. Antibodies for flow cytometry as needed. A possible panel of antibodies is: anti-mouse B220 Brilliant Violet 421 labeled; anti-mouse TCRβ APC-Cy7 labeled; anti-mouse CD38-Alexa 700 labeled, anti-mouse FAS/CD95 PE-Cy7 labeled.
7. Multiphoton microscope (*see* **Note 1**).
8. Flow cytometer/cell sorter (*see* **Note 2**)

3 Methods

The most challenging part of this method is determining the correct intensity of irradiation for optimal photoactivation without photodamage, and one should expect to spend some effort optimizing this process. This needs to be done empirically for each system, by altering parameters such as laser power and wavelength, pixel dwell-time and/or image resolution, and number of consecutive Z-stack passes. It is not uncommon for photoactivation of deeper structures (e.g., >150 μm below the capsule) to require 100% laser power with long dwell-times (e.g., 20 μs/pixel at 0.53 μm^2/pixel) to achieve the level of photoactivation required for detection by flow cytometry.

3.1 Induction of Germinal Centers (See Note 3)

1. Prepare a stock of NP-OVA in PBS at 1 μg/μl. In a 1.5 ml tube, combine 200 μl NP-OVA stock with 133 μl of PBS.
2. Resuspend alum thoroughly by vortexing.
3. Place tube with NP-OVA on a vortexer at slow speed and add 167 μl of alum dropwise while vortexing.
4. Close lid and vortex on maximum speed for an additional 10 s.
5. Incubate for 30 min at 4 °C on tube rotator.
6. To induce GCs in the pLN inject 25 μl per footpad (10 μg of antigen in alum).

3.2 Labeling FDC Networks

To locate GCs on the multiphoton microscope prior to photoactivation, the FDC network can be labeled using a non-blocking antibody to CD35 [8]. Inject each mouse intravenously with 10 μg of anti-CD35 monoclonal antibody (clone 8C12) conjugated to Texas Red X 1 day before analysis (*see* **Note 4**).

3.3 Lymph Node Preparation

1. Cover a flat surface with lens-cleaning paper.
2. Moisten paper with phosphate-buffered saline (PBS).
3. Euthanize the mouse according to your institution's guidelines.
4. Explant the popliteal lymph node (pLN) using mouse dissection tools, and transfer to the moistened lens-cleaning paper.
5. Remove fat and excess tissue from the cortical side of the pLN under a dissection microscope, using the Dumont forceps. Normally the medullary side is curved inward and displays visible blood vessels, whereas there are no visible vessels on the cortical side.
6. Apply vacuum grease to a coverslip as a central "ring" and four "posts," one in each corner (as shown in Fig. 1a).
7. Place the pLN in the central grease ring, with the cortical side facing upwards.
8. Add a drop of PBS to the central ring (Fig. 1b), then place a coverslip over the mounted node, pushing downwards until the cortical surface of the node is flat and the vacuum grease is touching both coverslips (Fig. 1c). Use the four grease "posts" in each corner to ensure that both coverslips are parallel.
9. Place the mounted pLN on a chilled aluminum block, using a droplet of water to attach the coverslip to the block by surface tension. Place the aluminum block in a tub of ice. The pLN is now ready for imaging (Fig. 1d).

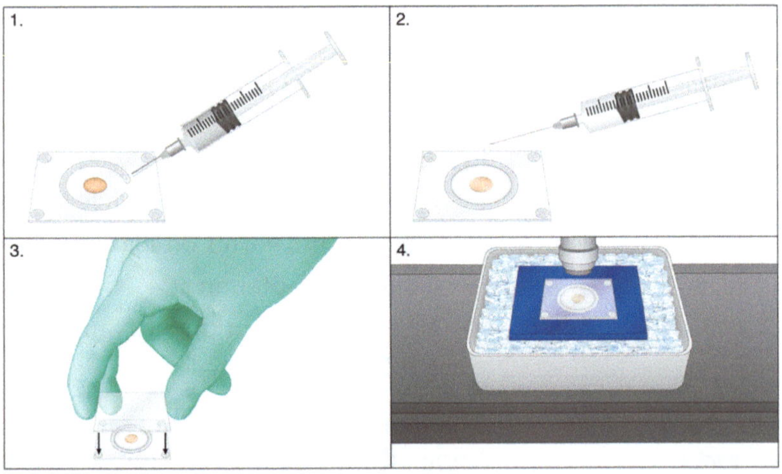

Fig. 1 LN preparation. (*1*) Place LN on coverslip and apply vacuum grease; (*2*) add a drop of PBS into the central vacuum grease ring, enough to cover the LN; (*3*) place a second coverslip over the LN; (*4*) place the mounted LN on an aluminum block in an ice bath, then onto the microscope stage

3.4 Microscopy and Photoactivation

1. Using the epifluorescence lamp, find the approximate plane of the LN cortex.
2. Set the multiphoton laser to 950 nm (the non-photoactivating wavelength) and scan the lymph node to locate a suitable GC (*see* **Note 5**). Once the GC is located, draw a region of interest (ROI) around the FDC network.
3. Locate the top and bottom of the volume to be photoactivated by scanning. Set a Z stack covering this volume with 5-μm steps. While the dense FDC staining can be used to determine the upper boundary of the GC, the DZ is more readily recognized based on autofluorescence from tingible body macrophages.
4. Set the appropriate scan speed and pixel density for photoactivation, then set the laser power at the upper and lower boundaries of the Z stack and turn on laser power interpolation (*see* **Note 6**).
5. Tune the multiphoton laser between 820 and 850 nm (exact wavelength should be determined empirically) and scan Z stack. More than one scan may be necessary.
6. To verify photoactivation, return the laser to 950 nm and turn off the ROI. A clearly distinguishable signal should be visible in the EGFP/YFP channel (Fig. 2). If a signal is not clearly visible, repeat **steps 2–5**.

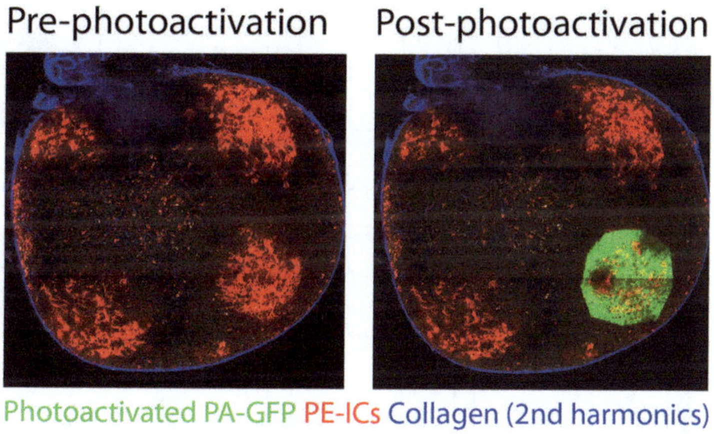

Fig. 2 GC photoactivation. In this example, one GC in a pLN of a NP-OVA-immunized mouse was photoactivated by drawing an ROI in the X/Y-plane and scanning a Z-stack with 5 μm steps, at a wavelength of 820 nm. Images show the same LN scanned at 950 nm wavelength prior to and after photoactivation. Tiled images of a single Z-plane are shown. In this specific case, FDCs were labeled using NP-tdTomato [8]

3.5 Flow Cytometry (See Note 7)

1. Remove pLN from coverslip chamber, place in 1.5-ml microfuge tube with 100 μl of PBE + 1 μl FcBlock and flow cytometry antibodies at 2× the final dilution. Our typical flow cytometry staining for photoactivated GC B cells is the following: anti-mouse B220 Brilliant Violet 421 labeled at 1/400 dilution; anti-mouse TCRβ APC Cy7 labeled at 1/200 dilution; anti-mouse CD38-Alexa 700 labeled at 1/200 dilution; and anti-mouse FAS/CD95 PE-Cy7 labeled at 1/800 dilution. Using this fluorochrome combination allows two channels to be left open for PAGFP: non-photoactivated (AmCyan channel on BD cytometers); compensated using unstained cells from PA-GFP-transgenic mouse; and photoactivated PAGFP (FITC channel on BD cytometers); compensated using a FITC-stained control (Fig. 3).
2. Carefully macerate pLN using a plastic pestle against the tube wall, and rinse the pestle with 100 μl PBE (so that flow cytometry antibodies are now 1×).
3. Incubate for 30 min at 4 °C in the dark.
4. Add 1 ml PBE, filter through a nylon mesh into a clean 1.5 ml tube, centrifuge on a benchtop microfuge at 7000 × *g* for 15 s.
5. Resuspend cell pellet in 400 μl of PBE, transfer to an appropriate tube, and run on flow cytometer/cell sorter.

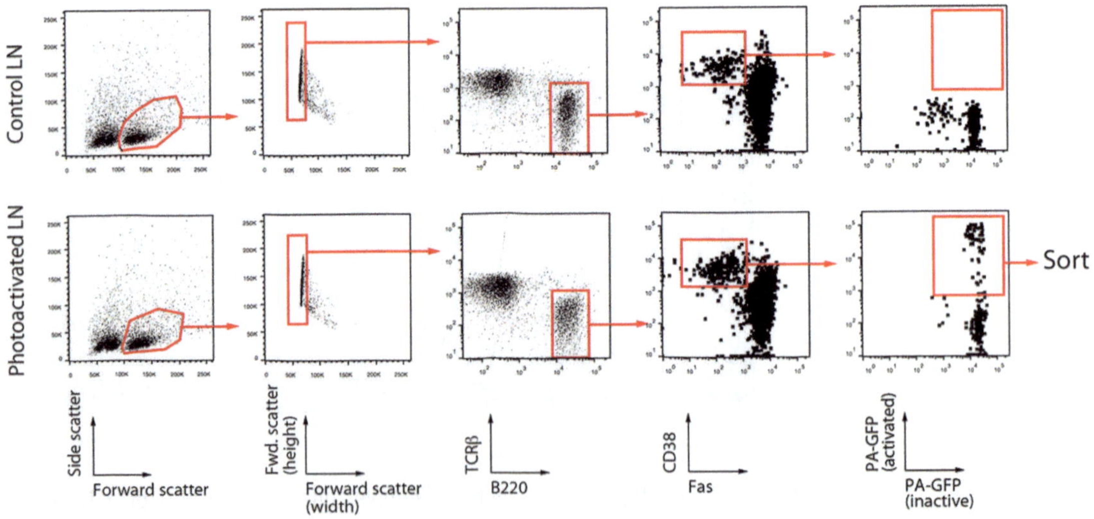

Fig. 3 Flow cytometry of photoactivated GCs. Figure shows gating strategy for sorting photoactivated cells from LN, gating sequentially on live cells, singlets, B cells, and GC B cells. The photoactivation gate (*rightmost panel*) is first set using a non-photoactivated control LN (*top*), then used to sort photoactivated cells (*bottom*)

4 Notes

1. Although different multiphoton setups can be used, ours consists of an Olympus FV1000 upright microscope fitted with a 25 × 1.05NA Plan water-immersion objective and a Mai-Tai DeepSee Ti-Sapphire laser (Spectraphysics). For photoactivation experiments, we use the following filters: a pair of CFP (480/40 nm) and YFP (525/50 nm) filters, separated by a 505-nm dichroic mirror and a third filter (605/70 nm) for RFP detection.
2. Most setups can be used to detect photoactivation. The minimum requirement is a 488 laser with a FITC/GFP setup. Availability of other channels will improve the identification of populations and measurement of different parameters. We recommend the standard configurations of the BD FACS Aria or FACS Fortessa.
3. The amounts of immunization reagents in this protocol is sufficient to inject ~20 mice. Even if lower numbers are needed, this is the minimum amount that can be generated reproducibly given the viscosity of the alum suspension. GCs induced using this protocol can first be observed ~6 days after immunization, lasting until at least ~20 days after immunization.
4. The GC LZ can be visualized as the area of densest FDC staining proximal to the LN collagen capsule. The DZ is the region of dim CD35 staining below the LZ (more distal to the capsule). Both zones are characterized by the presence of tingible body macrophages, which can be distinguished by their characteristic autofluorescence pattern (*see*, for example, ref. 4).
5. Because some photoactivation can occur even at 950 nm with higher laser power, start scanning with the lowest possible power, and increase to the minimum necessary to visualize the collagen capsule (second harmonics) and the FDC network (antibody stain).
6. Setting additional points at different depths may be required if the microscope software does not allow interpolation.
7. The yield of photoactivated cells from a single pLN is usually low, follow Subheading 3.5 to minimize cell loss.

References

1. Victora GD, Nussenzweig MC (2012) Germinal centers. Annu Rev Immunol 30:429–457. doi:10.1146/annurev-immunol-020711-075032
2. Allen CD, Okada T, Cyster JG (2007) Germinal-center organization and cellular dynamics. Immunity 27(2):190–202
3. Tas JM, Mesin L, Pasqual G, Targ S, Jacobsen JT, Mano YM, Chen CS, Weill JC, Reynaud CA, Browne EP, Meyer-Hermann M, Victora GD (2016) Visualizing antibody affinity maturation in germinal centers. Science 351(6277):1048–1054. doi:10.1126/science.aad3439

4. Victora GD, Schwickert TA, Fooksman DR, Kamphorst AO, Meyer-Hermann M, Dustin ML, Nussenzweig MC (2010) Germinal center dynamics revealed by Multiphoton microscopy with a photoactivatable fluorescent reporter. Cell 143(4):592–605. doi:10.1016/j.cell.2010.10.032
5. Patterson GH, Lippincott-Schwartz J (2002) A photoactivatable GFP for selective photolabeling of proteins and cells. Science 297(5588):1873–1877. doi:10.1126/science.1074952
6. Schneider M, Barozzi S, Testa I, Faretta M, Diaspro A (2005) Two-photon activation and excitation properties of PA-GFP in the 720-920-nm region. Biophys J 89(2):1346–1352. doi:10.1529/biophysj.104.054502
7. Shulman Z, Gitlin AD, Targ S, Jankovic M, Pasqual G, Nussenzweig MC, Victora GD (2013) T follicular helper cell dynamics in germinal centers. Science 341(6146):673–677. doi:10.1126/science.1241680
8. Fooksman DR, Schwickert TA, Victora GD, Dustin ML, Nussenzweig MC, Skokos D (2010) Development and migration of plasma cells in the mouse lymph node. Immunity 33(1):118–127. doi:10.1016/j.immuni.2010.06.015

Chapter 5

Fate Mapping and Transcript Profiling of Germinal Center Cells by Two-Photon Photoconversion

Imogen Moran and Tri Giang Phan

Abstract

The germinal center (GC) reaction is the key process for the generation of high affinity antibodies to foreign antigen. Standard experimental techniques such as fluorescence-activated cell sorting and histology have provided numerous insights into the composition and function of the GC. However, these approaches are limited to a "snapshot" in time and are unable to fully capture the dynamic nature of the GC. Intravital two-photon microscopy overcomes these disadvantages and has led to several major advances in the field but is restricted by practical and technical limits that prevent long-range mapping and molecular studies. Here we describe procedures for optical marking or "tagging" of cells in precise microanatomical compartments by two-photon photoconversion that can be used for long-term fate mapping and transcript profiling of GC T and B cells.

Key words Two-photon microscopy, Two-photon photoconversion, Fate mapping, Germinal center, Cell migration

1 Introduction

The germinal center (GC) is a transient structure that forms within the B cell follicle of secondary lymphoid organs where B cells undergo somatic hypermutation and antigen-based selection for BCR specificity and affinity [1, 2]. GCs were originally described by histology to consist of a proliferating dark zone (DZ) of centroblasts and nonproliferating light zone (LZ) of centrocytes. Within the LZ, GC B cells may interact with antigen trapped in the form of immune complexes on the surface of follicular dendritic cells (FDCs) that express high levels of complement and Fc receptors [1]. In the LZ, GC B cells may also interact with T follicular helper (Tfh) cells that provide critical cytokines and costimulatory molecules to support their differentiation, survival and growth [3]. In the classical model, GC B cells improve their affinity for antigen by repeatedly cycling between the DZ (where they proliferate and mutate their BCR), and the LZ (where they are selected to

Dinis Pedro Calado (ed.), *Germinal Centers: Methods and Protocols*, Methods in Molecular Biology, vol. 1623, DOI 10.1007/978-1-4939-7095-7_5, © Springer Science+Business Media LLC 2017

differentiate into long-lived plasma cells or memory cells). GC B cells that fail to compete for antigen and T cell help, or have acquired self-reactivity undergo apoptosis and are phagocytosed by tingible body macrophages (TBMs) located at the junction between the DZ and LZ. However, this complex cellular dynamics within the GC has meant that traditional experimental techniques such as fluorescence activated cell sorting (FACS) and histology can only provide snapshots of the underlying cellular processes.

Intravital two-photon microscopy (TPM) has become the gold standard for minimally invasive imaging of cell migration and interactions deep inside intact organs of live animals [4, 5] and provided key insights into the major events in B cell activation in the humoral immune response [6]. Since the first landmark studies that visualized cycling of GC B cells between LZ and DZ [7–9], there has been much progress in our understanding of GC dynamics. Thus, $CD169^{+}$ subcapsular sinus macrophages were shown to capture and present antigen to B cells to initiate the antibody response in the lymph node [10–14]. Activated B cells then relocalized to the T-B border and interfollicular zones where they interacted with T helper cells [15–17]. This interaction between T and B cells is dependent on the SLAM-associated protein (SAP) [18] and results in ICOS-mediated migration of Tfh cells [19] into the follicle and formation of the GC. Inside the GC, Tfh cells become "entangled" with GC B cells in multiple brief but extensive surface contacts [20, 21]. More recently, the migration pattern of primary [22, 23] and secondary Tfh cells [22, 24] have also been described.

Despite these advances, TPM does suffer from some practical limitations that make it difficult to achieve long-range fate mapping of cells over the course of the immune response [5]. In addition, while TPM is able to reveal dynamic changes in the location and behavior of the cells, it is unable to provide the molecular mechanisms that drive these changes. To overcome these limitations, investigators have developed strategies for in vivo optical marking ("tagging") of cells in precise microanatomical locations using two-photon photoactivation of PA-GFP [25] and two-photon photoconversion (TPP) of Kaede [22, 26] and Kikume [27]. Optically marked cells can be tracked for long distances to distant organs following recovery of mice from anesthesia [26] and cells can be isolated for microarray [25] or single cell transcript analysis [22]. Our lab has focused on the Kaede system [28] because the cells are visible prior to irreversible photoconversion making it possible to image photoconverted (red) and non-photoconverted (green) cells in the same experiment. Furthermore, the photoconverted Kaede protein is more stable than photoactivated GFP with an estimated half-life of >1 week for Kaede [26] compared to 30 h for PA-GFP [25]. Antigen-specific B and T cell responses can thus be tracked by breeding Kaede mice with immunoglobulin knockin mice such the

SW_{HEL} mouse [29, 30] and TCR transgenic mice such as OT2 mouse [31]. This protocol describes the procedure of TPP for fate mapping and transcript profiling of germinal center T and B cells.

2 Materials

2.1 Mice

1. Kaede transgenic mice (*see* **Note 1**).
2. BCR knockin SW_{HEL} B cells and OT2 TCR transgenic $CD4^+$ T cells (*see* **Note 2**).
3. SAP-deficient or wild-type mice as recipients (*see* **Note 3**).

2.2 Adoptive Cell Transfer

1. Dissecting instruments: dissecting scissors, forceps.
2. 2FD: 2% (v/v) fetal bovine serum in DMEM.
3. 15 mL polypropylene tubes (e.g., Falcon).
4. 50 mL polypropylene tubes (e.g., Falcon).
5. 1 mL syringe.
6. 70 μm cell strainer.
7. RBC Lysis Buffer: 0.15 M ammonium chloride (NH_4Cl), 10 mM potassium hydrogen carbonate ($KHCO_3$), 0.1 mM sodium EDTA (Na_2EDTA) in Baxter water, pH 7.2–7.4.
8. Fetal calf serum (FCS).
9. MACs Buffer: 0.5% bovine serum albumin, 2 mM EDTA, 0.09% azide in PBS, filter sterilized.
10. Biotinylated anti-B220 clone RA3-6B2, anti-CD11b clone M1/70, anti-CD11c clone HL3, anti-CD43 clone S7.
11. MACS anti-biotin magnetic beads (Miltenyi).
12. LS columns (Miltenyi).
13. MACs Separator (Miltenyi).
14. Hemocytometer.
15. Trypan Blue.
16. 5 mL round bottom flow cytometry tubes.
17. Lysozyme from chicken egg white (e.g., Sigma).
18. Fc-block anti-CD16/32 clone 2.4G2.
19. Fluorochrome conjugated anti-B220, anti-CD4, anti-$V_{\alpha 2}$ antibodies.
20. HyHEL9 monoclonal antibody.
21. Alexa Fluor 647 Monoclonal Antibody Labeling Kit (Invitrogen) or equivalent.
22. Flow cytometer.
23. FlowJo or equivalent software to analyze flow cytometry data.
24. 0.5 mL insulin syringe for intravenous injection.

2.3 Immunization

1. Antigen for immunization, (*see* **Note 4**).
2. Albumin from chicken egg white (e.g., Sigma), for OVA immunization.
3. $OVA_{323-339}$ peptide (CGGISQAVHAAHAEINEAGR), for conjugation to HEL to make HEL-OVA.
4. Succinimidyl-6-((ß-maleimidopropionamido) [5] hexanoate) (SMPH), for HEL-OVA conjugation.
5. Lysozyme from chicken egg white (e.g., Sigma), for HEL-OVA immunization.
6. Sigma Adjuvant System (SAS).
7. 0.5 mL insulin syringe for subcutaneous injection.

2.4 Labeling Germinal Centers

1. Anti-CD157 clone BP-3, (*see* **Note 5**).
2. Alexa Fluor 555, 647, and 680 labeling kit (Invitrogen) or CF680R kit (Biotium), (*see* **Note 6**).
3. 0.5-mL insulin syringe for subcutaneous injection.

2.5 Preparation of Mouse for Lymph Node Imaging

1. Anesthesia: Ketamine and xylazine diluted as described.
2. Bain circuit for gaseous anesthesia with isoflurane.
3. Custom Biotherm SmartStage (Cryologic).
4. Lacri-lube.
5. Electric razor.
6. Micropore Tape.
7. 70% ethanol.
8. Microdissecting Instruments: dissecting scissors, microscissors, forceps, sutures, needle holders.
9. Stereomicroscope.
10. Polydimethylsiloxane (PDMS) (e.g., Sylgard).
11. T-putty (Thermagon).
12. Vetbond tissue adhesive (3 M).
13. PBS.
14. Silicone grease.
15. Small O-ring.
16. Immersol Immersion Oil (Carl Zeiss).

2.6 Two-Photon Microscope

1. Software-tunable ultrafast Near Infra-Red (NIR) Laser. We use a Chameleon Vision II ultrafast Ti:Sa laser (Coherent Scientific) The laser wavelength and dispersion compensation settings are controlled by the real-time computer, which also controls and synchronizes data acquisition by the microscope.
2. Two-photon microscope. We use an upright Zeiss 7MP two-photon microscope (Carl Zeiss) with a W Plan-Apochromat

20×/1.0 DIC (UV) Vis-IR water immersion objective. Four external non-descanned detectors are used to detect blue (SP-485), green (BP 500–550), and far-red (BP 640–710) emissions. We use the ZEN software interface to enable real-time interactive control of photoconversion parameters.

2.7 Two-Photon Photoconversion

1. Plastic coverslips.
2. Vicryl sutures.
3. Temgesic diluted as described.

2.8 Fluorescence-Activated Cell Sorting (FACS)

1. FACS analyzer for data acquisition, e.g., LSR II SORP or FACS Canto (BD).
2. FACS sorter for single cell sorting, e.g., FACS ARIA (BD).

3 Methods

An overall experimental timeline is shown in (Fig. 1) and example experimental results is shown in (Fig. 2).

3.1 Transfer Fluorescent Cells to Recipient Mice

3.1.1 Harvest Tissue Donor Mice

1. Euthanize fluorescent antigen-specific reporter mice, such as Kaede OT2 and tdTomato SW_{HEL} mice according to local ethical guidelines, (*see* **Note** 7).
2. Harvest spleen into 2 mL 2FD media in 15-mL tube. If additional donor cells are required also harvest inguinal, axillary and brachial lymph nodes

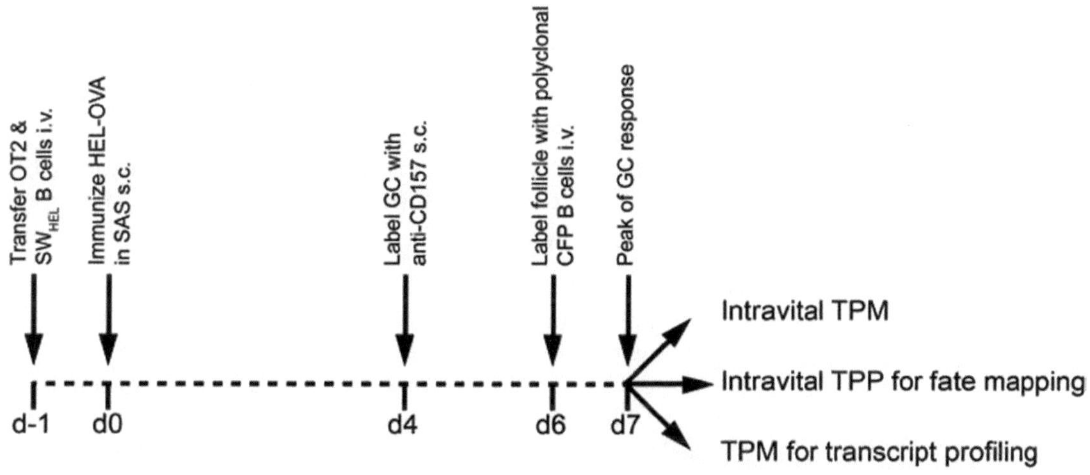

Fig. 1 Example experimental timeline for TPM or TPP of GCs in the inguinal lymph node. Fluorescent antigen-specific Kaede OT2 T cells and Tomato SW_{HEL} B cells are transferred into nonfluorescent recipient mice and immunized subcutaneously with antigen. The GC is labelled with fluorescently labeled anti-CD157 mAb injected 3 day before imaging, and the B cell follicle with polyclonal CFP B cells 1 day before imaging. GCs can then be imaged or photoconverted in the inguinal lymph node

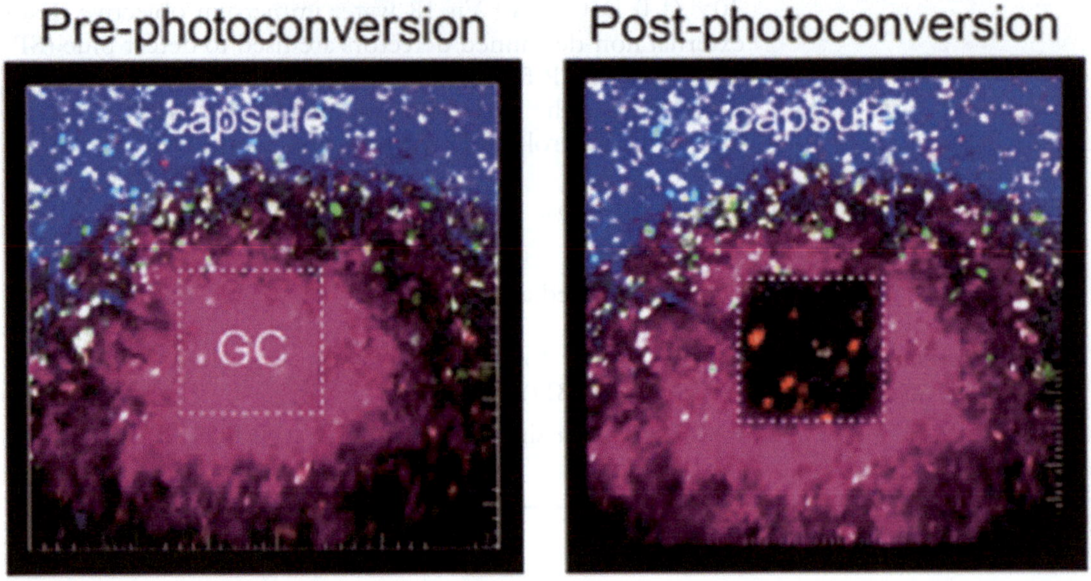

Fig. 2 Example of TPP of Tfh cells in the GC. Kaede OT2 T cells (*green*) were adoptively transferred into recipient mice and immunized with OVA in SAS. The GC was labeled by injection of anti-CD157 AF680 (*magenta*) on d4 and the follicle labeled by adoptive transfer of CFP B cells (*cyan*) on d6. Images of the draining inguinal lymph node were acquired on d7 (*left panel*) before TPP. The ROI (*dashed square*) targeted for photoconversion is shown. Images acquired following conversion (*right panel*) shows photobleaching of the GC label in the ROI which now contains photoconverted (*red*) Kaede OT2 T cells

3.1.2 Cell Preparation

All centrifugation steps for splenocytes in 50-mL tubes are performed at 453 × *g* for 5 min at 4 °C.

1. Prepare single cell suspension of splenocytes in 10 mL 2FD with 70 μm cell strainer and 1 mL syringe plunger.
2. Lyse RBC in 5 mL RBC Lysis buffer and underlay with 1 mL FCS.
3. Centrifuge and resuspend cells in 10 mL MACs buffer.
4. Repeat procedure for lymph nodes, if harvested, without RBC lysis step. Spleen and lymph node cells can then be combined.
5. Count cells and resuspend at 10^8 cells/mL.
6. To enrich donor population of interest, we utilize negative selection with biotinylated antibodies. Cells are stained with the following biotinylated antibodies typically at 5 μg/mL for 30 min on ice: To enrich for CD4 T cells in OT2 donor mice deplete with biotinylated anti-B220, anti-CD11b, anti-CD11c. To enrich for B cells in SW_{HEL} donor mice deplete with biotinylated anti-CD4, anti-CD11b, anti-CD11c, anti-CD43.
7. Wash cells twice in 10 mL MACs buffer

8. Resuspend cells at 10^8 cells/mL and stain with MACS anti-biotin magnetic beads for 30 min on ice. Wash cells twice in 10 mL MACs buffer.
9. Run filtered samples through MACs buffer prewashed LS columns on MACs Separator to isolate the negative fraction donor population of interest.
10. Determine the purity and yield of $CD4^+$ $V_{\alpha 2}{}^+$ OT2 population or $B220^+$ HEL-binding SW_{HEL} population by FACs analysis. Stain small volume of samples in FACs tube with following antibodies for 30 min on ice and wash twice in between staining steps at 453 × *g* for 5 min. For OT2 T cell cells stain: anti-CD4 (fluorochrome labeled) + anti-$V_{\alpha 2}$ (fluorochrome labeled). For SW_{HEL} population stain cells: Fc-block (anti-CD16/32) + HEL (200 ng/mL). Anti-B220 (fluorochrome labeled) + HyHEL9 monoclonal antibody labeled with A647.
11. Count cells after enrichment and determine total number of donor cell population from purity analysis, (*see* **Note 8**).
12. Transfer cells to recipient mice. 2.5×10^5 $CD4^+$ $V_{\alpha 2}{}^+$ fluorescent OT2 T cells and, if using, 2.5×10^5 $B220^+$ HEL-binding fluorescent SW_{HEL} B cells were injected i.v. into age and sex matched recipient mice.

3.2 Immunizations

1. Immunize mice the day after cell transfer.
2. If transferred OT2 T cells immunize subcutaneously with 20 μg of OVA in SAS in the lower flank and base of tail of mouse, (*see* **Note 9**)
3. If transferred both OT2 T cells and SW_{HEL} B cells, immunize subcutaneously with 20 μg HEL-OVA in SAS.
4. Time to generate GCs will depend on antigen utilized but OVA and HEL-OVA responses in this system typically peak around 7 days after primary immunization.

3.3 Labeling FDCs in the GC LZ

1. Label anti-CD157 clone BP-3 with fluorescent protein labeling kit such as Alexa Fluor 680 or CF680R.
2. Inject 40 μg fluorescent anti-CD157 s.c. on right hand imaging side of mouse in the lower flank and base of tail 3 days or more prior to imaging (*see* **Note 5**).

3.4 Labeling the B Cell Follicle

1. Isolate polyclonal B cells from different fluorescent donor color to transferred donor cells, as previously described in Subheading 3.1.
2. ~10 million injected i.v. 18–24 h prior to imaging.

3.5 Preparation of Mouse for Lymph Node Imaging

1. Induce anesthesia in the mouse with 100 mg/kg ketamine/ 5 mg/kg xylazine. Maintain anesthesia on 1–2% isoflurane supplemented with 100% oxygen.
2. Keep anesthetized mouse warm on a heating pad set to 37–38 °C.
3. Apply Lacri-Lube to eyes of mouse to prevent dryness.
4. Use an electric razor to shave hair on right flank and inguinal area. Use micropore tape to remove all hair from the area, (*see* **Note 10**).
5. Sterilize the skin with 70% ethanol and use sterile scissors to bluntly dissect a skin flap on the right hand side that exposes the inguinal lymph node and the inguinal ligament.
6. Apply T-putty to base of PDMS polymer (*see* **Note 11**) and stick to base of heating pad.
7. Stick skin flap, with skin to outside, down on PDMS, making sure the inguinal lymph node is in the middle of the PDMS not touching the VetBond glue.
8. Apply T-putty to seal the outside of skin flap. This will further fix the skin flap and also conduct heat from the heating pad.
9. Make a window in the skin overlying the inguinal lymph node using microscissors then apply PBS to keep the lymph node moist.
10. Carefully separate the lymph node from overlying fat and fascia layers to expose the cortical surface under a stereomicroscope with low-level illumination, (*see* **Note 12**).
11. Apply silicone grease to a small O-ring to adhere to skin around inguinal lymph node (*see* **Note 13**).
12. Put PBS in the O-ring to create a meniscus. Immersion oil may provide additional meniscus stability.

3.6 Two-Photon Microscopy

1. Transfer the mouse on heat pad to the microscope stage and reattach isoflurane nose cone to maintain anesthesia.
2. Set the microscope objective so that it is over the lymph node.
3. Tune the laser to the appropriate wavelength and adjust the power so that cells are visible, (*see* **Note 14**).
4. Set the upper and lower limits of the Z stack, typically 150 μm, and the number of images in the stack, typically step every 3 μm, (*see* **Note 15**).
5. Acquire image stacks at 30 s time intervals. Typically image for 30 min movies which is 60 cycles. Monitor drift and stop acquisition and check setup if large amount of drift is occurring.
6. Frequently monitor the mouse to check depth of anesthesia.

3.7 Real-Time Interactive Two-Photon Photoconversion for Fate Mapping

1. TPP is best performed with real-time interactive feedback to optimize the laser power and irradiation time for maximal photoconversion with minimal toxicity and photobleaching.
2. Locate cells of interest by scanning at 920 nm.
3. Set region of interest (ROI) to contain target cells. ROIs can be scaled from whole field of view down to single cell level by varying the laser power intensity (typically 20–60 mW) and duration of irradiation (typically 500–5000 cycles).
4. Tune laser to 840 nm and take a snap of the ROI. Note the mean pixel intensity in the green and red detectors. At 840 nm the Kaede green signal bleeds into the red channel.
5. Begin irradiating and note the changes in the mean pixel intensity in the green and red detectors. Initially the Kaede green signal and bleedthrough into the red channel will fall. Subsequently, the red signal will rise as Kaede is photoconverted. Adjust the laser power intensity if the green and red fluorescence is falling rapidly. Adjust stage position to correct for any lateral or axial shift during photoconversion.
6. Stop photoconverting when the green mean pixel intensity is less than half the original value and the rise in the red mean pixel intensity has plateaued, (*see* **Note 16**).
7. Determine the precision and level of photoconversion by scanning at 810 nm, (*see* **Note 17**).
8. Following photoconversion, clean and irrigate skin flap with PBS.
9. Suture skin window overlying the lymph node and close the skin flap with vicryl sutures.
10. Remove the mouse from the anesthetic machine and allow the mouse to recover from anesthesia on the heated stage (*see* **Note 18**).
11. Inject 75 μg/kg temgesic s.c. as postoperative analgesia once mice show signs of waking.
12. Provide hydration and nutritional gel packs as necessary and monitor mice and the surgical wound closely in the first 4–8 h and daily thereafter. It is best to house mice singly post-op to prevent wound dehiscence.

3.8 Real-Time Interactive Two-Photon Photoconversion for Transcript Analysis

1. TPP can also be performed on ex plant lymph nodes for FACS phenotyping and transcript analysis. This has advantage over intravital TPP that the lymph node preparation is relatively fast and the procedure is high throughput. Also the explant tissue is kept at room temperature at which cells are less motile and this improves the precision and efficiency of the photoconversion.

2. Euthanize mice and harvest the lymph node by separating the fascia and fat overlying it in the inguinal ligament.
3. Clean the lymph node under the stereomicroscope. Make sure all fat and connective tissue is removed.
4. Use 20 μm pipette tip to apply very small amount of VetBond to plastic coverslip and mount the lymph node on the coverslip with cortical side up.
5. Remove any excess VetBond under the stereomicroscope.
6. Photoconvert as above in Subheading 3.7.
7. Following photoconversion, cells can be isolated and FACs phenotyped or sorted to examine their cellular or molecular characteristics.

3.9 Image Analysis

1. Import raw image files into Imaris (Bitplane) or similar image analysis software that allows 3D visualization of cells and semi-automated tracking.
2. Adjust image intensities and thresholds and apply smoothing function such as Guassian filter used to reduce noise and improve contrast.
3. Drift correction is applied by utilizing stationary cells or landmarks.
4. Cells are detected using the spot detection function and tracks are then automatically generated which are then manually verified.
5. Surfaces and volumes can be applied to define structures such as the GC.
6. Motility parameters can be analyzed from the Imaris Statistics function such as average speed, arrest coefficient or displacement.

4 Notes

1. There are a number of genetically encoded fluorescent reporter mice that can be used for two-photon microscopy. We routinely use cyan fluorescent protein (CFP), green fluorescent protein (GFP), and tdTomato. For TPP we use Kaede transgenic mice crossed with SW_{HEL} (for B cells) or OT2 mice (for T cells). Kaede cells are green fluorescent and undergo photoconversion to red fluorescence on exposure to violet light or two-photon excitation with a NIR laser.
2. To examine antigen-specific and T and B cells in the GC, fluorescent reporter mice can be bred with SW_{HEL} (for B cells) or OT2 mice (for T cells). SW_{HEL} B cells recognize

hen-egg lysozyme (HEL) and OT2 CD4$^+$ T cells recognize a peptide derived from ovalbumin (OVA).

3. We use SAP-deficient mice [32] as recipients. Endogenous SAP-deficient CD4$^+$ T cells are less able to generate Tfh cells and this reduces competition with the adoptively transferred OT2 T cells.
4. HEL-OVA antigen is used when examining the response of both SW_{HEL} B cells and OT2 T cells. To prepare HEL-OVA, HEL protein is chemically conjugated to OVA peptide using the cross-linker SMPH as per the manufacturer's instructions. The HEL-OVA conjugation was checked by comparison to unconjugated HEL and a previous HEL-OVA conjugation on an SDS-PAGE gel with a Coomassie stain.
5. Several investigators have used different strategies for labeling the FDCs in the LZ of the GC such as fluorescent immune complexes and antigen. We have found that the anti-CD157 mAb BP-3 immediately labels the follicular stroma in resting lymph nodes but is redistributed to the GC after several days and labels the FDC network in immunized lymph nodes [22].
6. We have found that Alexa Fluor 647 rapidly bleaches compared to Alexa Fluor 680. More recently we have started using the CF680R labeling kits from Biotium and found them to be much more photostable for intravital imaging.
7. To examine Tfh cells only OT2 T cells are transferred and recipient mice immunized with OVA. To examine both Tfh and B cells, OT2 T cells and SW_{HEL} B cells are transferred and recipient mice immunized with HEL-OVA.
8. FACs purity is typically >80% for OT2 T cells and >97% for SW_{HEL} B cells.
9. It is important to warm the Sigma Adjuvant System for at least 30 min at 37 °C before use, typically in a water bath. After heating immediately mix the adjuvant 1:1 with OVA or HEL-OVA. Mix thoroughly and inject relatively quickly after preparation. Do not transport or store on ice.
10. It is necessary to remove as much hair as possible from the imaging area as hair is strongly autofluorescent.
11. We typically prepare the PDMS polymer in small petri dishes that can then be cut to an appropriate size and shape for the lymph node flap to sit on.
12. When exposing the lymph node it is important to be very careful to not disturb any of the blood vessels around the lymph node. A small capillary leak will obscure imaging from the hemoglobin in red blood cells. Also do not use bright light as this may inadvertently photoconvert the cells in the lymph node.

13. Silicone grease repels water so it is used both to stick the O-ring down to the skin flap and also to generate a hydrophobic barrier to help stabilize the meniscus.
14. Excitation wavelengths typically used are 810 nm (to optimally detect photoconverted Kaede red), 870 nm (to detect Alexa Fluor 555, Alexa Fluor 680, Kaede green), and 920 nm (to detect Kaede green, tdTomato). To save time we often reuse previous settings for different fluorescent proteins that give a rough guide of optimal settings for the current experiment.
15. Avoid oversampling by acquiring images at steps below the axial point-spread function and undersampling at steps larger than half the diameter of the cell of interest. Also avoid long time intervals as this will underestimate the instantaneous speed.
16. It is important to attenuate the laser power intensity correctly so that the cells are photoconverting without significant photobleaching. If the laser power is set too high it can result in photobleaching. High laser power can also be toxic to cells so it is important to check that TPP is nontoxic by comparing the migration velocities of the cells of interest both before and after photoconversion.
17. Photoconverted Kaede red is best detected at 810 nm. At 920 nm unphotoconverted Kaede bleeds into the red channel.
18. Anaesthetized mice are unable to maintain their body temperature and it is essential to keep mice warm at all times to improve surgical outcomes.

References

1. Allen CD, Okada T, Cyster JG (2007) Germinal-center organization and cellular dynamics. Immunity 27(2):190–202. doi:10.1016/j.immuni.2007.07.009
2. Victora GD, Nussenzweig MC (2012) Germinal centers. Annu Rev Immunol 30:429–457. doi:10.1146/annurev-immunol-020711-075032
3. Crotty S (2011) Follicular helper CD4 T cells (TFH). Annu Rev Immunol 29:621–663. doi:10.1146/annurev-immunol-031210-101400
4. Cahalan MD, Parker I (2008) Choreography of cell motility and interaction dynamics imaged by two-photon microscopy in lymphoid organs. Annu Rev Immunol 26:585–626. doi:10.1146/annurev.immunol.24.021605.090620
5. Phan TG, Bullen A (2010) Practical intravital two-photon microscopy for immunological research: faster, brighter, deeper. Immunol Cell Biol 88(4):438–444. doi:10.1038/icb.2009.116
6. Tangye SG, Brink R, Goodnow CC, Phan TG (2015) SnapShot: interactions between B cells and T cells. Cell 162(4):926–926 . doi:10.1016/j.cell.2015.07.055e921
7. Allen CD, Okada T, Tang HL, Cyster JG (2007) Imaging of germinal center selection events during affinity maturation. Science 315 (5811):528–531. doi:10.1126/science.1136736
8. Schwickert TA, Lindquist RL, Shakhar G, Livshits G, Skokos D, Kosco-Vilbois MH, Dustin ML, Nussenzweig MC (2007) In vivo imaging of germinal centres reveals a dynamic open structure. Nature 446(7131):83–87. doi:10.1038/nature05573
9. Hauser AE, Junt T, Mempel TR, Sneddon MW, Kleinstein SH, Henrickson SE, von

Andrian UH, Shlomchik MJ, Haberman AM (2007) Definition of germinal-center B cell migration in vivo reveals predominant intrazonal circulation patterns. Immunity 26 (5):655–667. doi:10.1016/j.immuni.2007.04.008

10. Junt T, Moseman EA, Iannacone M, Massberg S, Lang PA, Boes M, Fink K, Henrickson SE, Shayakhmetov DM, Di Paolo NC, van Rooijen N, Mempel TR, Whelan SP, von Andrian UH (2007) Subcapsular sinus macrophages in lymph nodes clear lymph-borne viruses and present them to antiviral B cells. Nature 450 (7166):110–114. doi:10.1038/nature06287
11. Carrasco YR, Batista FD (2007) B cells acquire particulate antigen in a macrophage-rich area at the boundary between the follicle and the subcapsular sinus of the lymph node. Immunity 27 (1):160–171. doi:10.1016/j.immuni.2007.06.007
12. Phan TG, Grigorova I, Okada T, Cyster JG (2007) Subcapsular encounter and complement-dependent transport of immune complexes by lymph node B cells. Nat Immunol 8(9):992–1000. doi:10.1038/ni1494
13. Phan TG, Green JA, Gray EE, Xu Y, Cyster JG (2009) Immune complex relay by subcapsular sinus macrophages and noncognate B cells drives antibody affinity maturation. Nat Immunol 10(7):786–793. doi:10.1038/ni.1745
14. Roozendaal R, Mempel TR, Pitcher LA, Gonzalez SF, Verschoor A, Mebius RE, von Andrian UH, Carroll MC (2009) Conduits mediate transport of low-molecular-weight antigen to lymph node follicles. Immunity 30 (2):264–276. doi:10.1016/j.immuni.2008.12.014
15. Okada T, Miller MJ, Parker I, Krummel MF, Neighbors M, Hartley SB, O'Garra A, Cahalan MD, Cyster JG (2005) Antigen-engaged B cells undergo chemotaxis toward the T zone and form motile conjugates with helper T cells. PLoS Biol 3(6):e150. doi:10.1371/journal.pbio.0030150
16. Kitano M, Moriyama S, Ando Y, Hikida M, Mori Y, Kurosaki T, Okada T (2011) Bcl6 protein expression shapes pre-germinal center B cell dynamics and follicular helper T cell heterogeneity. Immunity 34(6):961–972. doi:10.1016/j.immuni.2011.03.025
17. Kerfoot SM, Yaari G, Patel JR, Johnson KL, Gonzalez DG, Kleinstein SH, Haberman AM (2011) Germinal center B cell and T follicular helper cell development initiates in the interfollicular zone. Immunity 34(6):947–960. doi:10.1016/j.immuni.2011.03.024
18. Qi H, Cannons JL, Klauschen F, Schwartzberg PL, Germain RN (2008) SAP-controlled T-B cell interactions underlie germinal centre formation. Nature 455(7214):764–769. doi:10.1038/nature07345
19. Xu H, Li X, Liu D, Li J, Zhang X, Chen X, Hou S, Peng L, Xu C, Liu W, Zhang L, Qi H (2013) Follicular T-helper cell recruitment governed by bystander B cells and ICOS-driven motility. Nature 496(7446):523–527. doi:10.1038/nature12058
20. Liu D, Xu H, Shih C, Wan Z, Ma X, Ma W, Luo D, Qi H (2015) T-B-cell entanglement and ICOSL-driven feed-forward regulation of germinal centre reaction. Nature 517 (7533):214–218. doi:10.1038/nature13803
21. Shulman Z, Gitlin AD, Weinstein JS, Lainez B, Esplugues E, Flavell RA, Craft JE, Nussenzweig MC (2014) Dynamic signaling by T follicular helper cells during germinal center B cell selection. Science 345(6200):1058–1062. doi:10.1126/science.1257861
22. Suan D, Nguyen A, Moran I, Bourne K, Hermes JR, Arshi M, Hampton HR, Tomura M, Miwa Y, Kelleher AD, Kaplan W, Deenick EK, Tangye SG, Brink R, Chtanova T, Phan TG (2015) T follicular helper cells have distinct modes of migration and molecular signatures in naive and memory immune responses. Immunity 42(4):704–718. doi:10.1016/j.immuni.2015.03.002
23. Moriyama S, Takahashi N, Green JA, Hori S, Kubo M, Cyster JG, Okada T (2014) Sphingosine-1-phosphate receptor 2 is critical for follicular helper T cell retention in germinal centers. J Exp Med 211(7):1297–1305. doi:10.1084/jem.20131666
24. Shulman Z, Gitlin AD, Targ S, Jankovic M, Pasqual G, Nussenzweig MC, Victora GD (2013) T follicular helper cell dynamics in germinal centers. Science 341(6146):673–677. doi:10.1126/science.1241680
25. Victora GD, Schwickert TA, Fooksman DR, Kamphorst AO, Meyer-Hermann M, Dustin ML, Nussenzweig MC (2010) Germinal center dynamics revealed by multiphoton microscopy with a photoactivatable fluorescent reporter. Cell 143(4):592–605. doi:10.1016/j.cell.2010.10.032
26. Chtanova T, Hampton HR, Waterhouse LA, Wood K, Tomura M, Miwa Y, Mackay CR, Brink R, Phan TG (2014) Real-time interactive two-photon photoconversion of recirculating lymphocytes for discontinuous cell tracking in live adult mice. J Biophotonics 7(6):425–433. doi:10.1002/jbio.201200175
27. Suan D, Hampton HR, Tomura M, Kanagawa O, Chtanova T, Phan TG (2013) Optimizing fluorescence excitation and detection for intravital two-photon microscopy. Methods Cell

Biol 113:311–323. doi:10.1016/B978-0-12-407239-8.00014-8

28. Tomura M, Yoshida N, Tanaka J, Karasawa S, Miwa Y, Miyawaki A, Kanagawa O (2008) Monitoring cellular movement in vivo with photoconvertible fluorescence protein "Kaede" transgenic mice. Proc Natl Acad Sci U S A 105(31):10871–10876. doi:10.1073/pnas.0802278105
29. Phan TG, Amesbury M, Gardam S, Crosbie J, Hasbold J, Hodgkin PD, Basten A, Brink R (2003) B cell receptor-independent stimuli trigger immunoglobulin (Ig) class switch recombination and production of IgG autoantibodies by anergic self-reactive B cells. J Exp Med 197(7):845–860. doi:10.1084/jem.20022144
30. Brink R, Paus D, Bourne K, Hermes JR, Gardam S, Phan TG, Chan TD (2015) The SW (HEL) system for high-resolution analysis of in vivo antigen-specific T-dependent B cell responses. Methods Mol Biol 1291:103–123. doi:10.1007/978-1-4939-2498-1_9
31. Barnden MJ, Allison J, Heath WR, Carbone FR (1998) Defective TCR expression in transgenic mice constructed using cDNA-based alpha- and beta-chain genes under the control of heterologous regulatory elements. Immunol Cell Biol 76(1):34–40. doi:10.1046/j.1440-1711.1998.00709.x
32. Czar MJ, Kersh EN, Mijares LA, Lanier G, Lewis J, Yap G, Chen A, Sher A, Duckett CS, Ahmed R, Schwartzberg PL (2001) Altered lymphocyte responses and cytokine production in mice deficient in the X-linked lymphoproliferative disease gene SH2D1A/DSHP/SAP. Proc Natl Acad Sci U S A 98(13):7449–7454. doi:10.1073/pnas.131193098

Chapter 6

Intravital Microscopy of T–B Cell Interactions in Germinal Centers

Changming Shih and Hai Qi

Abstract

Dynamic imaging analyses of antigen-specific T–B interactions in germinal centers have advanced our understanding of the molecular mechanisms underlying affinity maturation and provided a wealth of information about how follicular helper T cells function in vivo. Here we describe a routine method to visualize fluorescence protein-expressing, antigen-specific T and B cells in germinal centers. The protocol for incorporating functional reporters or genetic perturbation of the T cells by retroviral transduction is also briefly described, using the FRET-based calcium reporter as an example.

Key words Germinal center, T–B cell interaction, Calcium signaling, Two-photon imaging, Intravital microscopy

1 Introduction

Germinal center (GC) is a highly dynamic microanatomical structure in which B cells undergo Darwinian competition and selection to effectuate affinity maturation of the antibody response against microbial pathogens [1, 2]. B cell receptors of varying affinities acquire antigen with different efficiencies, leading to different levels of surface presentation of peptide-major histocompatibility complexes (pMHC). This latter variation among competing GC B cells is detected by follicular helper T cells (TFHs) that, by titrating the help signals according to TCR stimulation received, determine how likely those B cells survive, expand, and differentiate into long-lived plasma cells and memory B cells. Owing to the nature of TCR–pMHC interactions and the requirement for additional surface-bound accessory molecules such as CD40 and CD40L, physical interactions between GC B and TFH cells are of paramount importance to the reaction [3, 4].

Antigen-specific T–B interactions initially take place at the T–B border [5], and they assume a form of stable and long-lived conjugates with activated B cells leading the pair [6, 7]. The quality of

Dinis Pedro Calado (ed.), *Germinal Centers: Methods and Protocols*, Methods in Molecular Biology, vol. 1623,
DOI 10.1007/978-1-4939-7095-7_6,

such interactions has been often evaluated with the duration of the contact. However, quite different from this period at the T–B border, antigen-specific T–B interactions during the GC phase are much more transient [8–10]. Recent findings suggest the surface contact area between the two types of cells reflect the quality of such interactions [6, 11, 12]. Because of the short-lived nature of GC B-TFH interactions, signaling reporters that can be visualized are important for understanding the functional importance of such interactions.

Because of its high spatial resolution and reduced tissue damage following prolonged imaging, two-photon microscopy is a method of choice when conducting intravital observation of highly dynamic T–B interactions within the GC. In this chapter, we provide detailed protocols for how to analyze short-lived T–B interactions inside GCs using two-photon microscopy. After briefly describing the surgical procedure, we emphasize parameters of adoptive cell transfer, viral transduction of signaling reporters, image acquisition and data analysis, all of which are critical aspects of an in-depth analysis of GC T–B interaction dynamics.

2 Materials

2.1 Virus Packaging

1. Cell culture facility: A biosafety level-2 tissue culture facility is recommended.
2. Retroviral constructs: for cells monitored by intravital microscopy, genetic perturbation such as overexpression and RNAi can be achieved by retroviral transduction. For experiments presented in this protocol, MSCV is chosen for gene overexpression in T cells. Specially, we use MSCV bearing genetically encoded calcium indicator YC-nano50^{CD} [11] to measure calcium signaling in T cells in vivo, (*see* **Note 1**).
3. Plat-E cell line: the cell line of choice for retrovirus production, (*see* **Note 2**).
4. Cell culture reagents: Dulbecco's Modified Eagle Medium (DMEM, with 10% v/v fetal bovine serum), 0.25% trypsin–EDTA. Plasmid transfection reagent.

2.2 Cell Isolation and Infection

1. Mice: To obtain antigen-specific T and B cells for intravital imaging, T cell receptor- and B cell receptor-transgenic mice are utilized to isolate T and B cells for adoptive transfer. In our exemplary experiments, OT-II T cells, which are specific to ovalbumin peptide 323–339 ($OVA_{323-339}$) presented by the I-A^b class II MHC molecule, and MD4 B cells, which are specific to hen egg lysozyme (HEL), are transferred to recipient mice for inducing antigen-specific GC reaction. These mice should also carry transgenes that express fluorescence

proteins (e.g., CFP-cyan fluorescence protein, GFP-green fluorescence protein, dsRed-discosoma red fluorescence protein), (*see* **Note 3**).

2. Cell culture facility: same as Subheading 2.1, **item 1**.
3. Cell isolation kit: 70 μm nylon cell strainer, 2 mL syringe, anti-CD4 magnetic microbeads, anti-CD19 magnetic microbeads, cell isolation magnetic columns, (*see* **Note 4**).
4. Cell culture reagent: phosphate-buffered saline (PBS pH = 7.4, autoclaved), MACS buffer (PBS with 1% v/v FBS and 5 mM EDTA pH = 8.0), RPMI-1640 medium (with 10% v/v FBS, 1 mM sodium pyruvate, 2 mM glutamax-1, 55 μM β-mercaptoethanol), interleukin-2 (IL-2, 4 μg/mL in RPMI-1640, 500×), Polybrene (4 mg/mL, 1000×), ACK erythrocyte lysing buffer (containing 155 mM NH_4Cl, 10 mM $KHCO_3$, 1 mM EDTA in water, pH = 7.4).

2.3 Cell Transfer and Immunization

1. Mice: C57BL/6 (B6) mice are used as recipients. All mice are maintained within the same condition as mentioned in Subheading 2.2, **item 1**.
2. Flow-cytometry cell sorter.
3. 1 mL insulin syringes with needle.
4. Reagent for immunization: HEL-OVA conjugating protein is used as antigens (*see* **Note 5**), alum plus lipopolysaccharide (LPS) are used as immunologic adjuvant. All the reagents are suspended in PBS.

2.4 Surgery

1. Surgery instruments: dissecting microscope, cautery pen, hemostatic forceps, thumb dressing forceps, curved forceps, sharp forceps, corneal scissors, dissecting scissors.
2. Anesthesia: isoflurane, oxygen, anesthesia vaporizer for isoflurane (with tubing to mix oxygen and isoflurane and connect inhaler for mice), heater for maintaining mouse body temperature, (*see* **Note 6**).
3. A custom-made platform for supporting the mouse and exposed lymph nodes (Fig. 2a, b).
4. Other reagents and tools: PBS, depilatory cream, high-vacuum grease, surgical tape, tissue adhesive, razor, liquid-repellent slide marker pen, peristaltic pump, vacuum pump (with tubing and spouts to maintain a stable water milieu as medium for microscope lens).

2.5 Image Acquisition

1. Microscope (e.g., Olympus FV1000) with 20× or 25× water-dipping lens.
2. One or two femtosecond lasers (e.g., MaiTai DeepSee laser, Spectra-Physics).

2.6 Data Analysis and Statistics

Software used for data analysis and statistics: Imaris (Bitplane), Microsoft Excel, ImageJ, Prism (Graphpad).

3 Methods

In this section, we describe a typical experiment for examining T–B interactions within GCs including visualization of T cell calcium signaling.

3.1 Packaging the Virus

The calcium indicator YC-nano50^{CD} construct is transduced into the T cells by a retroviral vector that is packaged in the Plat-E cell line.

1. Culture Plat-E cells in DMEM until 70–80% confluency.
2. Transfect Plat-E cells with pMSCV-YC-nano50^{CD} construct, (*see* **Note 7**).
3. Harvest viral supernatants 48 h and 72 h after transfection, remove cell debris by centrifugation, and store the virus at −80 °C until use.

3.2 T Cell Isolation and Activation

For cell isolation and culture, sterile techniques are to be used to avoid contamination.

1. Suspensions of single splenocytes of OT-II mice (*see* **Note 8**) are made by grinding spleens on a 70 μm cell strainer using the piston of a 2 mL syringe.
2. Erythrocytes are lysed by incubating in the ACK buffer (1 mL/ spleen) at the ambient temperature for 1–2 min.
3. Harvest cells by centrifugation at 500 × *g* for 5 min at 4 °C before isolation of CD4 T cells using the standard protocol of Miltenyi Microbeads or any other equivalent method, which should give >90% purity in the final CD4 T cell preparation.
4. Prepare the plate for T cell stimulation by coating each well of a 24-well plate with 300 μL anti-CD3/CD28 (8 μg/mL each) in PBS at 37 °C for 1–2 h, (*see* **Note 9**).
5. Suspend the CD4 T cells in RPMI-1640 supplemented with IL-2 (final concentration 8 ng/mL) and load into the coated well and culture for 48 h.

3.3 T Cell Infection and Sorting

1. To retrovirally transduce OT-II T cells thaw virus stock at 37 °C, and add IL-2 (optional) and Polybrene.
2. Gently aspirate medium from the 24-well plate in which T cells were activated by plate-bound anti-CD3 and anti-CD28.
3. Immediately add virus supernatants to the plate, and centrifuge the plate at 32 °C 1500 × *g* for 2 h, (*see* **Note 10**).

4. Expand infected T cells in RPMI-1640 supplemented with IL-2 as needed in T25 and T75 flasks.
5. FACS sort infected cells based on vector fluorescence, and rest the sorted cells in IL-2-supplemented RPMI-1640 for at least 2 h before adoptive transfer.

3.4 Cell Transfer and Immunization

1. While sorting infected OT-II T cells, dsRed-expressing MD4 B cells are isolated using similar protocols as Subheading 3.2 with anti-CD19 magnetic Microbeads.
2. Mix appropriate numbers of dsRed MD4 B cells and sorted OT-II T cells, and intravenously transfer into B6 recipients.
3. Immunize cutaneous sites that are drained by inguinal lymph nodes with HEL-OVA and adjuvant, and GC reactions appropriate for intravital imaging typically take place 5–7 days later (*see* **Note 11**).

3.5 Surgery

The ultimate goal is to adequately expose the lymph node while minimizing motion artifacts for visualizing cellular interactions within the lymphoid tissue. In principle, inguinal, popliteal and other cutaneous lymph nodes can be used for imaging, and there are a number of different surgical procedures to accomplish the goal. Here we describe one of the procedures that consistently produce the most stable view of the lymph node. A key during the surgery is to keep the mouse body as physiologically warm as possible by using a heater or heating pad.

1. Use a mixture of isoflurane (typically 1–2% v/v) and oxygen to anesthetize the mouse.
2. Shave the mouse to expose the necessary surgical area and remove residue hair with depilatory (Fig. 1a).
3. As shown in Fig. 1b, cut the skin in-between the midline and where the inguinal lymph node situates. Use PBS to moisturize the wound periodically.
4. Using cautery to isolate the fatty connective tissue around the inguinal fold (Fig. 1c).
5. Using forceps and dissecting scissors to remove connective tissues between the skin flap containing the inguinal lymph node and the peritoneal wall, until approximately a squared skin flap is isolated (Fig. 1d).
6. Glue the skin flap containing the lymph node onto the custom platform (Fig. 2a, 2b).
7. Use food-quality vacuum grease to build a water reservoir on top of the lymph node (Fig. 2c).
8. Gently remove fatty connective tissues surrounding the lymph node under the dissecting microscope, using the finest corneal

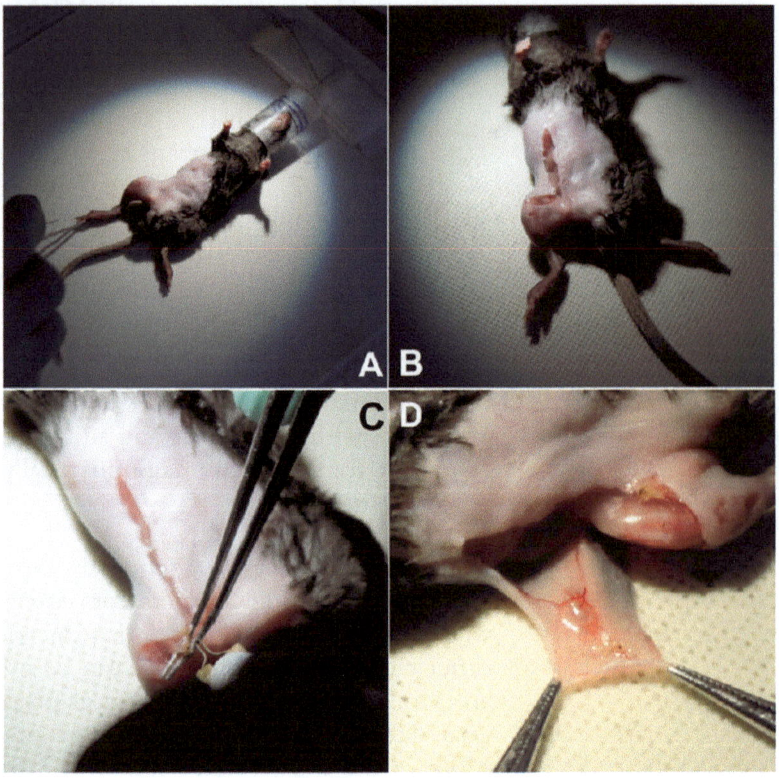

Fig. 1 Sequential steps of the surgical procedure to expose the inguinal lymph node

scissors and sharp forceps, periodically apply PBS to keep exposed tissues moisturized (Fig. 2d).

9. Once the capsule of the lymph node is largely exposed, apply a coverslip that has been marked with the liquid-repellent marker pen. The coverslip should barely touch the lymph node but amply pressed into the mass of vacuum grease (Fig. 2e).
10. Apply PBS on top of the coverslip, and set up inlet and outlet connected to peristaltic pump and vacuum, respectively, as shown in (Fig. 2f). By adjusting the flow rate and vacuum intensity, a constant flow of warm PBS over the coverslip should be achieved and the front glass of the lens is always immersed (Fig. 2f).

3.6 Intravital Imaging

Different microscope setups have different laser sources, different light paths, different filter sets, making it difficult to create universal applicable specific parameters to follow. Furthermore, different biological questions being asked may require adjustment of many parameters that are not necessarily of imaging per se. Several common principles should be followed as detailed below.

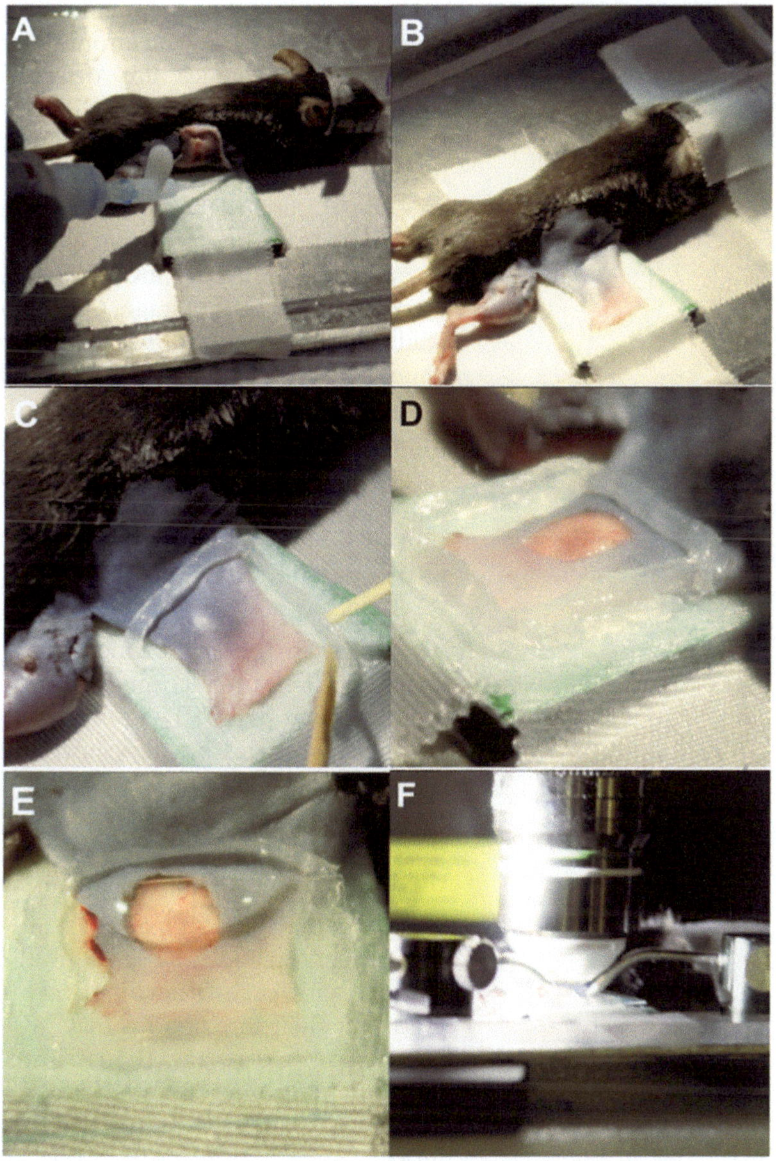

Fig. 2 Sequential steps of the surgical procedure to expose the lymph node for imaging

1. The microscope: it should be able to collect at least two channels of fluorescence signals, one for T cells and one for B cells. Ideally, a third channel would be available to collect information on tissue landmarks provided by secondary harmonic generation or other labeled cell populations (e.g., naïve B cells filling the follicular mantle).
2. The imaging wavelength: optimal excitation wavelengths for CFP (820–860 nm), GFP (900–940 nm), YFP (920–960 nm), and dsRed (920–960 nm) are different, and it can be

problematic if all these fluorescence proteins are to be excited together. However, commonly used fluorescence protein-expressing transgenes are driven by different promoters, which can express genes at very different levels, creating opportunities for convenient compromises, particularly when cell bodies and cellular processes instead of subcellular structures are to be differentiated. In our experience, the CAG promoter-driven CFP is the "weakest" but its expression level in the cytoplasm significantly increases as T or B cells become activated. Thus at the GC stage, CFP-expressing B or T cells can be readily imaged together with other much "brighter" FP transgene. Indeed, the ubiquitin promoter-driven GFP is extremely "bright," and it can be easily differentiated from CFP-expressing cells even though CFP and GFP have significant emission spectral overlap. We routinely use any two of the above four colors in our GC imaging, except for the combination of GFP and YFP, and an excitation wavelength between 840 nm and 860 nm is typically sufficient.

3. Imaging depth and orientation: Because T cells are almost exclusively in the light zone, which is closer to the capsule, analysis of T–B interactions inside GCs does not require a very large imaging depth. Typically, a range of 100–200 μm below the capsule would be sufficient. Because the light zone-dark zone orientation for any GC may be in parallel or perpendicular to the z axis of imaging, one may find GCs with T cells evenly distributed in the entire GC proper or skewed in one pole of the GC. This needs to be kept in mind when interpreting the dynamics of the T cells.
4. Imaging zoom and time resolution: Because T cells and B cells interact with each other in a very short-lived manner inside GCs, but their interactions may involve extensive surface contact, it is often advantageous to image the GC volume with higher repetition rates and digital magnification. For example, we typically use an XY pixel size below 0.5 mm, a Z-step no more than 1.5 mm, and a time resolution of 20–30 s per volume (*see* **Note 12**).
5. Cell density: GCs are clusters of B cells and T cells, and it is almost impossible to visualize and analyze in a quantitative manner individual T–B contacts if all the GC B cells are marked by the same fluorescence protein. Therefore, it is important to include unlabeled B cells of the same antigen specificity to "dilute" the labeled ones. Depending on the experimental need, a 19:1 ratio between the unlabeled and the labeled could be sufficient.
6. Detection of calcium flux: The Fluorescence Resonance Energy Transfer (FRET)-based calcium reporter YC-nano50^{CD} can be used to quantitatively measure calcium flux in T cells in the

GCs. T cells that express the reporter will be seen as CFP^+ and YFP^+, and the ratio YFP/CFP fluorescence intensity as calculated below reflects the calcium flux intensity. GC B cells express dsRed, and the optimal wavelength to image this combination is 830 nm, which excite well CFP but minimally YFP.

3.7 Data Analysis

Imaris is the software of choice for us, although other commercial and open-source software packages can be used for various aspects of the analyses.

1. T–B interactions: Historically, the duration of contact is the most widely used parameter to characterize antigen-specific T–B interactions [6, 7]. However, T–B contacts are universally short inside the GC, and the area of contacts is an important aspect of the interactions that is being regulated. The duration can be quantitated by counting the number of imaging frames that a T cell and a B cell in touch with each other. The area of contacts can be quantitated in two ways. For one, two surface objects can be created based on cytoplasmic fluorescence of each of the interacting partner cells, and the overlapping volume between the two reflects the extent of contact and can be normalized to either of the two objects. Alternatively, a z-projection image of the interacting cell partners can be used to outline the perimeters of the two cells, and the interface on the flattened image normalized to either of the two perimeters can also reflect the extent of the contact. These analyses are particularly suitable when genetically perturbed cells are compared in the same GC (e.g., two types of T cells interacting with one type of B cells), (Fig. 3).
2. Calcium signaling: Based on published methods [11, 13, 14], to quantitate calcium signaling by FRET, fluorescence intensity of both CFP and YFP should be quantitated. Imaris is used to render a 3D–surface for each cell according to cytoplasmic fluorescence signal. For each cell, the calcium signal is calculated as a FRET intensity according to the following formula:

$$R = \frac{I_{\mathrm{YFP}} - 0.44 \times I_{\mathrm{CFP}}}{I_{\mathrm{CFP}}}$$

Here, I represents intensity of the indicated channel. The 44% is the amount of spectral overlap that the YFP channel receives from the CFP emission even in the absence of FRET and, therefore, requires correction. Importantly, this is a parameter that needs to be estimated for individual microscope because it depends on the particular excitation wavelength and emission filters used. Because calcium flux transpires over time, one can define an arbitrary time 0,

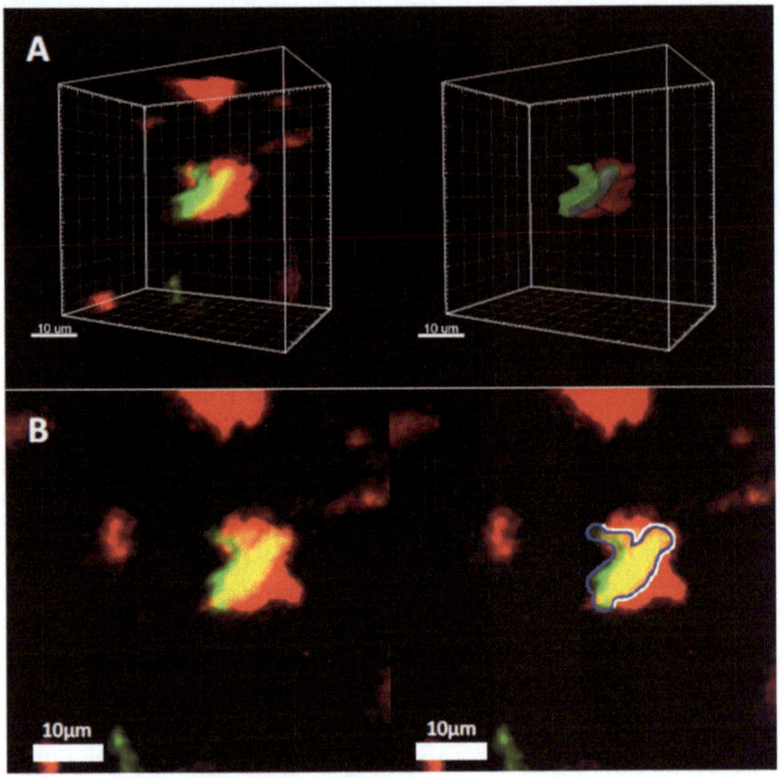

Fig. 3 Quantitation of GC T–B interaction area. (**a**) A GC T (*red*) and a GC B cell (*green*) are in contact (*left panel*). Surface objects for T and B cells are rendered according to their respective fluorescence intensities (*right panel*). The surface object in *blue* represents the overlap between the two volumes encapsulated by the two surfaces *blue*. The extent of T cell interaction with the B cell can be quantitated as the *blue* volume divided by the red volume. (**b**) The same T–B interacting pair as in (**a**) is presented as a maximum intensity projection image (*left panel*). Right panel shows the T cell perimeter (*blue line*) and the segment in contact (*white line*). Therefore, the extent of T cell interaction with the B cells can be quantitated as the fraction of the contacted segment in the perimeter

preferentially a condition in which one know the cell does not flux calcium as a priori, and one can then quantitate the calcium signal changes as ΔR_t,

$$\Delta R_t = \frac{R_t - R_0}{R_0}$$

In the context of T–B interactions, R_0 could represent basal FRET level of a given T cell, while R_t could represent FRET rate of the T cell in contact with a B cell at time-point t (Fig. 4).

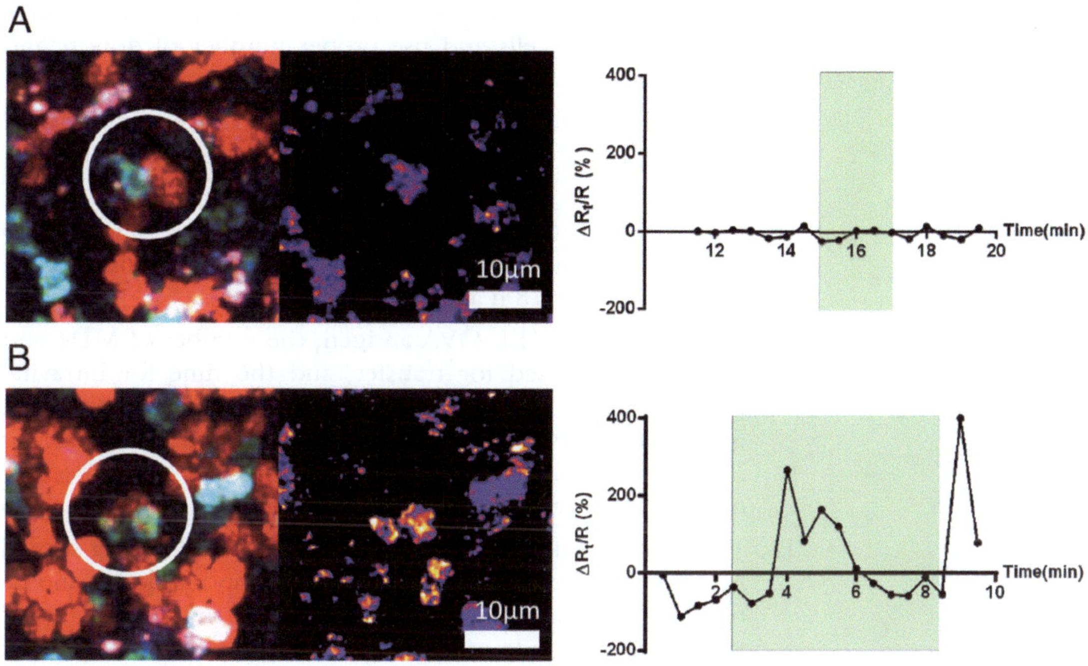

Fig. 4 Examples of T cell calcium signaling. T cells are in passing (**a**) or entangled (**b**) contacts with B cells, as defined in ref. [11]. The fluorescence images show the T cell (*cyan*), the B cell (*red*), and the calcium content (radiometric heat) in the T cell. The time series of relative calcium flux intensities are shown to the *right*, with *green* rectangles indicating durations of the two incidences of T–B contacts

4 Notes

1. Other virus constructs such as lentivirus can also be used.
2. For virus packaging, cell lines are alternative and should be chosen in accordance with virus construct.
3. Other antigen receptor transgenic mice can also be used.
4. We use magnetic cell isolation kit (Miltenyi Biotec), including antibody beads and magnetic column. Other protocols for enriching T or B cells from spleens or other lymphoid organs can also be used.
5. For a detailed protocol of making HEL-OVA, *see* [7].
6. To maintain mouse body temperature under anesthesia, a heating blanket can be used.
7. Any transient transfection protocols are in principle suitable.
8. In general, one naïve/unimmunized spleen can produce up to 0.8–1.3 × 10^7 T cells or 1–3 × 10^7 B cells. For adoptive transfer assay, each recipient requires around 2 × 10^5 naïve T cells or 5 × 10^5 in vitro activated T cell and 2 × 10^5 naïve B

cells. Therefore, before cell isolation, determine the total number of required cells and use proper number of donors and recipients.

9. Alternatively, the plate can be coated at 4 °C overnight. A total of 1–2 × 10^6 T cells are to be cultured in each well of the 24-well plate.
10. For each well, 0.5–2 mL virus are enough for infection. To increase infection rate, infection can be done twice, one at 24 h and the other at 48 h after T cell activation.
11. The amount of HEL-OVA antigen, the number of MD4 and OT-II T cells used for transfer, and the time for intravital imaging should be empirically optimized. As a starting point, 5 days post immunization are typically a good time point for imaging, and ~5 × 10^5 infected T cells together with ~2 × 10^5 naïve MD4 B cells are transferred per mouse.
12. The exact spatial and time resolution used for imaging should be adjusted empirically, as more spatiotemporal details can be obtained with smaller Z steps and shorter time intervals, but the necessary trade-off is a smaller tissue volume can be covered.

Acknowledgments

This work is supported in part by National Natural Science Foundation of China (grant No. 81330070 and 81425011). Photos of surgical procedures are courtesy of Jiacong Yan.

References

1. Victora GD, Nussenzweig MC (2012) Germinal centers. Annu Rev Immunol 30:429–457. doi:10.1146/annurev-immunol-020711-075032
2. MacLennan IC (1994) Germinal centers. Annu Rev Immunol 12:117–139. doi:10.1146/annurev.iy.12.040194.001001
3. Qi H, Liu D, Ma W, Wang Y, Yan H (2014) Bcl-6 controlled TFH polarization and memory: the known unknowns. Curr Opin Immunol 28:34–41. doi:10.1016/j.coi.2014.01.016
4. Crotty S (2011) Follicular helper CD4 T cells (T_{FH}). Annu Rev Immunol 29:621–663. doi:10.1146/annurev-immunol-031210-101400
5. Garside P, Ingulli E, Merica RR, Johnson JG, Noelle RJ, Jenkins MK (1998) Visualization of specific B and T lymphocyte interactions in the lymph node. Science 281 (5373):96–99
6. Okada T, Miller MJ, Parker I, Krummel MF, Neighbors M, Hartley SB, O'Garra A, Cahalan MD, Cyster JG (2005) Antigen-engaged B cells undergo chemotaxis toward the T zone and form motile conjugates with helper T cells. PLoS Biol 3(6):e150. doi:10.1371/journal.pbio.0030150
7. Qi H, Cannons JL, Klauschen F, Schwartzberg PL, Germain RN (2008) SAP-controlled T-B cell interactions underlie germinal centre formation. Nature 455(7214):764–769. doi:10.1038/nature07345
8. Allen CD, Okada T, Tang HL, Cyster JG (2007) Imaging of germinal center selection events during affinity maturation. Science 315(5811):528–531. doi:10.1126/science.1136736
9. Hauser AE, Junt T, Mempel TR, Sneddon MW, Kleinstein SH, Henrickson SE, von Andrian UH, Shlomchik MJ, Haberman AM (2007) Definition of germinal-center B cell

migration in vivo reveals predominant intrazonal circulation patterns. Immunity 26 (5):655–667. doi:10.1016/j.immuni.2007.04.008

10. Schwickert TA, Lindquist RL, Shakhar G, Livshits G, Skokos D, Kosco-Vilbois MH, Dustin ML, Nussenzweig MC (2007) In vivo imaging of germinal centres reveals a dynamic open structure. Nature 446(7131):83–87. doi:10.1038/nature05573
11. Liu D, Xu H, Shih C, Wan Z, Ma X, Ma W, Luo D, Qi H (2015) T-B-cell entanglement and ICOSL-driven feed-forward regulation of germinal centre reaction. Nature 517 (7533):214–218. doi:10.1038/nature13803
12. Gitlin AD, Shulman Z, Nussenzweig MC (2014) Clonal selection in the germinal centre by regulated proliferation and hypermutation. Nature 509(7502):637–640. doi:10.1038/nature13300
13. Horikawa K, Yamada Y, Matsuda T, Kobayashi K, Hashimoto M, Matsu-ura T, Miyawaki A, Michikawa T, Mikoshiba K, Nagai T (2010) Spontaneous network activity visualized by ultrasensitive Ca^{2+} indicators, yellow Cameleon-Nano. Nat Methods 7 (9):729–732. doi:10.1038/nmeth.1488. [pii]
14. Mues M, Bartholomaus I, Thestrup T, Griesbeck O, Wekerle H, Kawakami N, Krishnamoorthy G (2013) Real-time in vivo analysis of T cell activation in the central nervous system using a genetically encoded calcium indicator. Nat Med 19(6):778–783. doi:10.1038/nm.3180

Chapter 7

Identifying Follicular Regulatory T Cells by Confocal Microscopy

Ine Vanderleyden and Michelle A. Linterman

Abstract

Follicular regulatory T cells are a subset of Foxp3$^+$ regulatory T cells that migrate into the B cell follicle after infection or immunization and modulate the germinal center response. The anatomical positioning of follicular regulatory T cells within the germinal center is a defining characteristic of this subset of regulatory T cells; because of this, it is critical that studies of follicular regulatory T cells are able to identify them in situ. In this chapter we describe an immunofluorescence staining method to visualize follicular regulatory T cells in frozen secondary lymphoid tissue sections by confocal imaging.

Key words Follicular regulatory T cells, Regulatory T cells, Foxp3, Germinal center, Confocal microscopy, Immunofluorescence

1 Introduction

The germinal center (GC) response forms in secondary lymphoid tissues after infection or immunization, producing memory B cells and long-lived plasma cells that secrete high-affinity antibodies. The GC is a microenvironment of intense cellular collaboration of multiple cell types: GC B cells, follicular dendritic cells, tingible body macrophages, follicular helper T (Tfh) cells, and follicular regulatory T (Tfr) cells. Tfr cells are a subset of Foxp3$^+$ regulatory T cells (Treg) cells that are found within the GC where they act to modulate this response [1–4]. Tfr cells phenotypically resemble Tfh cells as they express the transcriptional repressor B-cell lymphoma 6 (Bcl-6), C-X-C chemokine receptor type 5 (CXCR5), and programmed cell death protein 1 (PD-1) among other typical proteins that have been implicated in Tfh cell biology. In addition to their phenotypic similarities to Tfh cells, Tfr cells also express typical Treg cell proteins such as Foxp3, the transcriptional regulator of the Treg subset and cytotoxic T-lymphocyte associated protein 4 (CTLA-4). Taken together, Tfr cells have a phenotype that has characteristics of both Treg and Tfh cells.

Dinis Pedro Calado (ed.), *Germinal Centers: Methods and Protocols*, Methods in Molecular Biology, vol. 1623,
DOI 10.1007/978-1-4939-7095-7_7, © Springer Science+Business Media LLC 2017

The confluence of Tfh and Treg cell features in Tfr cells allows them to be identified as $CXCR5^{+}PD\text{-}1^{+}Foxp3^{+}$ cells using flow cytometry. This is a very powerful technique for immunophenotyping, providing a high-throughput analysis of distinct immune cell types and making it possible to identify multiple distinct immune cell subsets in parallel. However, a major disadvantage of flow cytometry is the loss of information about the physical location of cells within a tissue. Imaging of stained frozen tissue sections by confocal microscopy allows for the in situ identification of immune cells. A key feature of Tfr cells, distinguishing them from other Treg cell subsets, is their localization within the GC [5]. It is therefore essential to be able to identify them in this highly specialized structure.

Herein we describe a method for the in situ visualization of Tfr cells in fixed/frozen secondary lymphoid tissues. The use of fixed/frozen tissue sections rather than paraffin embedded sections allows the parallel use of four different markers: Ki67, IgD, CD3, and Foxp3. This combination of markers makes it possible to identify Tfr cells ($CD3^{+}$ $Foxp3^{+}$) and Tfh cells ($CD3^{+}Foxp3^{-}$) within the GC area, which can be identified as the cluster of $Ki67^{+}$ GC B cells within the IgD^{+} B cell follicle (Fig. 1). A major disadvantage of using frozen tissue sections compared to paraffin sections is the poor morphology of frozen tissues. In this protocol we use periodate–lysine–paraformaldehyde (PLP) to fix the whole tissue prior to sectioning which improves tissue morphology, and improves the staining of Foxp3, which is essential for identifying Tfr cells.

2 Materials

2.1 Equipment

1. Cryomolds; these are commercially available but can also be made out of aluminum foil by wrapping the foil around a mold.
2. −80 °C freezer.
3. Cryostat.
4. PAP pen to encircle the tissue with a hydrophobic area to ensure reagents stay localized on the tissue and avoid spillage.
5. Superfrost plus slides.
6. Staining dishes.
7. Slide racks.
8. Humidified chamber.
9. Coverslips.
10. Confocal microscope.

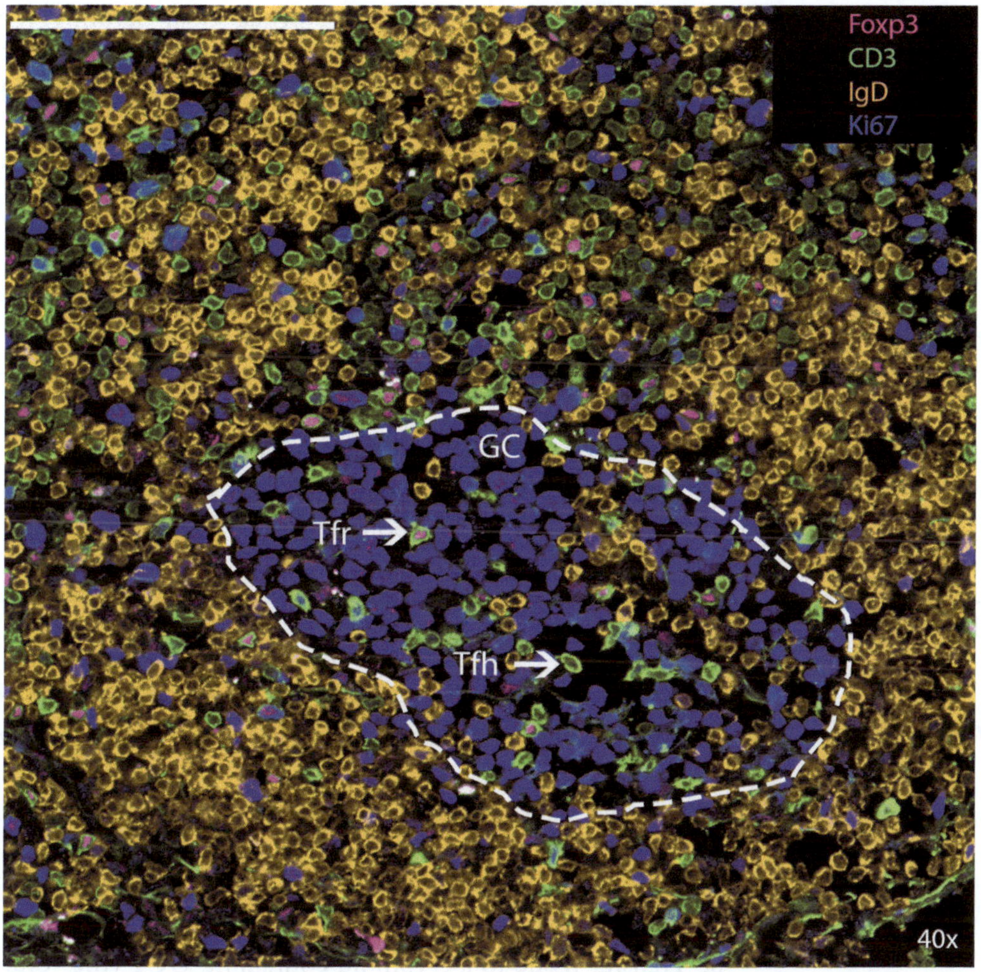

Fig. 1 GC in spleen of a wild-type C57BL/6 mouse 14 days after influenza A infection. This representative section was stained for Ki67 (*Blue*), IgD (*orange*), CD3 (*green*), and Foxp3 (*magenta*). Confocal image showing Tfr ($CD3^{+}Foxp3^{+}$) and Tfh ($CD3^{+}Foxp3^{-}$) cells (indicated by *arrows)* within the GC (indicated by the *white dashed line*). *Tfr cell* Follicular regulatory T cell, *Tfh* Follicular helper T cell, *GC* germinal center. Magnification 40×100 μm

2.2 Reagents and Solutions (See Note 1)

1. Periodate–lysine–paraformaldehyde (PLP): 1% PFA, 0.075 M L-lysine, 0.37 M sodium phosphate (pH 7.4), and 0.01 M $NaIO_4$. Make up fresh. Store at 4 °C (*see* **Note 2**).
2. 30% sucrose in dH_20 (*see* **Note 3**).
3. OCT®, Cryoprotective embedding medium. Store at room temperature.
4. 2-propanol.
5. Wash buffer: 0.5% (v/v) Tween 20 in PBS.
6. Blocking buffer: PBS+ 2% (w/v) BSA (bovine serum albumin) + 10% (v/v) normal goat serum (NGS), store aliquots at −20 °C (*see* **Note 4**).

7. Antibody diluents: 2% (v/v) Normal Goat Serum (NGS) in wash buffer (for antibodies generated in goat), store aliquots at −20 °C; 1% (w/v) BSA in wash buffer (for antibodies generated from other species), store aliquots at −20 °C.
8. Permeabilization buffer: PBS + 2% (v/v) Triton X. Make up fresh.
9. Aqueous mounting medium.

2.3 Antibodies (See Note 5)

1. Hamster anti-mouse CD3e purified (clone 500A2) and Alexa 568-conjugated goat anti-hamster IgG (polyclonal).
2. eFluor® 450-conjugated rat anti-mouse/rat Foxp3 (clone FJK16S).
3. Rat anti-mouse IgD Alexa 647 (clone 11-26c.2a).
4. Rabbit anti-mouse Ki67 (polyclonal) and Alexa-488 conjugated goat anti-rabbit (polyclonal).

3 Methods

3.1 Preparation of Frozen Tissue

1. Fix lymph node or spleen in 1–2 ml of PLP for 3–4 h at 4 °C.
2. Remove PLP and incubate with 1–2 ml of sucrose 30% overnight at 4 °C.
3. Remove sucrose and wash with 1–2 ml of sucrose 30%, five times for 15 min at 4 °C.
4. Remove all sucrose solution from sample and dab-dry using a tissue.
5. Transfer tissue to a cryomold containing OCT (the volume of OCT depends on the size of the mold but make sure the tissue is fully immersed) and place in a 2-propanol–dry ice cooling bath (−78 °C) (*see* **Note 6**).
6. Once the OCT is solid, store at −80 °C. Make sure the specimens are tightly wrapped.

3.2 Cryosectioning

1. Cut the frozen tissue into 10 μm serial sections at −20 °C using a cryostat.
2. Air-dry slides in slide box for 20 min.
3. Place on dry ice and store at −80 °C until required for the next step.

3.3 Immunofluorescence Staining

1. Remove slides from freezer and air-dry them overnight or dry under airflow for 30 min (e.g., hood, or fume cabinet).
2. Mark tissue sections with a PAP pen by drawing a circle around them, and leave for 10–30 min until the wax has completely dried.

3. Rehydrate slides: wash slides with wash buffer (transfer slides to a staining jar or dish) three times for 5 min (*see* **Note 7**).
4. Block nonspecific binding of the primary antigen by adding blocking buffer for 30 min at room temperature.
5. Wash slides with wash buffer three times for 5 min.
6. Permeabilization step: add Permeabilization buffer and incubate for 30–60 min at room temperature.
7. Wash slides with wash buffer three times for 5 min.
8. Primary antibody staining: Prepare an antibody cocktail and pipette this mix onto the slide making sure each section is completely covered. The antibody cocktail contains eFluor 450-conjugated rat anti-mouse/rat Foxp3 (4 μg/ml), hamster anti-mouse CD3e purified (2.5 μg/ml), rabbit anti-mouse Ki67 (10 μg/ml), and rat anti-mouse IgD Alexa 647 (2.5 μg/ml) in 1% (w/v) BSA in wash buffer. Incubate in the dark for 60–120 min at room temperature, or this step can be done overnight at 4 °C.
9. Wash slides with wash buffer three times for 5 min to remove unbound antibodies.
10. For secondary antibody staining, prepare an antibody cocktail and pipette this mix onto the slide making sure each section is completely covered. The antibody cocktail contains Alexa 568-conjugated goat anti-hamster IgG (4 μg/ml) and Alexa 488-conjugated goat anti-rabbit (5 μg/ml) in 2% (v/v) NGS in wash buffer for 60–120 min at room temperature or overnight at 4 °C.
11. Wash two times with wash buffer (this is a quick rinse, no incubation required).
12. Wash two times with PBS (this is a quick rinse, no incubation required).
13. Wipe off excess PBS without disrupting the tissue.
14. Mount the tissue with mounting medium and apply coverslip. Add a drop of mounting medium directly onto the tissue and gently place coverslip on top using forceps. Be careful to avoid the formation of air bubbles, as this will make imaging more difficult. Dry overnight at room temperature or until mounting medium has set (*see* **Note 8**).
15. Store at 4 °C for imaging (*see* **Note 9**).

4 Notes

1. For reagents stored at −20 °C, avoid freeze–thaw cycles by preparing aliquots of solutions.
2. Periodate oxidizes carbohydrates and forms aldehyde groups. The lysine, an amine, subsequently reacts with the aldehyde

group thereby cross-linking the carbohydrate groups. The low concentration of paraformaldehyde fixes and stabilizes the tissue by cross-linking proteins [6].

3. Sucrose removes water from the tissue and prevents the formation of water crystals in the tissue when it is frozen.
4. Blocking buffer: Adding blocking buffer to the slides is essential for the prevention of nonspecific binding of the primary antibody, thereby reducing background staining. Usually the blocking buffer is made up of PBS containing bovine serum albumin (BSA) and normal serum obtained from the species in which the secondary antibodies have been generated. It is crucial that the blocking buffer does not contain serum from any of the species in which the primary antibodies have been produced as this will cause the secondary antibody to bind to all the (nonspecific)-species non-epitope specific antibodies from that serum.
5. The antibodies used in this protocol are optimized for use with the Zeiss LSM 780 microscope. However, care must be taken to select antibodies conjugated to fluorescent probes compatible with the configuration of the available confocal microscope.
6. Prepare the dry ice–2-propanol cooling bath by filling a box with dry ice and placing a beaker on top of the dry ice. Fill the bottom of the beaker with a small volume of 2-propanol, making sure the 2-propanol does not come in contact with the OCT, as this will prevent the OCT from solidifying.
7. After rehydration it is important to keep the slides hydrated at all times. Therefore staining procedures should be performed in a covered humidified chamber (e.g., Add wet paper towels to the bottom of a slide box with lid).
8. The mounting medium used in this protocol does not contain DAPI, as it would interfere with imaging the eFluor 450-conjugated Foxp3.
9. Images are captured using a Zeiss LSM780 confocal microscope and image analysis (GC size, number of Tfh/Tfr per GC) is performed using Volocity.

References

1. Chung Y, Tanaka S, Chu F, Nurieva RI, Martinez GJ, Rawal S, Wang YH, Lim H, Reynolds JM, Zhou XH, Fan HM, Liu ZM, Neelapu SS, Dong C (2011) Follicular regulatory T cells expressing Foxp3 and Bcl-6 suppress germinal center reactions. Nat Med 17(8):983–988. doi:10.1038/nm.2426
2. Linterman MA, Pierson W, Lee SK, Kallies A, Kawamoto S, Rayner TF, Srivastava M, Divekar DP, Beaton L, Hogan JJ, Fagarasan S, Liston A, Smith KG, Vinuesa CG (2011) Foxp3+ follicular regulatory T cells control the germinal center response. Nat Med 17(8):975–982. doi:10.1038/nm.2425
3. Wollenberg I, Agua-Doce A, Hernandez A, Almeida C, Oliveira VG, Faro J, Graca L (2011) Regulation of the germinal center reaction by Foxp3+ follicular regulatory T cells. J

Immunol 187(9):4553–4560. doi:10.4049/jimmunol.1101328

4. Wu H, Chen Y, Liu H, Xu LL, Teuscher P, Wang S, Lu S, Dent AL (2016) Follicular regulatory T cells repress cytokine production by follicular helper T cells and optimize IgG responses in mice. Eur J Immunol 46(5):1152–1161. doi:10.1002/eji.201546094
5. Vanderleyden I, Linterman MA, Smith KG (2014) Regulatory T cells and control of the germinal centre response. Arthritis Res Ther 16 (5):471
6. McLean IW, Nakane PK (1974) Periodate-lysine-paraformaldehyde fixative. A new fixation for immunoelectron microscopy. J Histochem Cytochem 22(12):1077–1083

Chapter 8

Cytokine Expression by T Follicular Helper Cells

Christoph Jandl, Claudia Loetsch, and Cecile King

Abstract

T follicular helper (Tfh) cells are a specialized subset of $CD4^+$ T cells located within temporary structures known as germinal centers (GC) formed within B cell follicles of secondary lymphoid organs. In the GC, Tfh cells facilitate the production of high-affinity antibodies through secretion of effector cytokines, such as IL-21 and IL-4, and through cell-to-cell interactions. The flow cytometric-based method described here allows the detection of intracellular cytokines within the Tfh population of secondary lymphoid organs (e.g., spleen, lymph nodes, and lymphoid nodules such as Peyer's patches), enabling the study of Tfh responses to different stimuli in the context of immunity and autoimmunity.

Key words T follicular helper cells, Cytokines, IL-21, IL-4, Germinal center, Flow cytometry, Phenotyping

1 Introduction

Germinal centers (GCs) are specialized structures forming at the site of follicular dendritic cells (FDCs) within the B cell follicles of secondary lymphoid organs in response to infection or immunization [1]. Within the GC, antigen-specific B cells differentiate into high-affinity antibody-secreting plasma cells and memory B cells thereby providing long-term humoral immunity [2]. The process of GC formation and affinity maturation is crucially dependent on a specialized subset of CD4+ T cells that provide help to B cells through cell-to-cell interactions, as well as through the secretion of cytokines [1, 3, 4]. This subset of CD4+ T cells is known as T follicular helper (Tfh) cells, due to their localization within the B cell follicles of secondary lymphoid organs [5–7]. Tfh cells are classically identified by their combined surface expression of the chemokine receptor CXCR5 and the co-inhibitory molecule PD-1, as well as the intracellular expression of the transcription factor

Christoph Jandl and Claudia Loetsch have equally contributed.

Dinis Pedro Calado (ed.), *Germinal Centers: Methods and Protocols*, Methods in Molecular Biology, vol. 1623,
DOI 10.1007/978-1-4939-7095-7_8, © Springer Science+Business Media LLC 2017

Bcl-6 [5, 8–12]. Importantly, Tfh cells exert their function through the expression of CD40L, which stimulates CD40 expressed by GC B cells [13–15], and through the secretion of cytokines such as interleukin (IL)-21, IL-4, and interferon-γ (IFN-γ) [16–19]. In contrast to human Tfh cells, murine Tfh cells do not express IL-10, which is considered to be the most potent inducer of human plasma cell differentiation [20–22]. In mice, IL-10 is instead secreted by a FoxP3+ regulatory subset of follicular T cells called T follicular regulatory (Tfr) cells [23, 24] that has been shown to negatively regulate both Tfh cell number and function [25, 26]. Although the pro-inflammatory cytokine IL-17 is not produced by human tonsillar Tfh cells or murine Tfh cells induced by immunization [27], it has been shown to be produced by Tfh-like cells in autoimmune disease, where it is involved in the production of self-reactive antibodies and the formation of GCs [28–30]. In this chapter, we describe a method to identify Tfh cells ex vivo via their expression of both extracellular and intracellular markers, and a technique to assess the function of Tfh cells in terms of their ability to produce cytokines using multicolor flow cytometry.

2 Materials

2.1 Chemicals

1. Phorbol 12-myristate 13-acetate (PMA, e.g., Sigma).
2. Ionomycin (e.g., Sigma).
3. Brefeldin A (e.g., Sigma).

2.2 Reagents

1. Heat-inactivated fetal bovine serum (hiFBS): For complete complement inactivation, incubate FBS for 30 min at 56 °C.
2. Red blood cell (RBC) lysis buffer: 155 mM NH_4Cl, 10 mM $KHCO_3$, and 1 mM EDTA in dH_2O.
3. Cell processing buffer: RPMI 1640 Medium supplemented with 2% hiFBS.
4. Complete culture medium: RPMI 1640 Medium supplemented with 10% hiFBS, PMA (50 ng/ml), ionomycin (1 μg/ml), and brefeldin A (1 μl/ml).
5. FACS buffer: 0.5% BSA, 5 mM EDTA and 0.02% sodium azide in D-PBS.
6. BD Fixation/permeabilization buffer: 1× Cytofix/Cytoperm solution (BD Biosciences).
7. BD Permeabilization buffer: Prepare 1:10 dilution of 10× Perm/Wash stock solution (BD Biosciences) in dH_2O.
8. eBio Fixation/permeabilization buffer: Mix 1 part Fixation/Permeabilization Concentrate (eBioscience) and 3 parts Fixation/Permeabilization Diluent (eBioscience).

Table 1
Antibody summary

Antigen	Fluorophore	Dilution	Clone
PD-1	FITC	1:100	J43
IL-4	PE	1:100	11B11
CXCR5	Biotin	1:100	2G8
Streptavidin	Pe-Cy7	1:300	
IL21RFc	AF647	1:100	
TCRβ	APC-eF780	1:200	H57-597
Foxp3	eF450	1:100	FJK-16s
CD4	BV605	1:200	RM4-5
CD16/32 (Fc block)		1:300	2.4G2

9. eBio Permeabilization buffer: Prepare 1:10 dilution of 10× Perm/Wash stock solution (eBioscience) in dH_2O.
10. 70-μm cell strainers.
11. 96-well round bottom plates.
12. 15-ml conical tubes.
13. 6-well flat bottom plates.
14. 2-ml plastic syringes.

2.3 Antibodies

Label Recombinant Mouse IL-21 R Fc Chimera Protein (IL21RFc, R&D, #596-MR) with Alexa Fluor® 647 Antibody Labeling Kit (Thermo Fisher, # A20186) following the manufacturer's instructions. Alternatively, an AF647-conjugated anti-mouse IL21 antibody may be used (Table 1).

3 Methods

3.1 Cell Preparation (See Note 1)

1. Prepare 6-well plates containing 3 ml of cell processing buffer per well and add a 70-μm cell strainer to each well. Each organ should be processed in a separate well.
2. For LN, gently disrupt the tissues between the frosted ends of microscope slides and filter cell suspension through the mesh of a 70-μm cell strainer.
3. For spleen, inject 2 ml of media into tail of spleen and gently tease cells out of spleen sac using the end of a 2 ml syringe plunger, then filter cell suspension through the 70-μm cell strainer. Rinse the strainer with another 10 ml of cell processing media and transfer the single cell suspension into a 15-ml conical tube.

4. Centrifuge the tubes at 350 × *g* for 5 min at 4 °C to pellet the cells.
5. For lymph nodes: Aspirate the supernatant and resuspend the cell pellet in 2–5 ml of cell processing media depending on the size of the pellet. Cells obtained from lymph nodes are ready for counting at this point.
6. For spleens: Aspirate supernatant and resuspend the cell pellet in 1 ml of RBC lysis buffer.
7. Incubate splenocytes for 1 min on ice (*see* **Note 2**).
8. Add 4 ml of cell processing media to the splenocyte suspension and centrifuge at 350 × *g* for 5 min at 4 °C.
9. Aspirate the supernatant and resuspend the splenocytes in 10 ml of processing media. Proceed to cell counting.

3.2 Stimulation of Cytokine Production (See Notes 3 and 4)

1. Preheat cell culture medium to 37 °C (100 μl required/sample).
2. Plate 2×10^7/ml cells of each sample per well into a 24-well round bottom plate and centrifuge the plate at 350 × *g* for 5 min at 4 °C to pellet the cells.
3. Aspirate the supernatant and resuspend cells at 2×10^7 cells/ml in complete culture medium containing 50 ng/ml PMA and 1 μg/ml ionomycin.
4. Incubate the cells for 4–6 h at 37 °C, 5% CO_2.
5. After stimulation, add 100 μl of ice-cold FACS buffer to each well and centrifuge the plate at 350 × *g* for 5 min at 4 °C to wash the cells.
6. Aspirate supernatant and repeat washing step. Aspirate supernatant.

3.3 Cell Surface Staining

1. Resuspend cells in 50 μl FACS buffer containing anti-mouse CD16/32 (Fc block) and incubate for 10 min on ice.
2. Add 50 μl FACS buffer containing anti-CXCR5 antibody at a 1:50 dilution on top of Fc block (for a final concentration of 1:100) and 1 μl/ml brefeldin A, incubate for 30 min at room temperature (*see* **Note 5**).
3. Add 100 μl of ice-cold FACS buffer to each well and centrifuge at 350 × *g* for 5 min at 4 °C to wash cells.
4. Aspirate the supernatant.
5. Add 100 μl of FACS buffer containing appropriate concentrations of anti-CD4, PD-1, TCRβ, CD25, streptavidin-Pe-Cy7. Incubate for 30 min on ice in the dark.
6. Add 100 μl ice-cold FACS buffer and centrifuge at 350 × *g* for 5 min at 4 °C to wash cells.
7. Aspirate supernatant and repeat washing step. Aspirate supernatant.

3.4 Fixing and Permeabilization (See Note 6)

1. Resuspend surface-stained cells in 100 μl of Cytofix/Cytoperm fixing solution and incubate for 30 min on ice.
2. Add 100 μl of ice-cold 1 × Perm/Wash buffer and centrifuge at 350 × *g* for 5 min at 4 °C to wash cells.
3. Aspirate supernatant and repeat washing step using Perm/Wash buffer. Aspirate supernatant.

3.5 Intracellular Immunostaining (See Note 6)

1. To stain for intracellular markers and cytokines, resuspend each sample in 100 μl of 1 × Perm/Wash buffer containing Fc block and anti-FoxP3 antibody at the appropriate concentrations, as well as antibodies against your cytokine of interest (*see* **Notes** 7 and **8**).
2. Incubate your samples for 30 min at room temperature in the dark.
3. Add 100 μl ice-cold 1 × Perm/Wash buffer and centrifuge at 350 × *g* for 5 min at 4 °C to wash cells.
4. Aspirate supernatant and repeat washing step using Perm/Wash buffer. Aspirate supernatant.
5. Resuspend the cells in 100 μl FACS buffer and filter them through a mesh top FACS tube to remove any cell clumps or debris.
6. Keep the cells in the dark until acquiring them on a cytometer.

3.6 Gating Strategy for Flow Cytometric Analysis (Figs. 1 and 2)

1. Gate out cell doublets.
2. Gate on the lymphocyte population.
3. Gate on $CD4^+$ $TCR\beta^+$ cells.
4. Gate on $CXCR5^+$ $PD\text{-}1^+$ cells.
5. Gate on $FoxP3^-$ Tfh cells.
6. Gate on cytokine containing Tfh cells.

4 Notes

1. All procedures should be carried out on ice or at 4 °C, unless specified otherwise.
2. The optimal duration of the RBC lysis depends in the size of the spleen and on the strength of the individual RBC buffer, and should be optimized. Generally, durations between 45 s and 1 min yield good results. Incubation with RBC lysis buffer for an extended period of time may lead to unwanted lysis of cells of interest.
3. Different cytokines might require a different stimulation regime, but for IL-2, IL-4, IL-10, IL-17 and IL-21 stimulation with 50 ng/ml PMA and 1 μg/ml ionomycin for 4–6 h yields good results.

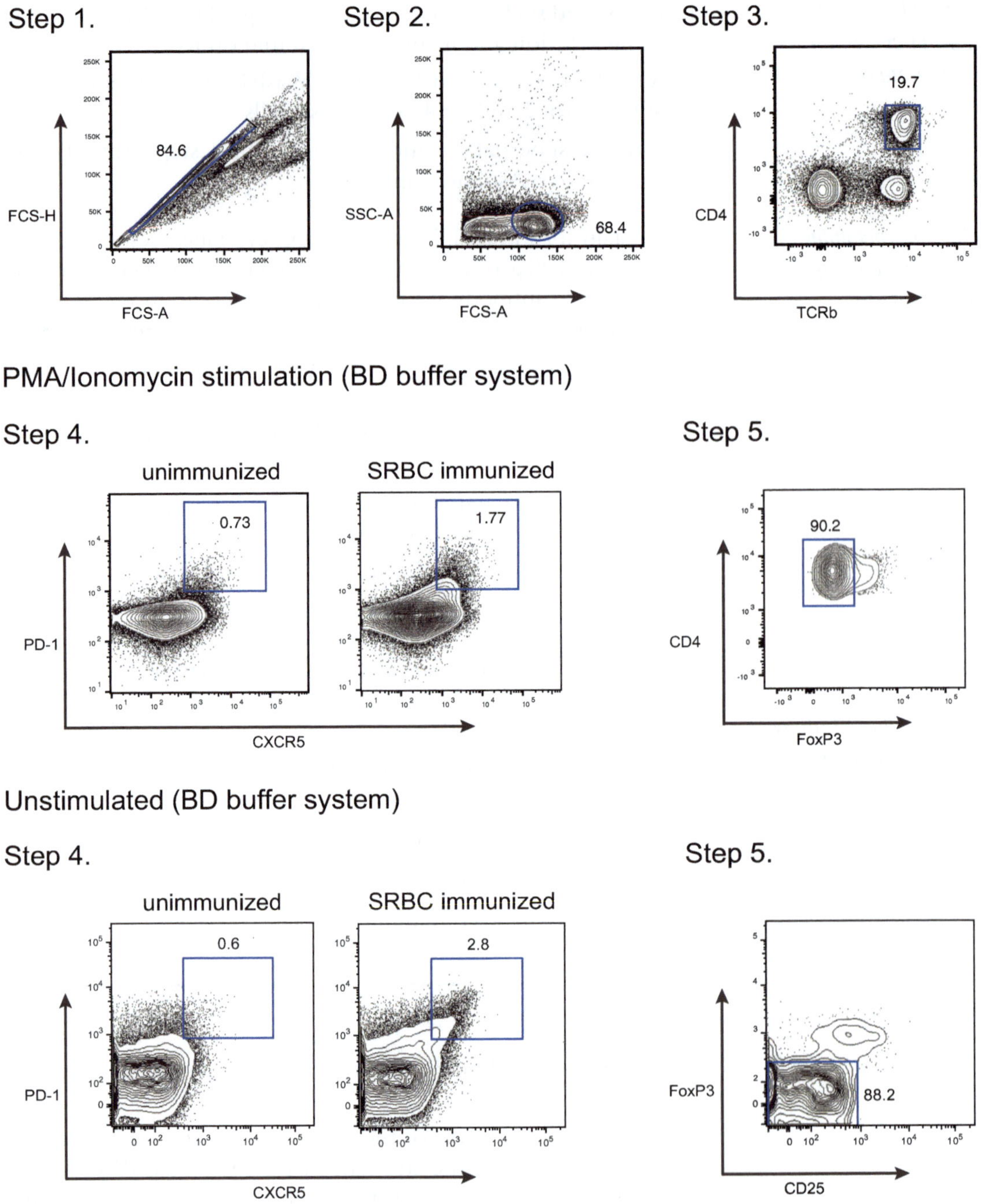

Fig. 1 Gating strategy for Tfh cells. Splenocytes isolated from C57BL/6 mice immunized with 2×10^8 sheep red blood cells i.v., 7 days prior. The cells were isolated and prepared as described in Subheading 3.1. All figures following step 1 are a depiction of the events selected within the previous gate. *Step 1*: exclusion of doublets. *Step 2*: Lymphocyte gate. *Step 3*: CD4+ T cell gate. *Step 4*: CXCR5hi PD-1hi. *Step 5*: exclusion of FoxP3 + Tfr cells. Tfh cell gate: PMA/ionomycin stimulation can reduce the fluorescence intensity of CXCR5 and PD1 as shown in *Step 4* Tfh cells after PMA/ionomycin compared with unstimulated Tfh cells

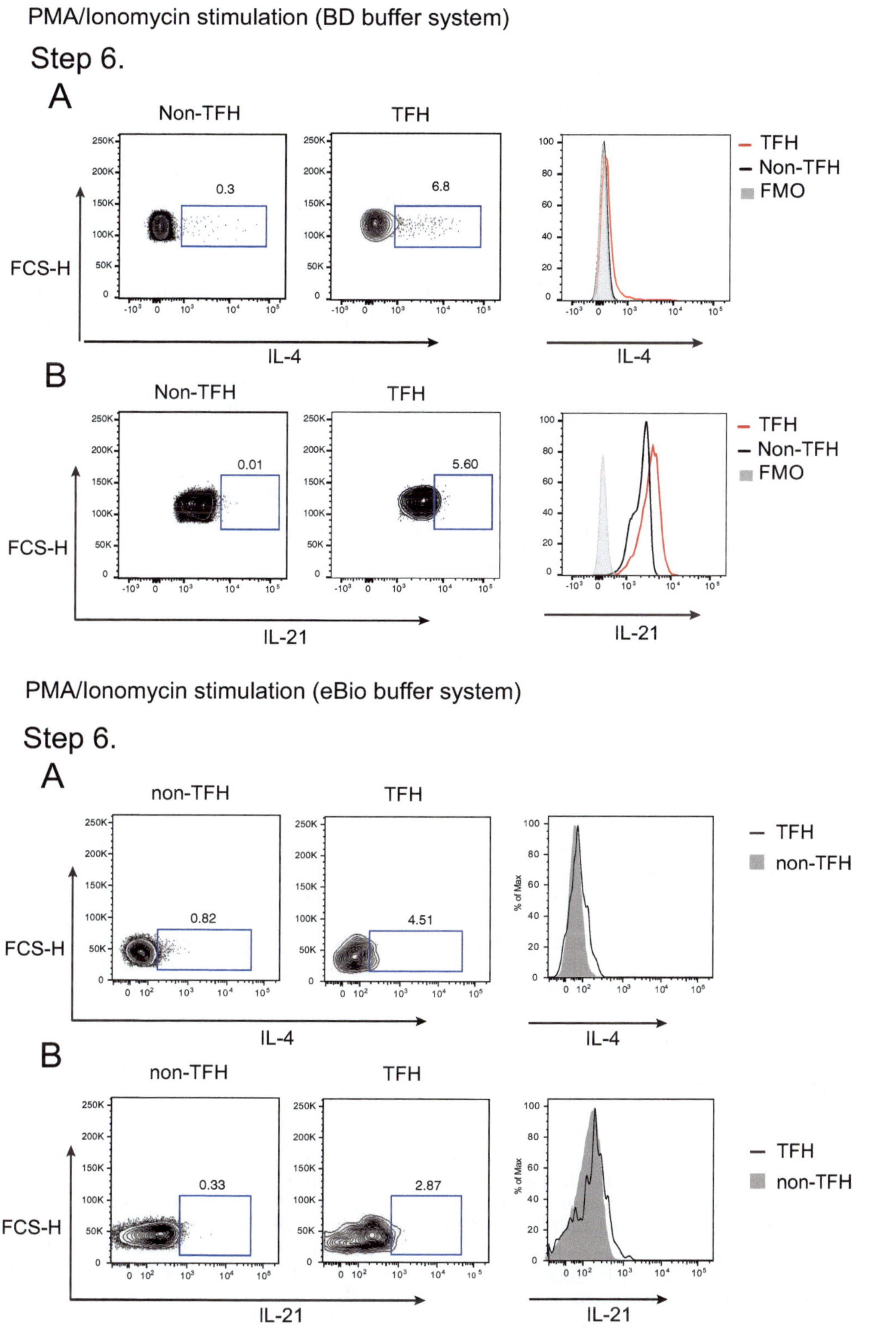

Fig. 2 Intracellular immunostain for cytokines in Tfh cells. Intracellular immunostaining for IL-4 and IL-21 using the BD Cytofix/Cytoperm Kit and the eBioscience FoxP3/Transcription Factor Staining Buffer Set (*Step 6*). The representative dot plots and histogram overlays of CXCR5hi PD-1hi FoxP3- TCRb+ CD4+ T cells (Tfh cells) and CXCR5- PD-1- CD4+ TCRb+ T cells (non-Tfh cells) shown in *Step 6* follow the *Steps 1–5* shown in Fig. 1

4. Depending on the experimental setting, cytokines may be detected directly ex vivo; however, in vitro stimulation greatly increases the availability of intracellular cytokines.
5. Performing this step at room temperature enhances the detection of CXCR5. Alternatively, this step can be performed at 4 °C for 1 h.
6. Both the BD Cytofix/Cytoperm Kit and the eBioscience FoxP3/Transciption Factor Staining Buffer Set can be used for intracellular cytokine immunostaining. For procedures requiring the nuclear detection of Bcl-6 or FoxP3, the eBiosciences protocol is preferable, but reduces the fluorescence intensity of surface markers including CXCR5 and PD-1, which requires optimization.
7. For detection of IL-21 we use IL-21 receptor chimera (IL-21rFc) (R&D Systems) conjugated in-house to the fluorochrome A647 using the Molecular Probes Alexa Fluor 647 antibody labeling kit.
8. The best negative control is cytokine deficient T cells. In the absence of cytokine deficient T cells (to minimize false-positive results) we highly recommend including appropriate isotype controls for each cytokine included in your immunostaining panel.

References

1. MacLennan IC (1994) Germinal centers. Annu Rev Immunol 12:117–139. doi:10.1146/annurev.iy.12.040194.001001
2. Liu YJ, Malisan F, de Bouteiller O, Guret C, Lebecque S, Banchereau J, Mills FC, Max EE, Martinez-Valdez H (1996) Within germinal centers, isotype switching of immunoglobulin genes occurs after the onset of somatic mutation. Immunity 4(3):241–250
3. Claman HN, Chaperon EA, Triplett RF (1966) Thymus-marrow cell combinations. Synergism in antibody production. Proc Soc Exp Biol Med 122(4):1167–1171
4. Miller JF, Mitchell GF (1968) Cell to cell interaction in the immune response. I. Hemolysin-forming cells in neonatally thymectomized mice reconstituted with thymus or thoracic duct lymphocytes. J Exp Med 128(4):801–820
5. Breitfeld D, Ohl L, Kremmer E, Ellwart J, Sallusto F, Lipp M, Förster R (2000) Follicular B helper T cells express CXC chemokine receptor 5, localize to B cell follicles, and support immunoglobulin production. J Exp Med 192 (11):1545–1552
6. Schaerli P, Willimann K, Lang AB, Lipp M, Loetscher P, Moser B (2000) CXC chemokine receptor 5 expression defines follicular homing T cells with B cell helper function. J Exp Med 192(11):1553–1562
7. Kim CH, Rott LS, Clark-Lewis I, Campbell DJ, Wu L, Butcher EC (2001) Subspecialization of CXCR5+ T cells: B helper activity is focused in a germinal center-localized subset of CXCR5+ T cells. J Exp Med 193(12):1373–1381
8. Hardtke S, Ohl L, Forster R (2005) Balanced expression of CXCR5 and CCR7 on follicular T helper cells determines their transient positioning to lymph node follicles and is essential for efficient B-cell help. Blood 106 (6):1924–1931. Epub 2005 May 1917
9. Choi YS, Kageyama R, Eto D, Escobar TC, Johnston RJ, Monticelli L, Lao C, Crotty S (2011) ICOS receptor instructs T follicular helper cell versus effector cell differentiation via induction of the transcriptional repressor Bcl6. Immunity 34(6):932–946. doi:10.1016/j.immuni.2011.03.023
10. Fazilleau N, McHeyzer-Williams LJ, Rosen H, McHeyzer-Williams MG (2009) The function of follicular helper T cells is regulated by the strength of T cell antigen receptor binding. Nat Immunol 10(4):375–384. doi:10.1038/ni.1704

11. Bauquet AT, Jin H, Paterson AM, Mitsdoerffer M, Ho IC, Sharpe AH, Kuchroo VK (2009) The costimulatory molecule ICOS regulates the expression of c-Maf and IL-21 in the development of follicular T helper cells and TH-17 cells. Nat Immunol 10(2):167–175. doi:10.1038/ni.1690
12. Chtanova T, Tangye SG, Newton R, Frank N, Hodge MR, Rolph MS, Mackay CR (2004) T follicular helper cells express a distinctive transcriptional profile, reflecting their role as non-Th1/Th2 effector cells that provide help for B cells. J Immunol 173(1):68–78
13. Foy TM, Laman JD, Ledbetter JA, Aruffo A, Claassen E, Noelle RJ (1994) gp39-CD40 interactions are essential for germinal center formation and the development of B cell memory. J Exp Med 180(1):157–163
14. Renshaw BR, Fanslow WC, Armitage RJ, Campbell KA, Liggitt D, Wright B, Davison BL, Maliszewski CR (1994) Humoral immune responses in CD40 ligand-deficient mice. J Exp Med 180(5):1889–1900
15. van Essen D, Kikutani H, Gray D (1995) CD40 ligand-transduced co-stimulation of T cells in the development of helper function. Nature 378(6557):620–623. doi:10.1038/378620a0
16. Ozaki K, Spolski R, Feng CG, Qi C-F, Cheng J, Sher A, Morse HC, Liu C, Schwartzberg PL, Leonard WJ (2002) A critical role for IL-21 in regulating immunoglobulin production. Science 298(5598):1630–1634. doi:10.1126/science.1077002
17. Vogelzang A, McGuire HM, Yu D, Sprent J, Mackay CR, King C (2008) A fundamental role for interleukin-21 in the generation of T follicular helper cells. Immunity 29(1):127–137. doi:10.1016/j.immuni.2008.06.001
18. Reinhardt RL, Liang HE, Locksley RM (2009) Cytokine-secreting follicular T cells shape the antibody repertoire. Nat Immunol 10(4):385–393
19. Crotty S (2014) T follicular helper cell differentiation, function, and roles in disease. Immunity 41(4):529–542. doi:10.1016/j.immuni.2014.10.004
20. Kim CH, Lim HW, Kim JR, Rott L, Hillsamer P, Butcher EC (2004) Unique gene expression program of human germinal center T helper cells. Blood 104(7):1952–1960. doi:10.1182/blood-2004-03-1206
21. Banchereau J, Rousset F (1992) Human B lymphocytes: phenotype, proliferation, and differentiation. Adv Immunol 52:125–262
22. Rousset F, Garcia E, Defrance T, Péronne C, Vezzio N, Hsu DH, Kastelein R, Moore KW, Banchereau J (1992) Interleukin 10 is a potent growth and differentiation factor for activated human B lymphocytes. Proc Natl Acad Sci U S A 89(5):1890–1893
23. Linterman MA, Pierson W, Lee SK, Kallies A, Kawamoto S, Rayner TF, Srivastava M, Divekar DP, Beaton L, Hogan JJ, Fagarasan S, Liston A, Smith KGC, Vinuesa CG (2011) Foxp3+ follicular regulatory T cells control the germinal center response. Nat Med 17(8):975–982. doi:10.1038/nm.2425
24. Chung Y, Tanaka S, Chu F, Nurieva RI, Martinez GJ, Rawal S, Wang Y-H, Lim H, Reynolds JM, X-h Z, H-m F, Z-m L, Neelapu SS, Dong C (2011) Follicular regulatory T cells expressing Foxp3 and Bcl-6 suppress germinal center reactions. Nat Med 17(8):983–988. doi:10.1038/nm.2426
25. Cai G, Nie X, Zhang W, Wu B, Lin J, Wang H, Jiang C, Shen Q (2012) A regulatory role for IL-10 receptor signaling in development and B cell help of T follicular helper cells in mice. J Immunol 189(3):1294–1302. doi:10.4049/jimmunol.1102948
26. Wu HY, Quintana FJ, Weiner HL (2008) Nasal anti-CD3 antibody ameliorates lupus by inducing an IL-10-secreting CD4+ CD25- LAP+ regulatory T cell and is associated with down-regulation of IL-17+ CD4+ ICOS+ CXCR5+ follicular helper T cells. J Immunol 181(9):6038–6050
27. Yu D, Rao S, Tsai LM, Lee SK, He Y, Sutcliffe EL, Srivastava M, Linterman M, Zheng L, Simpson N, Ellyard JI, Parish IA, Ma CS, Li Q-J, Parish CR, Mackay CR, Vinuesa CG (2009) The transcriptional repressor Bcl-6 directs T follicular helper cell lineage commitment. Immunity 31(3):457–468. doi:10.1016/j.immuni.2009.07.002
28. Hsu H-C, Yang P, Wang J, Wu Q, Myers R, Chen J, Yi J, Guentert T, Tousson A, Stanus AL, Le T-vL, Lorenz RG, Xu H, Kolls JK, Carter RH, Chaplin DD, Williams RW, Mountz JD (2008) Interleukin 17-producing T helper cells and interleukin 17 orchestrate autoreactive germinal center development in autoimmune BXD2 mice. Nat Immunol 9(2):166–175. doi:10.1038/ni1552
29. Moudgil KD, Choubey D (2011) Cytokines in autoimmunity: role in induction, regulation, and treatment. J Interf Cytokine Res 31(10):695–703. doi:10.1089/jir.2011.0065
30. Wichner K, Stauss D, Kampfrath B, Krüger K, Müller G, Rehm A, Lipp M, Höpken UE (2015) Dysregulated development of IL-17- and IL-21-expressing follicular helper T cells and increased germinal center formation in the absence of RORγt. FASEB J. doi:10.1096/fj.15-274001

Chapter 9

Follicular Dendritic Cell Isolation and Loading of Immune Complexes

Balthasar A. Heesters, Cees E. Van der Poel, and Michael C. Carroll

Abstract

Follicular dendritic cells (FDCs) are stromal cells that are centrally located within B cell follicles of lymph nodes and other lymphoid organs such as the spleen. Due to their relative low abundance and difficulty to isolate, FDCs are still largely an enigma. Here we describe how to isolate FDCs for ex vivo cell culture, sorting by flow cytometry and how to load them in vivo or in vitro with immune complexes.

Key words Cell isolation, FACS, Follicular dendritic cell, Cell culture, Immune complex

1 Introduction

Follicular dendritic cells (FDCs) are stromal-derived cells that are essential for maintenance of the architecture of B cell follicles within lymph nodes (LNs) and other secondary lymphoid organs such as the spleen [1–3]. They are unique in that they retain antigen for extensive periods, i.e., up to months after immunization, and are critical for clonal selection of B cells within germinal centers (GC) [4–6]. Due to their low abundance, i.e., less than 1% of total LN stromal cells, and difficulty in isolation, much of our understanding of FDC is derived from studies in vivo [1–7]. However, recent studies of FDC in ex vivo cultures reveal a complex biology. Intriguingly, complement-opsonized immune complexes are taken up by FDC via the CD21 receptor into a cycling endosomal compartment where they periodically are cycled to the cell surface [8, 9]. In order to understand the biology of FDC at a more cellular level, it is important to isolate them to relatively high purity. Here, we describe a relatively straightforward approach to isolate FDCs for study in cell culture or for sorting by flow cytometry and how to load them in vivo or in vitro with immune complexes.

Dinis Pedro Calado (ed.), *Germinal Centers: Methods and Protocols*, Methods in Molecular Biology, vol. 1623,
DOI 10.1007/978-1-4939-7095-7_9, © Springer Science+Business Media LLC 2017

2 Materials

Prepare all buffers and medium under sterile conditions. Culture cells under 10% CO_2 in an incubator at 37 °C.

2.1 Buffers and Medium

1. Collection buffer: supplement Hanks' Balanced Salt Solution (HBSS) with 20 mM HEPES (4-(2-hydroxyethyl)-1-piperazineethanesulfonic acid) and adjust pH to 7.4.
2. Digestion mix: supplement Collection buffer with 100 μg/ml DNase I, 800 μg/ml dispase, and 200 μg/ml collagenase.
3. Harvest buffer: supplement Collection buffer with 2 mM ethylenediaminetetraacetic acid, EDTA, and 1% bovine serum albumin (BSA).
4. FDC medium: supplement Dulbecco's modified Eagle's medium (DMEM) with 10% fetal calf serum (FCS, do not heat-inactivate [10]), 20 mM HEPES, 2 mM L-glutamine, 200 μM Nonessential amino acids (NEAA), 1 μg/ml gentamycin, and 55 μM β-mercaptoethanol (2-ME, 14.83 M stock). Adjust pH to 7.4.
5. Sort buffer: supplement Collection buffer with 0.5% BSA and 2 mM EDTA.
6. Gelatin veronal buffer supplemented with Ca^{2+} and Mg^{2+} (GVB++): 0.1% gelatin, 5 mM Veronal, 145 mM NaCl, 0.025% NaN_3, 0.15 mM calcium chloride, and 0.5 mM magnesium chloride, pH 7.3.
7. Phosphate-buffered saline (PBS).

2.2 Reagents

1. Glass Pasteur pipettes: fire polish glass Pasteur pipettes into three sizes of decreasing order for trituration of LNs. Suggested sizes are approximately 75%, 50%, and 25% of the original size.
2. Coverslips: dissolve type I collagen in a 0.2% acetic acid solution to get a final concentration of 50.8 μg/ml. Use sterile 18 mm glass coverslips and coat them with 250 μl of the collagen solution and air-dry overnight. This will result in a coating of approximately 5 μg/cm^2.
3. 5-ml and 15-ml polystyrene conical centrifuge tubes.
4. 50-ml polypropylene conical centrifuge tubes.
5. A column-free based cell separation magnet system (e.g., STEMCELL Technologies).

2.3 Antibodies (See Note 1)

1. Anti-mouse CD45-coated magnetic microbeads.
2. Streptavidin-coated magnetic microbeads.
3. Streptavidin-conjugated Alexa 488 (SA-Alexa 488).

4. Biotin-conjugated anti-mouse CD35, clone 8C12 (CD35/8C12-Bio).
5. Phycoerythrin-cyanine 7-conjugated anti-mouse CD45, clone 30-F1 (CD45-PE/Cy7).
6. Phycoerythrin-cyanine 7-conjugated anti-mouse Ter119, clone TER-119 (Ter119-PE/Cy7).
7. Allophycocyanin-conjugated anti-mouse podoplanin, clone 8.1.1 (Podoplanin-APC).
8. Peridinin chlorophyll protein-cyanine 5.5 anti-mouse CD31, clone 390 (CD31-PerCPCy5.5).
9. Fixable dead cell stain efluor780 (LIVE/DEAD-e780).
10. Pacific Blue-conjugated anti-mouse CD21/CD35, clone 7E9 (CD21/CD35-Pacific Blue).
11. Unconjugated anti-mouse VLA-4, clone PS2 (α-VLA4).
12. Unconjugated anti-mouse LFA, clone 2E6 (α-LFA).
13. PE conjugated anti-mouse CD16/32, clone 93 (CD16/32-PE).
14. B-phycoerythrin (B-PE).
15. Anti-phycoerythrin antibody raised in rabbit (Rabbit α-PE).

2.4 Fluorescent Activated Cell Sorting (FACS)

To use the combination of specified fluorochrome conjugated antibodies a flow cytometer equipped with a Blue, Red, and Violet laser is required.

3 Methods

3.1 Isolation of Follicular Dendritic Cells

3.1.1 Initial Steps for the Isolation of Follicular Dendritic Cells

1. On the day before cell isolation; inject 10 μg of CD35/8C12-Bio in 50 μl 1× PBS intravenous (i.v.) per mouse (~25 g weight) 24 h before harvest. For a typical experiment 4–8 mice are used.
2. Coat sterile glass coverslips 24 h before cell isolation.
3. Collect the Popliteal, Inguinal, Axillary, Brachial, Superficial Cervical, Deep Cervical, and Auricular LNs (Fig. 1) in collection buffer using blunt dissection and clean off as much fat as possible (*see* **Note 2**).
4. Transfer LNs to 1 ml of digestion mix in a 50-ml conical tube.
5. Incubate LNs at 37 °C for 5 min, swirl and incubate for another 5 min.
6. Gently triturate the LNs using fire-polished Pasteur pipettes (decreasing size each round), 10–20 times until LNs easily pass though (*see* **Note 3**).

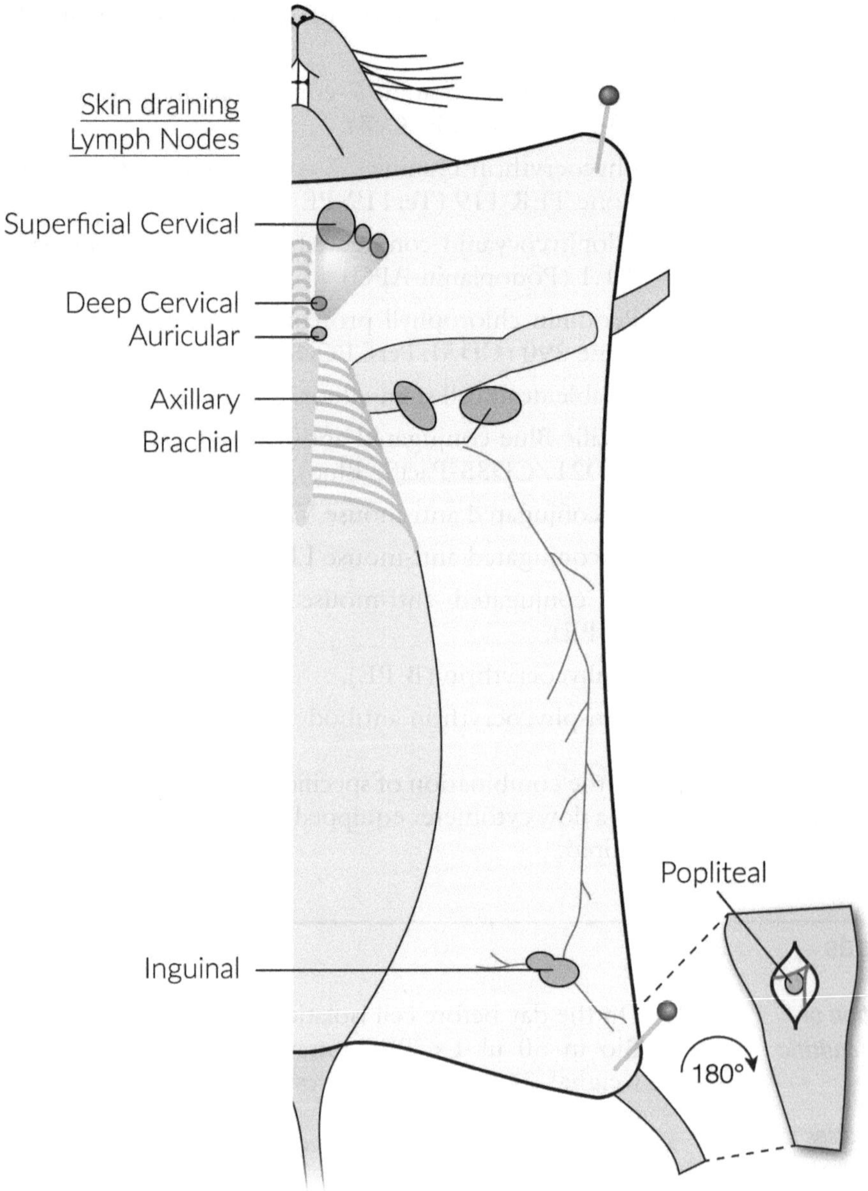

Fig. 1 Schematic of mouse skin draining lymph node anatomical location. Diagram indicates lymph nodes typically harvested for FDC isolation. From *top* to *bottom*: superficial cervical, deep cervical, auricular, axillary, brachial, inguinal, and popliteal lymph nodes

7. Take supernatant and transfer to 50 ml tube with 15 ml of harvest buffer on ice.
8. Add new digestion mix to LNs.
9. Repeat **steps 5–8** until LNs are fully digested.
10. Proceed for isolation for flow cytometry in Subheading 3.1.2 or for cell culture in Subheading 3.1.3.

3.1.2 Follicular Dendritic Cell Flow Cytometry

1. Spin single cell suspension in harvest buffer at 200 × *g* for 5 min.
2. Resuspend single cell suspension in 350 μl harvest buffer and add 50 μl of CD45-microbeads. Simultaneous release B cells by incubation with blocking antibodies for VLA-4 and LFA: 4 μl of PS2 (~3 mg/ml, α-VLA-4) and 4 μl of 2E6 (~750 μg/ml, α-LFA) [1:100].
3. Incubate for 30 min on ice.
4. Add 7 ml of harvest buffer into a 15-ml polystyrene tube.
5. Place tube on a cell separation magnet.
6. Incubate on ice for 5 min.
7. Pour supernatant into clean tube 15-ml polystyrene tube.
8. Repeat **steps 5–7** three times.
9. Spin down supernatant at 200 × *g* for 5 min.
10. Resuspend in 200 μl of the antibody mastermix in sort buffer: SA-Alexa488 [1:400], CD45-PE/Cy7 [1:400], Ter119-PE/Cy7 [1:400], Podoplanin-APC [1:200], CD31-PerCPCy5.5 [1:400], LIVE/DEAD-e780 [1:2000], CD21/CD35-Pacific blue [1:200], CD16/CD32-PE [1:300], α-VLA-4) [1:100],α-LFA [1:100] (*see* **Note 1**).
11. Incubate for 30 min on ice.
12. Wash, spin at 200 × *g* for 5 min.
13. Resuspend in 500 μl sort buffer.
14. Acquire cells on a flow cytometric sorter machine (Fig. 2).

3.1.3 Follicular Dendritic Cell Culture

1. Spin single cell suspension in harvest buffer at 200 × *g* for 5 min.
2. (**Steps 2–9** are optional) Resuspend single cell suspension in 350 μl harvest buffer and add 50 μl of CD45-microbeads. Simultaneous release B cells by treating with blocking abs; 4 μl (~3 mg/ml, α-VLA-4) and 4 μl (~750 μg/ml, α-LFA).
3. Incubate for 30 min on ice.
4. Add 7 ml of harvest buffer into a 15-ml polystyrene tube.
5. Place tube on a cell separation magnet.
6. Incubate on ice for 5 min.
7. Pour supernatant into clean tube 15-ml polystyrene tube.
8. Repeat **steps 5–7** three times.
9. Spin down supernatant at 200 × *g* for 5 min.
10. Suspend pellet in 500 μl harvest buffer.
11. Add 50 μl SA-microbeads, incubate for 30 min on ice.
12. Fill the polystyrene tube till the height of the magnet.

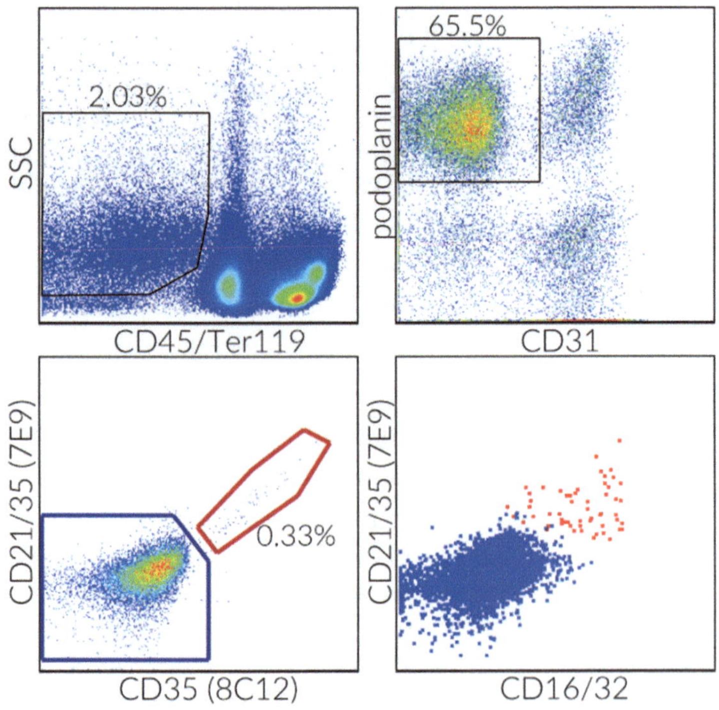

Fig. 2 Gating strategy for flow-cytometric analysis of FDC. Cell suspensions were prepared as described in the protocol and stained using the indicated antibodies. After FSC/SSC gating, singlet gating and gating for cells negative for live/dead staining (not shown), stromal cells can be defined as CD45/ter119 negative (*top left panel*). The main stromal cell populations can be further discriminated using podoplanin and CD31 staining (*top right panel*). Within the FRC/FDC gate (podoplanin positive, CD31 negative) FDC are defined as CD21/35 (7E9) and in vivo labeled CD35 (8C12) double positive cells (*lower left panel*). The lower right panel shows a dotplot overlay with Fc receptor positive FDC in *red* and FRC in *blue*

13. Incubate on ice for 5 min.
14. Decant tube.
15. Wash tube with 1 ml harvest buffer.
16. Repeat **steps 12–15** three times.
17. Resuspend cells in FDC media and count cells.
18. Plate 100 k cells in 500 μl FDC media per collagen coated coverslip (12 mm circular coverslips in 24 well tissue culture plate).

3.2 Loading of Immune Complexes onto Follicular Dendritic Cells

3.2.1 Preparation of PE-Immune Complex (PE-IC) for In Vitro Studies

1. Bleed a mouse for fresh serum.
2. Incubate blood for 30 min at RT, then spin down at 9400 × *g* for 5 min.
3. Collect the serum (supernatant).
4. Prepare PE-IC by mixing 10 μl of 1 mg/ml of B-PE, 1 μl of 10 mg/ml of rabbit α-PE, 10 μl of fresh mouse serum, and 79 μl of GVB++ and incubate the mixture for 30 min at 37 °C.

3.2.2 In Vitro PE-IC Binding to Naïve B Cells

1. Prepare a single cell suspension from a red blood cell lysed mouse spleen.
2. Mix PE-IC with 10^6 splenocytes in 500 μl FDC media (*see* **Note 4**).
3. Incubate for 30 min at 37 °C.
4. Wash with 10 ml FDC media and suspend in 100 μl prewarmed FDC media.

3.2.3 In Vitro Transfer of PE-IC from B Cell to FDC

1. Change FDC media with fresh prewarmed FDC media to remove dead cells.
2. Incubate 10^6 B-cells with the cultured FDC for 30 min at 37 °C.
3. Wash four times with 2 ml prewarmed FDC media.

3.2.4 Loading of FDC with Immune Complexes In Vivo

1. Mix 10 μl of 1 mg/ml of B-PE with 1 μl of 10 mg/ml of rabbit α-PE in 200 μl PBS.
2. Inject 10 μl in the mouse footpad.
3. Alternatively to the above two first steps inject 100 μg of 10 mg/ml of rabbit α-PE i.v.
4. Twenty four hours after **step 3**, inject 10 μl of 1 mg/ml of B-PE solution (10 μg B-PE in 200 μl PBS) in the mouse footpad.

4 Notes

1. Antibodies, viability dye and their dilutions have been titrated in our laboratory. Please be aware that optimal concentrations are likely to vary depending on the company, product lot, and experimental conditions.
2. Viability drastically decreases with time, the time from mouse to single cell suspension should be minimized.
3. FDCs are extremely sensitive and gentle trituration cannot be emphasized enough.
4. Negative controls are required, i.e., B-cells without PE-IC; and PE-IC without serum.

Acknowledgments

We thank members of the Carroll lab for helpful comments and experimental results that were important in working out the conditions for FDC isolation and maintenance in cell culture. We gratefully acknowledge research support from NIH grants AI110164 and AI039246-19.

References

1. Krautler NJ, Kana V, Kranich J, Tian Y, Perera D, Lemm D, Schwarz P, Armulik A, Browning JL, Tallquist M, Buch T, Oliveira-Martins JB, Zhu C, Hermann M, Wagner U, Brink R, Heikenwalder M, Aguzzi A (2012) Follicular dendritic cells emerge from ubiquitous perivascular precursors. Cell 150(1):194–206. doi:10.1016/j.cell.2012.05.032
2. Tew JG, Kosco MH, Burton GF, Szakal AK (1990) Follicular dendritic cells as accessory cells. Immunol Rev 117:185–211
3. Cyster JG (2010) B cell follicles and antigen encounters of the third kind. Nat Immunol 11 (11):989–996. doi:10.1038/ni.1946
4. Tew JG, Mandel TE (1979) Prolonged antigen half-life in the lymphoid follicles of specifically immunized mice. Immunology 37(1):69–76
5. Cyster JG, Hartley SB, Goodnow CC (1994) Competition for follicular niches excludes self-reactive cells from the recirculating B-cell repertoire. Nature 371(6496):389–395. doi:10.1038/371389a0
6. Heesters BA, Myers RC, Carroll MC (2014) Follicular dendritic cells: dynamic antigen libraries. Nat Rev Immunol 14(7):495–504. doi:10.1038/nri3689
7. Jarjour M, Jorquera A, Mondor I, Wienert S, Narang P, Coles MC, Klauschen F, Bajenoff M (2014) Fate mapping reveals origin and dynamics of lymph node follicular dendritic cells. J Exp Med 211(6):1109–1122. doi:10.1084/jem.20132409
8. Heesters BA, Chatterjee P, Kim YA, Gonzalez SF, Kuligowski MP, Kirchhausen T, Carroll MC (2013) Endocytosis and recycling of immune complexes by follicular dendritic cells enhances B cell antigen binding and activation. Immunity 38(6):1164–1175. doi:10.1016/j.immuni.2013.02.023
9. Heesters BA, Lindqvist M, Vagefi PA, Scully EP, Schildberg FA, Altfeld M, Walker BD, Kaufmann DE, Carroll MC (2015) Follicular dendritic cells retain infectious HIV in cycling endosomes. PLoS Pathog 11(12):e1005285. doi:10.1371/journal.ppat.1005285
10. Look N (1996) Heat inactivation – are you wasting your time? Art to Science (HyClone Newsletter) 15:1–5

Chapter 10

Isolation and Characterization of Mouse and Human Follicular Dendritic Cells

Mohey Eldin M. El Shikh, Riham El Sayed, and Costantino Pitzalis

Abstract

Follicular dendritic cells (FDCs) reside in the B cell follicles of secondary and tertiary lymphoid tissues where they play key roles in the development and maintenance of lymphoid tissue architecture and function. FDCs trap native antigens for extended periods of time in the form of immune complexes which critcally regulate germinal center reactions in health and disease. Here, we describe how to isolate and characterize FDCs from murine and human lymphoid tissues.

Key words Follicular dendritic cells, FDC isolation, Germinal center, GCR, Lymph nodes, Spleen, Tonsil

1 Introduction

Follicular dendritic cells (FDCs) are unique accessory immune cells that help maintain the structure and function of primary and secondary B cell follicles [1]. The cardinal feature of FDCs is the long-term retention of unprocessed antigens in the form of immune complexes (ICs) via Fc and complement receptors (Figs. 1 and 2) [2]. This unique ability to trap unlimited arrays of molecules in the presence of specific antibodies and complement is a double-edged sword in regulated and dysregulated immunity [3]. On the one hand, induction of protective germinal center reactions (GCRs) against microbial antigens retained on FDCs is essential for long-standing protection and maintenance of immunological memory [4]. On the other hand, retention of autoantigens [5], viral particles [6], and prion proteins [7] on FDCs has been implicated in the pathogenesis of chronic inflammatory autoimmune diseases, chronic viral infections including AIDS, prion diseases, and infection-driven B cell lymphomas in humans and animals. In addition to antigen presentation, FDCs regulate the GCRs by providing secreted and membrane-bound B cell co-stimulatory signals including C4b-binding protein (C4bBP), CD21L, 8D6, BAFF, IL-1β,

Dinis Pedro Calado (ed.), *Germinal Centers: Methods and Protocols*, Methods in Molecular Biology, vol. 1623,
DOI 10.1007/978-1-4939-7095-7_10, © Springer Science+Business Media LLC 2017

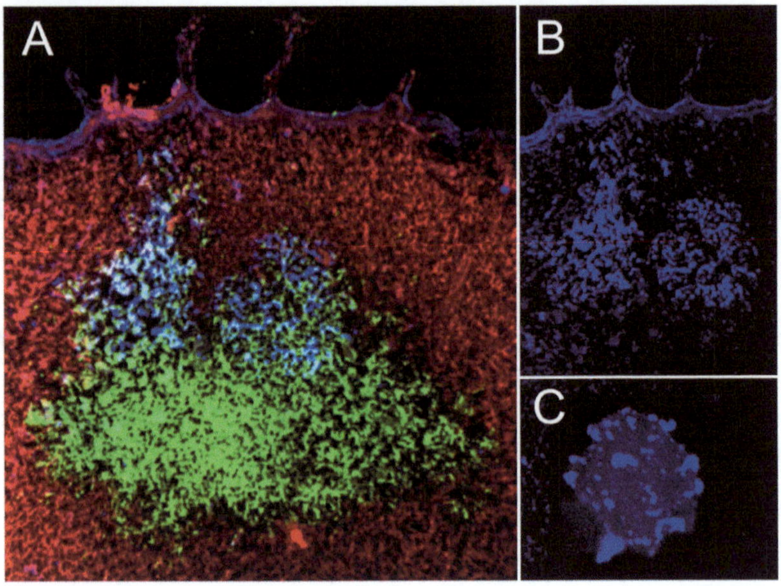

Fig. 1 Murine FDCs retaining HEL-ICs in a secondary B cell follicle of a mouse LN. C57BL/6 mice were passively immunized with 1 mg rabbit anti-HEL IP followed 24 h later by 100 μg HEL SC in the back of the neck. Five days later, mice were sacrificed and the axillary LNs were collected. Ten μm thick sections were labeled with PE-conjugated anti-IgD (resting B cell marker) and FITC-conjugated GL-7 (GC B cell marker) then visualized by scanning confocal microscopy. (**a**) AMCA-labeled HEL (*blue*) is retained in the FDC reticulum and GC B cells are labled green with GL-7. (**b**) Single channel showing AMCA-HEL on FDCs. (**c**) FDCs isolated by magnetic cell separation from the draining axillary LNs show surface-retained AMCA-HEL (*blue*)

IL-6, IL-7, IL-10, and IL-15 and these co-signals are upregulated on activated FDCs [1–3].

FDCs form reticular networks that intemitly interact with germinal center B cells in vivo. The (1) fragile nature of FDCs after dissociation of such networks during isolation, (2) the relatively small numbers of FDCs in lymphoid tissues (0.1% of cells in secondary lymphoid tissues and <2% of germinal center cells), and (3) the strong synapses holding FDCs and B cells are all important factors that should be considered in FDC isolation protocols.

FDC enrichment protocols [8–12] generally start with lymphocyte depletion by irradiation and preparation of single cell suspension from lymphoid tissues using enzymatic digestion. Several approaches can then be used to isolate FDCs from the single cell suspension. These approaches are mainly based on (1) Percoll or bovine serum albumin (BSA) density gradients cell fractionation, (2) fluorescence-activated cell sorting (FACS), (3) magnetic cell separation, or (4) a combination of more than one technique. Here we describe protocols for the isolation of murine and human FDCs from secondary lymphoid tissues, together with methods of identification of purified FDCs in vitro.

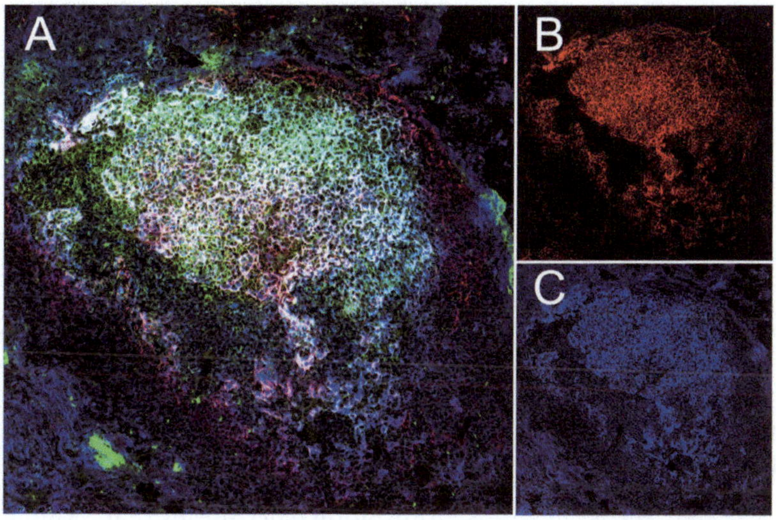

Fig. 2 IC retention on human FDCs in a secondary B cell follicle of a human tonsil. Ten μm thick sections of surgically removed tonsil were labeled with CNA-42 (FDC, *red*), CD20 (B cells, *green*) and anti human IgG (*blue*, ICs) then visualized by scanning confocal microscopy (**a**) An overlay showing co-localization of ICs (*blue*) on human FDCs (*red*) with GC B cells (*green*). The fine FDC meshwork in the follicular mantle (*red only*) does not show IC retention. (**b, c**) Single color channels showing FDCs (CNA.42, *red*) and IgG in ICs (*blue*) in (**b**) and (**c**) respectively

2 Materials

2.1 Human Tissue or Mice and Irradiation

1. Human lymphoid tissues: freshly obtained palatine tonsils, lymph nodes, splenic fragments, or malignant lymphomas.
2. Naïve or immunized mice, 8–10 weeks old (*see* **Note 1**).
3. Sterile ventilated irradiation container.
4. Cesium-137 (^{137}Cs) irradiator.
5. 70 μm cell strainer.

2.2 Medium and Solutions

1. Dulbecco's Modified Eagle Medium (DMEM) [High glucose—No glutamine base medium]. For human tissue use Roswell Park Memorial Institute (RPMI) 1640 instead of DMEM.
2. Complete DMEM (cDMEM): Add 20 mM HEPES buffer; 2 mM L-glutamine; 0.2 mM nonessential amino acids; and 10 μg/mL of gentamicin.
3. Preparation of cDMEM with FCS (cDMEM with 10% FCS): Heat-inactivate FCS at 56 °C for 45 min in a water bath. Add heat-inactivated FCS to cDMEM [final concentration 10%].

Make sure media, sera, and supplements are sterile before use. Ready-made media and supplements can be used straight without sterilization. Media prepared in-house has to be filter-sterilized before use. Store media at 4 °C.

4. Enzyme cocktail for tissue digestion: 22 mg/mL of Collagenase D (*see* **Note 2**), and 5 mg/mL of DNase I.

2.3 Tissue Preparation for Digestion

1. Petri dishes (35 × 10 mm).
2. Sterile disposable needles (26 G).
3. Sterile disposable scalpels, scissors, and tweezers.

2.4 Labelling of Mouse Tissue and Human Tissue

1. Purified rat anti-mouse FDC M1 (clone FDC-M1).
2. Biotinylated mouse anti rat Ig kappa light chain (clone MRK-1).
3. Anti-biotin MicroBeads (e.g., Miltenyi Biotec).
4. Purified mouse anti human FDC mAbs (IgM clone HJ2 or CNA.42).
5. Anti-mouse IgM MicroBeads (e.g., Miltenyi Biotec).

2.5 Magnetic Cell Separation

1. Filter-sterile degassed MACS Buffer: Phosphate buffered saline containing 0.5% bovine serum albumin and 2 mM EDTA.
2. MACS LS column (Miltenyi Biotec).
3. Magnet and stand (Miltenyi Biotec).

2.6 Immunofloure-scent Staining and Flow Cytometry

1. FACS buffer: Phosphate-buffered saline containing 2 mM EDTA, 2% heat-inactivated FCS and 0.5% bovine serum albumin (BSA).
2. For mouse tissue use purified rat anti-mouse CD16/CD32 (Mouse Fc Block) and for human tissue human BD Fc Blocker (BD Biosciences).
3. For mouse tissue use flourochrome-conjugated mAbs against mouse CD21/CD35, ICAM-1, VCAM-1, CD40, CD16/32, and fluorophore-matched isotype controls; and for human tissue use Biotin-conjugated CNA.42, HJ2, and isotype-matched controls.
4. Streptavidin-conjugated flourophores.
5. For mouse tissue make biotinylated FDC-M1: conjugate purified rat anti mouse FDC M1 (FDC-M1) to biotin-xx Succinimidyl Ester following the manufacturer's instructions.
6. Avidin/biotin blocking kit.

3 Methods

3.1 Isolation of Mouse Follicular Dendritic Cells

1. Irradiate mice with 1000–1500 Rads (1 Rad = 0.01 Gy) 24–48 h prior to sacrificing mice (*see* **Note 3**).
2. On the day of isolation, sacrifice mice by cervical dislocation.
3. Dissect the popliteal, inguinal, axillary, brachial, submental, mandibular, cervical, and mesenteric lymph nodes from four irradiated mice and place in a petri dish containing cDMEM with 10% FCS on ice. Remove the lymph nodes with as little fat and connective tissue as possible (*see* **Note 4**).
4. Gently open the capsule of each lymph node using two 26-gauge needles mounted on plastic syringes. Teasing the capsule facilitates entry of the enzyme cocktail into the lymph node, digestion of the stroma, and release of the cells.
5. Digest the opened lymph nodes in a cocktail of 2 mL cDMEM with 10% FCS + 1.5 mL of 22 mg/mL collagenase D + 0.5 mL of 5 mg/mL DNase I [= total of 4 mL with collagenase D at a final concentration of 8 mg/mL and DNase I at a final concentration of 625 μg/mL].
6. Incubate for 45 min at 37 °C in a humified 5% CO_2 incubator.
7. After incubation, gently release the cells by pipetting the digested lymph node fragments using wide-bore Pasteur pipette.
8. Run the digestion mix through a 70 μm cell strainer and collect the released cells in 10 mL ice-cold cDMEM with 10% FCS.
9. Repeat **steps 5–8** to digest any residual tissue fragments and harvest the released cells from the second round of digestion in 10 mL ice-cold cDMEM with 10% FCS.
10. Combine the released cells from the two rounds of digestion and pellet the suspended cells by centrifugation at 300 × *g* for 10 min.
11. Resuspend the cell pellet in 1 mL cDMEM with 10% FCS.
12. Incubate the suspended cells with 0.7 μg/mL FDC-M1 [per 20×10^6 viable cells] for 1 h on ice (*see* **Note 5**).
13. Wash the cells by adding 5 mL cDMEM and centrifuge at 300 × *g* for 10 min. Aspirate supernatant completely.
14. Resuspend the cell pellet in 1 mL cDMEM with 10% FCS and incubate with 5 μg/mL biotinylated anti-rat kappa antibody (clone MRK-1) for 45 min on ice.
15. Wash the cells by adding 5 mL MACS buffer and centrifuge at 300 × *g* for 10 min. Aspirate supernatant completely.
16. Resuspend the cell pellet in 1 mL MACS buffer and incubate with 40 μL anti-biotin microbeads for 15 min on ice.

17. Wash the cells by adding 5 mL MACS buffer and centrifuge at 300 × *g* for 10 min. Aspirate supernatant completely.
18. Resuspend the cell pellet in 1 mL MACS buffer.
19. Place LS column in the magnetic field of a MACS separator and rinse it with 3 mL MACS buffer.
20. Layer the cell suspension onto the column and collect the flow-through (non-FDC fraction).
21. Wash the column with 3 mL MACS buffer 3 times and collect unlabeled cells that pass through (non-FDC fraction).
22. Remove the column from the separator and place it on a suitable collection tube.
23. Add 5 mL MACS buffer onto the column and flush out the magnetically labeled cells (FDCs) by firmly pushing the plunger into the column.
24. Collect the eluted FDCs by centrifugation at 300 × *g* for 10 min.
25. Resuspend the FDC pellet in 1 mL cDMEM with 10% FCS.

3.2 Flow Cytometry of Mouse Follicular Dendritic Cells

1. FDCs express high levels of FcRs. To reduce Fc receptor-mediated binding of antibodies of interest, preincubate the cell suspension with mouse FcR blocker [purified anti-mouse CD16/CD32 mAb 2.4G2] at a concentration of 1 μg/million cells in 100 μL at 4 °C for 5 min.
2. In the presence of the mouse Fc blocker, add flourophore-conjugated mAbs (1 μg/million cells) against mouse CD21/CD35, ICAM-1, VCAM-1, CD40, CD16/32, or matched isotype controls for 45 min on ice.
3. Wash cells and resuspend in FACS buffer [final cell concentration 1×10^6 cell/mL].
4. Analyze using flow cytometry.
5. To label FDCs with the FDC-specific marker FDC-M1, first block any biotin remaining from the isolation procedure using avidin/biotin blocking kit; then incubate FDCs with 1 μg/million cell biotinylated FDC-M1 followed by streptavidin-conjugated flourophores. Analyze by flow cytometry.
6. To assess the capacity of purified FDCs to retain immune complexes (ICs) in vitro, first prepare ovalbumin (OVA) ICs by incubating OVA with anti-OVA IgG at a ratio of 1 μg OVA to 8 μg anti-OVA for 30 min. Load 100 ng OVA in the form of ICs per million FDC and incubate for 2 h at 37 °C in a 5% CO_2 incubator. Wash the cells twice in FACS buffer to get rid of unbound ICs. Detect membrane-retained ICs with anti IgG or anti-OVA at 1 μg/million cells for 45 min. Wash and analyze by flow cytometry (*see* **Note 6**).

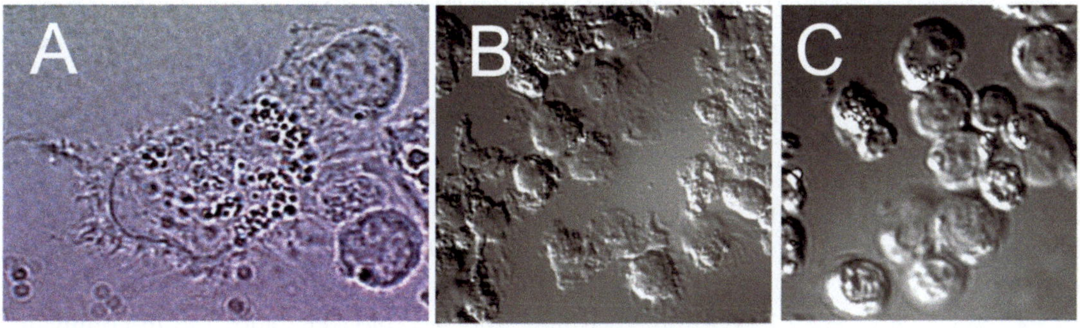

Fig. 3 Purified murine FDCs. (**a**) Unstained murine FDC freshly isolated from mouse LNs by Percoll gradient separatrion. The FDC dendrites are in contact with round lymphocytes that remained attached during the isolation process (**b**, **c**) FDCs isolated by magnetic cell separation as seen by the Nomarski optics (**b**). The dendritic morphology of FDCs is absent in a control preparation of lymphocytes (**c**)

7. Typically, 0.5–1 million FDCs can be obtained from the lymph nodes of unprimed mouse with 85–90% purity (Fig. 3). Isolated FDCs express CD32, CD21/35, CD40, CD54, FDC-M1 and can retain ICs in vitro.

3.3 Isolation of Human Follicular Dendritic Cells

1. Place the tissue(s) in petri dishes containing cRPMI-1640 with 10% FCS and dissect all connective and non-lymphoid tissue from the sample(s).
2. Cut the tissues into 2 × 2 mm using scaples and/or scissors.
3. Digest the tissues as described for the murine lymph nodes (**steps 5–8** in Subheading 3.1); however the total volume of the digestion cocktail must be increased to 20–30 mL to accommodate the larger masses of human tissues.
4. Pre-enrich the FDC population on 50% continious Percoll gradient (*see* **Note 5**). Human FDCs are collected from the 1.052 to 1.060 density interface.
5. Minimize B cell binding to FDCs by incubating the single cell suspention with 2–10 μg/mL mouse anti human CD11a/LFA-1 in 10 mL cRPMI with 10% FCS at 4 °C overnight (*see* **Note 5**).
6. Deplete macrophages by incubation on plastic Petri dishes (*see* **Note 5**). After macrophage depletion, collect the nonadherent cells by centrifugation.
7. Incubate the cells with 1 μg/million cell anti-human FDC mAb (clone HJ2 or CNA.42) in MACS buffer for 2 h at room temperature on a rocker (*see* **Note 7**).
8. Wash, resuspend the cells, and use anti mouse IgM MicroBeads to magnetically label the FDCs according to the manufacturer instructions (**steps 15–18** in Subheading 3.1).

9. Positively select the FDCs on LS columns as described in the murine FDC isolation protocol [**steps 19–25** in Subheading 3.1] (*see* **Notes 8** and **9**).
10. The typical yield contains 80–90% FDCs as determined by flow cytometry using the FDC-specific mAb HJ2 or CNA.42.

3.4 Flow Cytometry of Human Follicular Dendritic Cells

1. Block the FDC-FcRs by incubation with human BD Fc Block (2.5 μg/million cell in 100 ul FACS buffer for for 30 min at room temperature.
2. Without washing the Fc blocker, add biotin-conjugated CNA.42 or HJ2 (1 μg/million cells) or isotype matched controls for 45 min on ice in FACS buffer. Wash and incubate with streptavidin-conjugated flourophores. Analyze by flow cytometry.
3. To label for the capacity of purified human FDCs follow the steps described in murine FDC isolation (**step 6** in Subheading 3.2).

4 Notes

1. Immunization of mice: although FDCs can be isolated from naïve and immunized mice; the number of isolated FDCs significantly increases with immunization. Three immunization approaches can be used to increase the yield of FDCs: Passive immunization with 1 mg (per mouse) antigen-specific antibody intra-peritoneally (IP) or subcutaneously (SC) (e.g., OVA-specific polyclonal rabbit IgG), followed 24 h later by antigen administration at a dose of 100 μg/mouse (e.g., OVA) IP or SC. Injection of preformed immune complexes is also possible, but it may precipitate at the injection site. Mice are left for 5–7 days then FDCs are isolated. Active immunization with alum-precipitated antigens with *B. pertussis* at a dose of 200 μg/mouse SC followed 2 weeks later by a booster dose of 200 μg/mouse alum-precipitated antigens with *B. pertussis.* Mice can be used after 3–7 days after secondary immunization. Active immunization with 2×10^8 washed sheep red blood cells (SRBCs) IP followed 2 weeks later by a booster dose of 1×10^8 SRBCs. Mice can be used after 3–7 days after the booster dose.
2. Purity of collagenases: contamination of crude collagenases with trypsin or trypsin-like proteases can degrade cell surface molecules and alter the phenotype and function of FDCs. The presence of proteases can be tested by incubating T cells with collagenase for 30 min at 37 °C and then screening for the loss of the trypsin-sensitive cell surface markers CD4 and CD8 by

flow cytometry. If crude collagenases are contaminated with proteases, the use of more pure preparations like Liberase Blendzyme at a final concentration of 0.5 Wünsch units/mL is recommended.

3. Irradiation of mice: irradiation with 1000 Rads is typical for elimination of lymphocytes in this protocol; however, doses between 600 and 3500 Rads have been used. Higher doses of irradiation (>1000 Rads) eliminate more lymphocytes and thus reduce lymphocyte contamination. The high irradiation doses do not interfere with FDC phenotype or functions and this may be due to a high level of thiol compounds in FDCs that protect them against radiation injury [13]. Alternatively, lymphocytes can be eliminated by intra-peritoneal injection of cyclophosphamide monohydrate ISOPAC at a dose of 300 mg/kg 3 days before FDC isolation.

4. Other tissues: FDCs can be obtained from other lymphoid tissues (e.g., spleen, or Peyer's patches) or other species using the same approach. Tissues need to be cut into small pieces (2–3 mm) with scalpels or needles before digestion in the enzyme cocktail.

5. Alternative approach: fluorescence-activated cell sorting (FACS) is an alternative approach to magnetic cell separation of murine FDCs: (1) Murine FDCs in single cell suspensions (**step 11** in Subheading 3.1) are labeled with 1 μg/million cells biotinylated FDC-M1 then flow sorted using fluorophore-conjugated streptavidin. One or more sorting cycles can be used with an average purity of 93–95% after the second sorting [14]. Isolation of FDCs by flow sorting using FDC-M2 has also been reported [15]. (2) Prior to sorting, one or more of the following optional steps can be performed (after **step 11** in Subheading 3.1) to increase the purity of the FDC preparations. (1) Percoll gradient pre-enrichment: FDC-enriched preparations of 30–50% purity can be prepared from the single cell suspension obtained from enzymatic digestion of lymphoid tisses using density gradient separation (Fig. 3). These FDC-enriched fractions can then be labeled with FDC-M1 and FDCs either magnetically separated (*see* Subheading 3.1) or flow sorted (*see* above on this note). Fifty per cent continuous Percoll gradient is generated by combining 163 mL of Percoll solution, 114 mL of distilled water, 33 mL of 1.5 M NaCl, and 20 mL of HEPES buffer. Twenty-five milliliters are then dispensed in Beckman ultraclear cellulose nitrate centrifuge tubes and centrifuged at 30,000 × *g* for 20 min at 4 °C, using a fixed-angle rotor. These gradients can then be stored for several weeks at 4 °C where they will not be disturbed. Percoll gradients are calibrated in one of the tubes by adding 10–20 μL of density marker beads (equivalent to density levels from 1.033

to 1.090 g/mL) to ensure proper formation of gradients and determine the appropriate band for each density required. Two to five milliliters cell suspension in medium is then layered on top of the Percoll gradient and the tubes are centrifuged at 400 × *g* for 10 min at 4 °C. FDCs are collected from the layer between 1.060 and 1.074 g/mL, washed and resuspended in cDMEM with 10% FCS. (2) Interference with the FDC-B cell synapse to minimize B cell contamination: this can be achieved by incubating the single cell suspentions in cDMEM 10% FCS with 2–10 μg/mL rat anti-mouse CD54/ICAM-1 mAb before FDC labeling with biotinylated FDC-M1 for flow sorting. (3) Incubation on plastic culture plates to minimize macrophage contamination: macrophages are depleted from gradient-enriched FDC fractions by incubation in a plastic culture dish for 60 min at 37 °C, 5% CO_2 and collection of the nonadherent fraction.

6. Retention of directly labeled antigens: Flourophore-tagged antigens (e.g., phycoerythrin-conjugated OVA) can be loaded in the form of immune complexes on isolated FDCs and directly analyzed by flow cytometry without anti-OVA or anti IgG Abs.
7. Human FcR blockade: FDC-FcRs blockade by incubation with aggregated human IgG for 30 min at 4 °C can be performed before the addition of mouse anti numan FDC mAb.
8. Alternative approach: fluorescence-activated cell sorting (FACS) is an alternative approach to magnetic cell separation of human FDCs: Human FDC-enriched fractions obtained by Percoll gradient separation (**step 4** in Subheading 3.3) are labled with 1 μg/million cell anti-CD21 and anti-CD14 for 45 min on ice in FACS buffer, washed and flow sorted. FDCs are $CD21^{Hi}$ $CD14^{+}$, whereas B cells ($CD21^{+}CD14^{-}$) and monocytes ($CD21^{-}CD14^{+}$) are excluded [12]. Instead, FDCs in Percoll enriched fractions are labeled with mouse IgM anti-human FDC mAb (HJ2) followed by flourophore-conjugated anti-mouse IgM Ab, then sorted using flow cytometry [16].
9. Irradiation of purified FDCs: because FDCs are radiation resistant, human FDC preparations can be irradiated (3000 Rads) before use in downstream applications.

References

1. El Shikh ME, El Sayed RM, Tew JG et al (2009) Follicular dendritic cells (B lymphocyte stimulating). In: Encyclopedia of life sciences. John Wiley & Sons, Ltd, Chichester
2. El Shikh ME, El Sayed RM, Sukumar S et al (2010) Activation of B cells by antigens on follicular dendritic cells. Trends Immunol 31:205–211

3. El Shikh ME, Pitzalis C (2012) Follicular dendritic cells in health and disease. Front Immunol 3:292
4. Tew JG, Phipps RP, Mandel TE (1980) The maintenance and regulation of the humoral immune response: persisting antigen and the role of follicular antigen-binding dendritic cells as accessory cells. Immunol Rev 53:175–201
5. Pitzalis C, Jones GW, Bombardieri M et al (2014) Ectopic lymphoid-like structures in infection, cancer and autoimmunity. Nat Rev Immunol 14:447–462
6. Burton GF, Masuda A, Heath SL et al (1997) Follicular dendritic cells (FDC) in retroviral infection: host/pathogen perspectives. Immunol Rev 156:185–197
7. Aguzzi A, Krautler NJ (2010) Characterizing follicular dendritic cells: a progress report. Eur J Immunol 40:2134–2138
8. Sukumar S, El Shikh ME, Tew JG et al (2008) Ultrastructural study of highly enriched follicular dendritic cells reveals their morphology and the periodicity of immune complex binding. Cell Tissue Res 332:89–99
9. Sukumar S, Szakal AK, Tew JG (2006) Isolation of functionally active murine follicular dendritic cells. J Immunol Methods 313:81–95
10. Szakal AK, Gieringer RL, Kosco MH et al (1985) Isolated follicular dendritic cells: cytochemical antigen localization, Nomarski, SEM, and TEM morphology. J Immunol 134:1349–1359
11. Fakher M, Wu J, Qin D et al (2001) Follicular dendritic cell accessory activity crosses MHC and species barriers. Eur J Immunol 31:176–185
12. Liu YJ, Xu J, de Bouteiller O et al (1997) Follicular dendritic cells specifically express the long CR2/CD21 isoform. J Exp Med 185:165–170
13. Tew JG, Wu J, Qin D et al (1997) Follicular dendritic cells and presentation of antigen and costimulatory signals to B cells. Immunol Rev 156:39–52
14. Burton GF, Conrad DH, Szakal AK et al (1993) Follicular dendritic cells and B cell costimulation. J Immunol 150:31–38
15. Garin A, Meyer-Hermann M, Contie M et al (2010) Toll-like receptor 4 signaling by follicular dendritic cells is pivotal for germinal center onset and affinity maturation. Immunity 33:84–95
16. Zhou X, Shapiro L, Fellingham G et al (2011) HIV replication in CD4+ T lymphocytes in the presence and absence of follicular dendritic cells: inhibition of replication mediated by alpha-1-antitrypsin through altered IkappaBalpha ubiquitination. J Immunol 186: 3148–3155

Chapter 11

In Vitro-Induced Germinal Center B Cell Culture System

Kei Haniuda, Takuya Nojima, and Daisuke Kitamura

Abstract

In germinal centers (GCs), B cells undergo repeated cycles of proliferation and affinity-based selection, and differentiate into memory B cells or long-lived plasma cells. It has been difficult to elucidate regulatory mechanisms for the dynamic GC B cell maturation and differentiation, partly because experimental manipulation of GC B cells has been limited. Here we describe a culture system in which we can induce massive expansion of naive B cells that exhibit GC B cell-like phenotype and acquire abilities to differentiate into memory B cells or bone marrow plasma cells depending on cytokine conditions. This system will allow us to elucidate the molecular mechanisms of GC B cell differentiation.

Key words iGB cell, iMB cell, 40LB cell, Germinal center, Memory B cell, Long-lived plasma cell, IL-4, IL-21

1 Introduction

In the field of T cells, in vitro culture systems that allow long-term proliferation of naive T cells and their differentiation toward specific effector/memory T cell subsets have been widely used to study molecular mechanisms for the T-cell differentiation. However, there has been no practical in vitro culture system to effectively induce expansion of B cells with high viability and differentiation towards memory B cells or long-lived plasma cells. Since B cell division/proliferation and differentiation are thought to be tightly connected processes [1–5], such systems would enable elucidation of molecular basis for the GC B cell differentiation that remains unclear.

We have developed a novel B-cell culture system that mimics GC B-cell development [6]. For this system, we generated a feeder cell line termed 40LB, a fibroblast cell line expressing exogenous CD40-ligand (CD40L) and B-cell activating factor (BAFF) that promote B cell proliferation and survival. When cultured with IL-4 for the first 4 days and then with IL-21 for the second 4 days on the feeder cells, naive B cells proliferate logarithmically up to 10,000

Dinis Pedro Calado (ed.), *Germinal Centers: Methods and Protocols*, Methods in Molecular Biology, vol. 1623, DOI 10.1007/978-1-4939-7095-7_11, fold, efficiently undergo isotype switching to IgG1 and IgE, and express GC B-cell markers such as GL7 and Fas-ligand, and Bcl6, a well-known master regulator of GC B cells. Thus, we termed the B cells propagated in this system in vitro-induced GC B (iGB) cells, and this system iGB cell culture system. When iGB cells after the primary culture with IL-4 are transferred into irradiated recipient mice, they survive in the secondary lymphoid tissues of the recipients for months and acquire the phenotype of memory B cells, which we termed induced memory B (iMB) cells. Alternatively, when those after the secondary culture with IL-21 are transferred, they differentiate into plasma cells in the bone marrow. Thus, we can recapitulate normal GC B cell development in vitro to some extent. This system is now widely used to study the late B cell development, and several groups have reported their studies utilizing this system [7–14]. Here, we describe a detailed protocol of the iGB cell culture system.

2 Materials

1. All reagents are stored at 4 °C. Cells are stored in liquid nitrogen.
2. 40LB cells: BALB/c3T3 cells (provided by RIKEN BRC) stably expressing mouse CD40L and mouse BAFF (*see* **Note 1**).
3. Cell culture plates: 10-cm dish (e.g., Falcon #353003), 6-well plate (e.g., Costar #3516), 12-well plate (e.g., Costar #3513).
4. 3T3 medium: D-MEM (high glucose) supplemented with 10% fetal bovine serum (FBS), 100 units/mL penicillin, and 100 μg/mL streptomycin.
5. B cell medium: RPMI-1640 supplemented with 10% FBS (*see* **Note 2**), 10 mM HEPES, 1 mM sodium pyruvate, 5.5×10^{-5} M 2-ME, 100 U/mL penicillin, and 100 μg/mL streptomycin.
6. Phosphate-buffered saline (PBS; 20×): Dissolve 160 g NaCl, 4 g KCl, 23 g Na_2HPO_4, and 4 g KH_2PO_4 in 1 L H_2O.
7. MACS buffer: PBS supplemented with 0.5% bovine serum albumin (BSA) and 2 mM EDTA.
8. Antibody cocktail for B cell isolation (for one spleen): 100 μL of MACS buffer containing 3 μg/mL of biotinylated anti-CD43 (BD Pharmingen, clone: S7), 2.5 μg/mL of biotinylated anti-Ter119 (Biolegend, clone: TER-119), 1.5 μg/mL of biotinylated anti-CD4 (Biolegend, clone: GK1.5), 1.5 μg/mL of biotinylated anti-CD8 (Biolegend, clone: 53-6.7), and 1.5 μg/mL of biotinylated anti-CD49b (Biolegend, clone: DX5).
9. Antibody cocktail for feeder cell depletion (for cells from one 10-cm dish): 260 μL of MACS buffer containing 2.9 μg/mL of

biotinylated anti-H-2K[d] (Biolegend, clone: SF1-1.1). To deplete plasma cells simultaneously, biotinylated anti-CD138 (BD Pharmingen, clone: 281-2) is also added at 1.15 μg/mL to the cocktail.

10. Streptavidin Particles Plus-DM (BD IMag #557812).
11. Recombinant mouse IL-4: PeproTech #214-14.
12. Recombinant mouse IL-21: PeproTech #210-21.

3 Methods

3.1 Handling and Irradiation of 40LB Cells

40LB feeder cells are typically cultured and maintained with 10 mL of 3T3 medium in a 10 cm dish at 37 °C, 5% CO_2. We usually passage the 40LB cells every 3 days with sixfold dilutions and keep on culture up to 1 month, as described below. After 1 month or so, we restart the culture with a frozen stock (*see* **Note 3**). The following protocol is for the culture using 10 cm dishes. Change the volume of solutions in proportion to the bottom area of the plate (well) to use, as shown by examples in Table 1.

1. Thaw a frozen 40LB cells, dilute the content with warm 3T3 medium, centrifuge the cell suspension, resuspend the cell pellet with warm 3T3 medium, and seed into culture dishes. Typically, we initially seed the cells from one frozen tube (containing ~10^7 cells) into two dishes, which become confluent after 2 days.
2. Remove the medium from the cell culture dish, and rinse it with 2 mL of PBS. Add 1 mL of trypsin–EDTA and leave the dish at room temperature until cells get dispersed (takes about 1–3 min). To inactivate trypsin, add 3 mL of 3T3 medium and gently pipette up and down to completely detach the cells. Transfer the cell suspension to a tube and centrifuge for 5 min at 290 × *g*.

Table 1
Plate type, cell number to seed, and volume of solutions

Plate type	10 cm	6 well	12 well
No. 40LB cells to seed (× 10^4)	300	50	20
Vol. PBS to rinse a dish or a well (mL)	2	0.5	0.25
Vol. Trypsin–EDTA to harvest cells (mL)	1	0.5	0.25
Vol. 3T3 medium to inactivate trypsin (mL)	3	1	0.5
No. B cells to seed (×10^4)	50	5	2
Vol. B cell medium to seed (mL)	40	8	4

3. Discard the supernatant and disrupt the cell pellet by gentle shaking. Add the desired volume of fresh 3T3 medium and pipette up and down 3–4 times to resuspend the cells. Seed appropriate number of 40LB cells to desired plate (*see* Table 1), and allow cells at least 12 h to attach the plate (*see* **Note 4**).
4. Irradiate the cells with 80 Gy of γ-ray to stop proliferation (*see* **Note 5**).
5. To freeze the 40LB cells for storage, resuspend the pellet from **step 2** in 1 mL of D-MEM supplemented with 20% FBS and 10% DMSO.

3.2 Isolation of B Cells from Mouse Spleen

Keep cells cold and use prechilled solutions to prevent damaging cells. All centrifugation steps are performed at 4 °C for 7 min, 394 × *g*. Here we describe the protocol for one spleen.

1. Push out cells from a mouse spleen using MACS buffer with 26G needle and syringe, and then gently disrupt the remaining spleen capsular using a pair of frosted glasses. Transfer the cell suspension to a tube and centrifuge. Resuspend the cell pellet with 5 mL of MACS buffer, pass through a 70-μm mesh and centrifuge.
2. Discard the supernatants and disrupt the cell pellet with gentle shaking. Add 100 μL of antibody cocktail for B cell isolation and gently pipette up and down to completely suspend the cells. Incubate the cells for 20 min on ice.
3. Wash the cells twice with 10 mL of MACS buffer with a centrifugation at each step.
4. Discard the supernatants and disrupt the cell pellet with gentle shaking. Add 50 μL of Streptavidin Particles Plus-DM and gently pipette up and down to completely suspend the cells. Incubate the cells for 20 min on ice.
5. Place two 5 mL round-bottom tubes onto the iMag magnet.
6. Resuspend cells with 2 mL of MACS buffer and transfer into the first tube on the magnet. Wait for 5 min until the labeled cells attach to the wall of the tube beside the magnet.
7. Collect unbound cells (negative fraction: B cells) and transfer it to the second tube. Resuspend the attached cells in the first tube with a new 2 mL of MACS buffer and place the tube back onto the magnet and leave it for 5 min as in **step 6**.
8. Repeat **step 7** once more and centrifuge the whole collected cells in the second tube.
9. Place a LS column onto the MACS magnet and place 40 μm mesh on top of the column and attach a 22G needle to an outlet of the column. Rinse the column with 2 mL of MACS buffer.

10. Resuspend the cell pellet from **step 8** with 1 mL of MACS buffer and apply the cell suspension onto the LS column through the mesh. Wash the column three times with 1 mL of MACS buffer and once with 3 mL of MACS buffer. Collect all the effluent (negative fraction: B cells) and centrifuge it.
11. Resuspend the cell pellet with 10 mL of B cell medium and count the cell number (*see* **Note 6**).
12. Check the purity of isolated B cells using flow cytometry (*see* **Note 7**).

3.3 iGB Cell Culture

Use prewarmed buffer and medium. All centrifugation steps are performed at room temperature for 5 min, 394 × *g* (*see* **Note 8**). Here we describe a protocol for the culture using 10-cm dishes.

1. Remove culture medium of irradiated feeder cells and add 35 mL (*see* **Note 9**) of B cell media supplemented with IL-4 (final concentration is 1 ng/mL). Seed 5 mL of cell suspension containing appropriate number of the isolated B cells (*see* Table 1) and culture at 37 °C, 5% CO_2 for 4 days (*see* **Note 10**).
2. To harvest cells, carefully remove medium from the surface down to the level about 0.5 cm above the bottom and transfer the remaining medium to a collection tube. Add 4 mL of MACS buffer and leave dish at room temperature until cells begin to get dispersed (takes about 3–5 min). Gently shake the dish to detach the cells and collect them to the tube. Rinse the plate with 5 mL of MACS buffer and collect the medium into the tube. Repeat the same once more and centrifuge all the collected cells.
3. Discard the supernatants and disrupt the cell pellet by gentle shaking. Resuspend the cell pellet with 10 mL of B cell medium and count the cell number. iGB cells are distinguishable as the cells smaller than the 40LB cells.
4. To culture the cells sequentially, seed appropriate number of the iGB cells onto a new irradiated feeder layer and culture them as described in **step 1** with an appropriate cytokine (*see* **Note 11**).

3.4 Feeder Cell Depletion

It is desirable to remove feeder cells from harvested cells of the iGB cell culture when they are transferred into mice or analyzed at a molecular level. Here we describe a protocol to deplete feeder cells from iGB cells cultured on a 10 cm dish. Prewarmed solutions are used. All centrifugation steps are performed at room temperature for 5 min, 394 × *g*.

1. Harvest cells as in Subheading 3.3, **step 2**.
2. Discard the supernatants and disrupt the cell pellet with gentle shaking. Add 260 μL of an antibody cocktail containing

biotinylated anti-H-2K^d (*see* Subheading 2, **item 8** and **Note 12**) and gently pipette up and down to completely suspend the cells. Incubate for 20 min at room temperature.

3. Wash the cells twice with 10 mL of MACS buffer with a centrifugation at each step.
4. Discard the supernatants and disrupt the cell pellet with gentle shaking. Add 50 μL of Streptavidin Particles Plus-DM and 100 μL of MACS buffer, and then gently pipette up and down to completely suspend the cells. Incubate for 20 min at room temperature.
5. Apply to iMag and MACS procedure as described in Subheading 3.2, **steps 5–10**.
6. Resuspend the collected cells in a 10 mL of appropriate medium or buffer and count the cell number.

3.5 Generation of iMB Cells and Bone Marrow Plasma Cells

Use mice maintained under specific pathogen free conditions. To distinguish donor iMB cells from B cells of recipient mice with flow cytometry, choose a combination of mice with different congenic markers, such as Ly5.1 and Ly5.2. Transfer iGB cells primarily cultured with IL-4 into the recipient mice to generate iMB cells, and iGB cells secondarily cultured with IL-21 to generate bone marrow plasma cells.

1. Irradiate mice with 6.5 Gy of γ-ray (sublethal does, *see* **Note 13**).
2. Resuspend purified iGB cells (deprived of feeder cells; 4×10^7/mL) in PBS and inject 500 μL of the cell suspension into the irradiated mice (2×10^7 per mouse).
3. We typically analyze the peripheral lymphoid organs in the recipient mice after 30 days (*see* **Note 14**).

4 Notes

1. To make the 40LB cell line, BAFF expression vector containing a neomycin-resistant gene (pCA-neo BAFF) and CD40L expression vector containing a puromycin-resistant gene (pApuro2 CD40L) were transfected into Balb/c 3T3 cells [6]. To increase BAFF expression, the original 40LB cells were transfected with the same BAFF-expression vector again, selected with a higher concentration of G418 (2 mg/mL) for 1 week, and then survived colonies were bulk cultured and expanded. The resultant cell line was named 40LB-NB. B cells grow better on the 40LB-NB than the original 40LB. We usually use the 40LB-NB which is now simply called 40LB.
2. The lot of FBS affects the cell growth and class switching. Indeed, only ~30% of the lots we tested worked well.

When we select the FBS lot for iGB cell culture, we usually use the following parameters: (1) Cell expansion factor during the primary culture with IL-4 (more than 60-fold in 96 h) and the secondary culture with IL-21 (more than 100-fold in 96 h). (2) FACS profiles after the primary and the secondary culture: efficient class switching to IgG1 and IgE (but no more than 50% IgE$^+$ cells) and homogeneous expression of GL7 and Fas, and minimal differentiation to CD138$^+$ cells after the primary culture. Most of the FBS lots should work for the culture of 40LB cells.

3. We do not usually use G418 or puromycin during expansion of 40LB cells, but the expression of CD40L and BAFF is homogeneous and apparently stable. For those who prefer using these antibiotics, rather low concentrations would be recommended (puromycin at 2.5 μg/mL and G418 at 0.5 mg/mL). The expression of CD40L is quite low by flow cytometry, but it remained the same level for at least 6 months, in a culture without puromycin. BAFF expression on 40LB cells could be monitored by intracellular staining (e.g., with anti-BAFF-PE, clone 121808, R&D) but not by cell surface staining.
4. Alternatively, we sometimes irradiate the harvested 40LB cells in a tube, then seed them onto plates, and culture them for at least 14 h (until the 40LB cells attach to the plates in a good shape) before seeding B cells onto them.
5. In our original article, we described the irradiation dose as 120 Gy [6]. Recently, however, we have found that 80 Gy is enough to arrest the growth of 40LB cells and to support the growth of B cells on them.
6. We usually obtain 1~4 × 10^7 B cells from one spleen.
7. Use fluorochrome-conjugated antibodies against B cell makers such as CD19 and B220. Purity of B cells is usually more than 97%. If B cells with a lower purity are used for iGB cell culture, feeder cells would become crumbled due to contaminated CD8$^+$ T cells.
8. After the B cells start proliferation and become "iGB cells," the iGB cells should be treated at room temperature during procedures such as centrifugation, antibody staining, iMag, and MACS or flow cytometry. iGB cells tend to die on ice.
9. Less volume of medium with more frequent replacement would also be acceptable.
10. Alternatively, positively sorted B cells using techniques such as MACS or flow cytometry could be used for the culture. B cells from peripheral blood or other lymphoid organs such as lymph nodes could also be used.
11. Primary cultured iGB cells can be re-expanded on the new feeder layer with IL-21 (or IL-4, which will promote further

switching to IgE, with a slower growth rate than IL-21), as described in the original article. At this time, you can put the whole harvested cells including the feeder cells onto the new feeder layer, adjusting the number of iGB cells as desired. For example, primary iGB cells on day 4 can be seeded, to culture with IL-21, at 5×10^5 (for 3 more days) or 5×10^4 (for 4 more days) per 10 cm dish. In the secondary culture with IL-21, the growth rate of iGB cells is difficult to predict, thus we change the medium when it begins to turn yellowish. The secondary culture can be continued for as long as 6 days without feeder renewal; however, the longer iGB cells are in culture, the more they become $CD138^+$ plasmablasts, and the cells eventually stop growing around day 10.

12. For cells from 1 well of 6-well or 12-well plate, use 1/6 or 1/12 volume, respectively, of diluted antibody with the same concentration.
13. We have confirmed that 6.5 Gy of γ-ray is sublethal dose for a C57BL/6 mouse in our facility. After the irradiation, lymphocytes can hardly be detected in the spleen, but mice recover after about 2 weeks.
14. The frequency of the iMB cells in the spleens of the recipient mice is about 0.3% (of $CD19^+$ cells) after 30 days. The more IgE^+ cells and/or $CD138^+$ cells are included in the transferred iGB cells, the less frequent iMB cells will be generated in the recipient mice. Much more frequency of donor cells can be detected at earlier days after transfer, for example, 5–10% at day 7, but most of them retain iGB cell phenotype, not having differentiated to iMB cells. This higher frequency of donor cells is likely due to the lower number of host lymphocytes yet under recovery from irradiation at the early days.

Acknowledgment

This work was supported by Grant-in-Aid for Scientific Research (B) to D.K. The authors declare no financial conflict of interest.

References

1. Hasbold J, Corcoran LM, Tarlinton DM, Tangye SG, Hodgkin PD (2004) Evidence from the generation of immunoglobulin G-secreting cells that stochastic mechanisms regulate lymphocyte differentiation. Nat Immunol 5 (1):55–63. doi:10.1038/ni1016
2. Duffy KR, Wellard CJ, Markham JF, Zhou JH, Holmberg R, Hawkins ED, Hasbold J, Dowling MR, Hodgkin PD (2012) Activation-induced B cell fates are selected by intracellular stochastic competition. Science 335(6066):338–341. doi:10.1126/science.1213230
3. Barnett BE, Ciocca ML, Goenka R, Barnett LG, Wu J, Laufer TM, Burkhardt JK, Cancro MP, Reiner SL (2012) Asymmetric B cell division in the germinal center reaction. Science 335(6066):342–344. doi:10.1126/science.1213495

4. Thaunat O, Granja AG, Barral P, Filby A, Montaner B, Collinson L, Martinez-Martin N, Harwood NE, Bruckbauer A, Batista FD (2012) Asymmetric segregation of polarized antigen on B cell division shapes presentation capacity. Science 335(6067):475–479. doi:10.1126/science.1214100
5. Nutt SL, Hodgkin PD, Tarlinton DM, Corcoran LM (2015) The generation of antibody-secreting plasma cells. Nat Rev Immunol 15 (3):160–171. doi:10.1038/nri3795
6. Nojima T, Haniuda K, Moutai T, Matsudaira M, Mizokawa S, Shiratori I, Azuma T, Kitamura D (2011) In-vitro derived germinal centre B cells differentially generate memory B or plasma cells in vivo. Nat Commun 2:465. doi:10.1038/ncomms1475
7. Caganova M, Carrisi C, Varano G, Mainoldi F, Zanardi F, Germain PL, George L, Alberghini F, Ferrarini L, Talukder AK, Ponzoni M, Testa G, Nojima T, Doglioni C, Kitamura D, Toellner KM, Su IH, Casola S (2013) Germinal center dysregulation by histone methyltransferase EZH2 promotes lymphomagenesis. J Clin Invest 123(12):5009–5022. doi:10.1172/JCI70626
8. Fukao S, Haniuda K, Nojima T, Takai T, Kitamura D (2014) gp49B-mediated negative regulation of antibody production by memory and marginal zone B cells. J Immunol 193(2):635–644. doi:10.4049/jimmunol.1302772
9. Wu L, Parekh VV, Hsiao J, Kitamura D, Van Kaer L (2014) Spleen supports a pool of innate-like B cells in white adipose tissue that protects against obesity-associated insulin resistance. Proc Natl Acad Sci U S A 111(43):E4638–E4647. doi:10.1073/pnas.1324052111
10. Kitabatake M, Soma M, Zhang T, Kuwahara K, Fukushima Y, Nojima T, Kitamura D, Sakaguchi N (2015) JNK regulatory molecule G5PR induces IgG autoantibody-producing plasmablasts from peritoneal B1a cells. J Immunol 194(4):1480–1488. doi:10.4049/jimmunol.1401127
11. Purwada A, Jaiswal MK, Ahn H, Nojima T, Kitamura D, Gaharwar AK, Cerchietti L, Singh A (2015) Ex vivo engineered immune organoids for controlled germinal center reactions. Biomaterials 63:24–34. doi:10.1016/j.biomaterials.2015.06.002
12. Webb LM, Datta P, Bell SE, Kitamura D, Turner M, Butcher GW (2016) GIMAP1 is essential for the survival of naive and activated B cells in vivo. J Immunol 196(1):207–216. doi:10.4049/jimmunol.1501582
13. Kuraoka M, Schmidt AG, Nojima T, Feng F, Watanabe A, Kitamura D, Harrison SC, Kepler TB, Kelsoe G (2016) Complex antigens drive permissive clonal selection in germinal centers. Immunity 44(3):542–552. doi:10.1016/j.immuni.2016.02.010
14. Domeier PP, Chodisetti SB, Soni C, Schell SL, Elias MJ, Wong EB, Cooper TK, Kitamura D, Rahman ZS (2016) IFN-gamma receptor and STAT1 signaling in B cells are central to spontaneous germinal center formation and autoimmunity. J Exp Med 213(5):715–732. doi:10.1084/jem.20151722

Chapter 12

CRISPR/Cas9-Mediated In Vitro Mutagenesis in GC-Like B Cells

Van Trung Chu, Robin Graf, and Klaus Rajewsky

Abstract

The CRISPR/Cas9 technology has developed into a powerful tool for genome editing, both in terms of gene silencing and the insertion of precise mutations. However, the application of CRISPR/Cas9-mediated mutagenesis in primary immune cells, in particular in B cells, is still in its infancy because of the difficulty to deliver the CRISPR/Cas9 system into these cells. Here, we describe a new method to use CRISPR/Cas9 for manipulating genes in germinal center (GC)-like B cells in vitro. We isolated Cas9-expressing B cells from R26-Cas9iGFP/+ mice (expressing Cas9 constitutively from the Rosa26 locus) and mixed them with control B cells. Primary B cells were cultured on CD40L- and BAFF-expressing feeder cells and transduced with retroviral particles expressing the sgRNAs of interest. Using this system, we have achieved complete gene knockouts in up to 92% of activated B cells.

Key words B cells, 40LB feeder cells, CRISPR/Cas9, Genome editing, In vitro

1 Introduction

The CRISPR/Cas9 technology has been applied to genome editing. The system utilizes the type II CRISPR/Cas9 adaptive immune system derived from *Streptococcus pyogenes* which targets the Cas9 nuclease to any given genetic locus through a 20-nt guide sequence cloned upstream of the "protospacer adjacent motif" (PAM), leading to DNA double-strand breaks (DSBs) [1, 2] (Fig. 1). In mammalian cells, the CRISPR/Cas9-induced DSBs are mostly repaired via the nonhomologous end-joining (NHEJ) pathway introducing insertions and deletions that can lead to loss-of-function alleles, generating genetic knockouts [3]. Alternatively and at lower frequency, DSBs are repaired via homology-directed repair (HDR) [4]. HDR can be utilized for the introduction of precise genetic modifications if suitable DNA templates are provided. To date, CRISPR/Cas9-mediated gene silencing through NHEJ pathway has worked efficiently in mouse embryonic stem cells, mouse zygotes as well as mouse and human cell lines [3, 5–7]. However,

Dinis Pedro Calado (ed.), *Germinal Centers: Methods and Protocols*, Methods in Molecular Biology, vol. 1623,
DOI 10.1007/978-1-4939-7095-7_12,

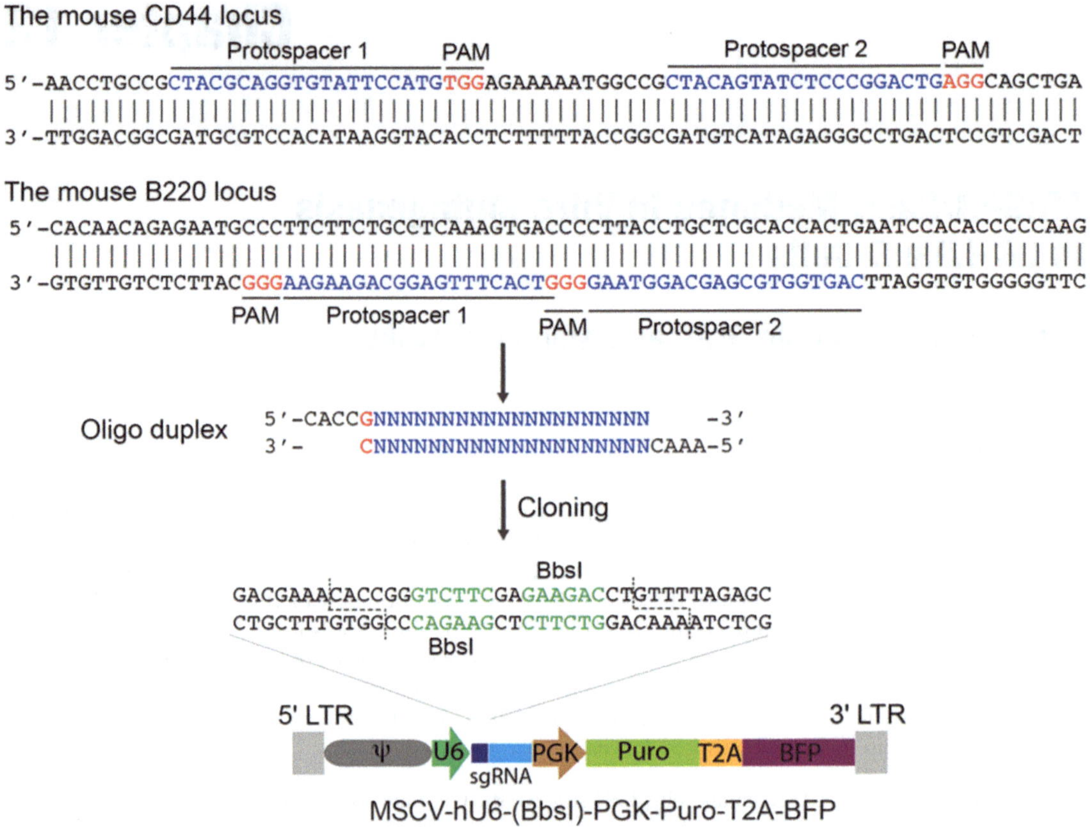

Fig. 1 Generation of mouse CD44- and B220-specific sgRNA expression vectors. Schematic of the sgRNA expression vectors. Two sgRNAs specific for the mouse CD44 and B220 loci (*blue*) were selected. The PAM (NGG) signal is shown in *red*. Because the hU6 promoter preferably initiates transcription with a G nucleotide as starting nucleotide, an additional G is added 5′ of protospacers (*red*). Duplex oligonucleotides with compatible overhangs were cloned into the BbsI cloning sites (*green*) of the MSCV-hU6-PGK-Puro-T2A-BFP expression vector. The 20-nt gRNA sequences are shown as N

precise gene correction via the HDR pathway has remained inefficient. Recently, it has been reported that HDR efficiency in cell lines and mouse zygotes can be enhanced up to eightfold by suppressing the NHEJ pathway [8, 9].

Using Cas9 knockin mice, Flatt et al. have shown CRISPR/Cas9-mediated genome editing in mouse bone marrow-derived dendritic cells [10]. Furthermore, a genome-wide CRISPR screen has been performed in bone marrow-derived dendritic cells isolated from Cas9 knockin mice [11]. However, CRISPR/Cas9-mediated gene editing in primary B cells has not yet been reported. Upon stimulation by T cell dependent or T cell-independent antigens, naïve B cells proliferate and differentiate into antibody-secreting plasma cells [12, 13]. Similarly, in vitro stimulation of naïve B cells with anti-CD40 antibodies and IL-4/IL-21 leads to cell proliferation, antibody class switching and terminal differentiation [14, 15].

To mimic T cell dependent activation, Nojiama et al. have generated feeder cells expressing CD40L and the B cell survival factor BAFF, which are both supporting B cell activation and survival. On this feeder cell system, naïve B cells differentiate into GC-like B cells and proliferate over prolonged times [16]. We previously generated Rosa26-LSL-Cas9iGFP mice by knocking a conditional Cas9-IRES-GFP cassette into the mouse *Rosa26* locus, with the Cas9 transgene expression under the control of a loxP-flanked Stop cassette [17]. Cas9 transgenic mice were generated by crossing these R26-LSL-Cas9iGFP with Cre-deleter mice.

Here, we describe a protocol for CRISPR/Cas9-mediated gene inactivation in activated mouse B cells, exemplifying this method for the surface markers B220 and CD44 (*see* Figs. 1 and 2). We achieved gene silencing in 68–88% of the cells for B220, and in 90–92% for CD44 (*see* Fig. 3). We observed these high knockout-frequencies targeting many additional genes (not shown). For reasons yet to be clarified, the pattern of mutagenesis differed between the two loci (*see* Fig. 4).

2 Materials

2.1 Generation of sgRNA Expressing MSCV Vector

1. Forward and reverse oligonucleotides targeting mouse B220 and CD44 loci (Eurofins) (*see* **Note 1**) (*see* Fig. 1).
2. Fast digest BbsI and Fast Alkaline Phosphatase (Thermo Scientific).
3. T4 PNK and T4 ligase (NEB).
4. LB agar plate and liquid medium (Roth).
5. DH5α bacteria (NEB) (*see* **Note 2**).
6. NucleoSpin® Miniprep/Maxiprep kits (Macherey-Nagel).

2.2 Isolation of Mouse Splenic B Cells

1. PBS, pH 7.2 (Gibco).
2. BD Falcon Cell strainer (BD Bioscience).
3. Anti-FcγR antibody clone 2.4G2 (BioLegend) (*see* **Note 3**).
4. 50-ml and 15-ml Falcon tubes (Greiner Bio-One).
5. BSA (Sigma).
6. Erythrocyte lysis buffer (pH 7.5): 155 mM NH_4Cl, 10 mM $KHCO_3$, 0.1 mM EDTA (*see* **Note 4**).
7. Mouse CD43 microbeads, LS column, and magnetic separator (Miltenyi Biotec).

2.3 Cell Culture

1. B cell medium: DMEM−/− (Gibco), 10% FBS (Gibco), 2 mM Hepes (Gibco), 2 mM L-glutamine (Gibco), 2 mM sodium pyruvate (Gibco), 1× NAA (Gibco), beta-mercaptoethanol (Sigma), and 10 μg/ml gentamicin (Lonza) (*see* **Note 5**).

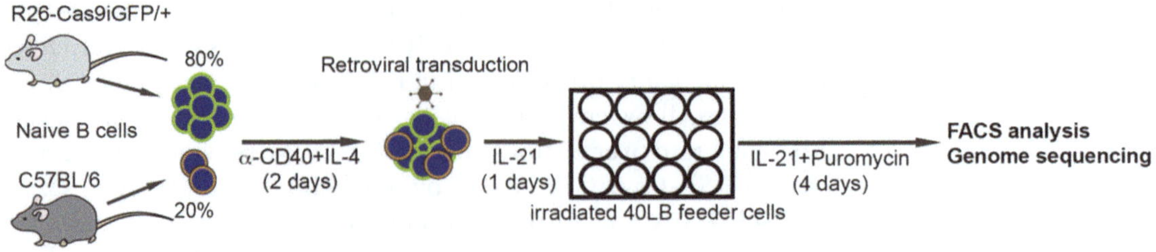

Fig. 2 In vitro B cell stimulation and transduction. B cells are isolated from spleens of R26-Cas9iGFP/+ and C57BL/6 mice. 80% of Cas9-expressing B cells are mixed with 20% of wild-type B cells, and subsequently stimulated with anti-CD40 antibodies and IL-4 cytokines for 2 days. On day 2, the mixed B cells are spin-transduced with sgRNA expression retroviral particles in the presence of IL-21. One day later, the transduced B cells are transferred to the irradiated 40LB feeder cell layers supplied with IL-21. Transduced B cells are then selected with Puromycin for 4 days and analysed by FACS or sorted for genomic sequencing

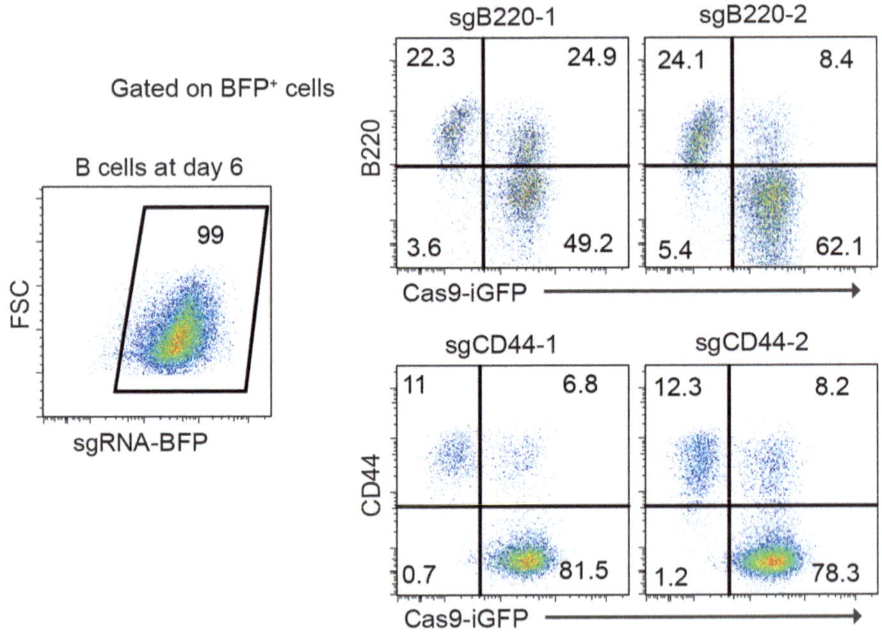

Fig. 3 CRISPR-mediated knockout efficiency of mouse B220 and CD44 genes. Puromycin selection for 4 days was sufficient to eliminate non-transduced B cells. The knockout efficiency of Cas9-expressing (GFP+) B cells is shown in the BFP+ gate. B220-specific sgRNAs inactivated both alleles of the B220 gene in 68% (sgB220-1) and 88% (sgB220-2) of transduced GFP+ cells. The inactivation efficiency of the mouse CD44 gene was 92% (sgCD44-1) and 90% (sgCD44-2)

2. 40LB feeder cell medium: DMEM−/− (Gibco), 10% FBS (Gibco), 2 mM L-glutamine (Gibco), 2 mM sodium pyruvate (Gibco) (*see* **Note 6**).
3. Plat-E retroviral packaging cell medium: DMEM+/+ (Gibco), 10% FBS (Gibco) (*see* **Note 6**).
4. Anti-mouse CD40 antibody (BioLegend) (*see* **Note 7**).
5. IL-4 and IL-21 cytokines (Peprotech) (*see* **Note 7**).

Fig. 4 Micro-indel pattern of CRISPR-targeted B220 and CD44 sequences. B220⁻GFP⁺ and CD44⁻GFP⁺ B cells were sorted to extract genomic DNA. The targeted sequences in B220 and CD44 were amplified by high-fidelity PCR. The PCR products were subcloned into pSTBlue-1 sequencing vector. Insertions of nucleotides are shown in *green*, and deletions of nucleotides are indicated as *dash* (–)

2.4 Retroviral Production and Transduction

1. FuGENE HD® transfection reagent (Promega) (*see* **Note 8**).
2. Opti-MEM medium (Gibco) (*see* **Note 8**).
3. Amicon® Ultra-15 (Millipore) (*see* **Note 9**).
4. Polybrene (Sigma).
5. Puromycin (Thermo Scientific) (*see* **Note 9**).

2.5 FACS, PCR and Sequencing

1. FACS buffer: PBS–1%BSA.
2. Antibodies: B220, CD19, and CD44 (BioLegend).
3. PI to exclude dead cells (Sigma).
4. DNA extraction solution (Epicentre) (*see* **Note 10**).
5. PCR Herculase II Fusion DNA Polymerase kit (Agilent) (*see* **Note 10**).
6. pSTBlue-1 blunt cloning kit (Novagen) (*see* **Note 10**).

3 Methods

3.1 Generation of sgRNA Vectors to Target Mouse B220 and CD44 Loci

1. Design sgRNAs based on specificity and activity (*see* Fig. 1). See note added in proof.
2. Add a G nucleotide at the 5′ end of protospacers (*see* **Note 1**).
3. Generate forward and reverse oligonucleotides as in Fig. 1.
4. Anneal forward and reverse oligonucleotides and phosphorylate them using T4 PNK.
5. Dilute the oligonucleotide-duplexes and clone them into the BbsI sites of the retroviral MSCV-hU6-PGK-Puro-T2A-BFP vector (*see* **Note 11**) (*see* Fig. 1).

6. Transform the plasmids into DH5α bacteria. Pick bacterial colonies and isolate the DNA using Nucleospin® miniprep kit. Confirm the correct insertion by sequencing (*see* **Note 11**).
7. Amplify the correct plasmids using the Nucleospin® Maxi kit.

3.2 In Vitro B Cell Stimulation

1. To isolate resting B cells, isolate spleens from R26-Cas9iGFP/+ or C57BL/6 mice and place them on BD cell strainers, process the spleen to a single cell suspension by mechanical disruption, and centrifuge at 290 × *g* for 7 min at 4 °C. Discard the supernatant and lyse erythrocytes by resuspending in 5 ml erythrocyte lysis buffer for 5 min on ice. Add 30 ml of FACS buffer (PBS–1%BSA) to stop the reaction (*see* **Note 12**).
2. Centrifuge the cells as before and resuspend in 1 ml FACS buffer, block with anti-FcγR antibody for 10 min at 4 °C (*see* **Note 3**). Subsequently, add 50 μl mouse anti-CD43 microbeads. Mix and incubate for 15 min at 4 °C (*see* **Note 13**).
3. Wash the cells by centrifugation at 290 × *g* for 7 min, remove the supernatant, and resuspend the pellets with 1 ml FACS buffer. Load the cell suspension on magnetic LS column. The resting ($CD43^-$) B cells are collected as negative fraction, i.e., the flow-through (*see* **Note 13**).
4. Count the B cells and resuspend them at a density of 1×10^6 cells/ml. Mix 80% of Cas9-expressing B cells with 20% of wild-type B cells and stimulate with 5 μg/ml of anti-CD40 antibody and 10 ng/ml of IL-4. Place the cells into a 37 °C cell incubator (*see* **Note** 7) (*see* Fig. 2).

3.3 Retroviral Production

1. Twenty-four hours before transfection, plate the Plat-E retroviral packaging cells at a density of 3×10^6 cells per 10 cm dish in 15 ml completed 40LB medium (DMEM+10%FBS).
2. The next day, mix 19 μg of the sgRNA-expressing MSCV plasmid with Opti-MEM medium to obtain a total volume of 879 μl (*see* **Note 8**).
3. Add 56 μl of FuGENE HD® reagent to the DNA solution and mix carefully by pipetting up and down. Incubate the DNA complex for 15 min at room temperature (RT) (*see* **Note 8**).
4. Add 850 μl of the DNA complex per 100 mm dish and mix by slow shaking. Put the cell dishes into the 37 °C incubator.
5. The next day, remove the supernatant and add 15 ml of new Plat-E medium. Transfer the dishes to the **32 °C** incubator (*see* **Note 14**).
6. Collect the supernatant 48 h and 72 h after transfection. Keep the 48 h supernatant at 4 °C and mix the two supernatants together (*see* **Note 14**).

7. To increase the infection rate, concentrate the retroviral supernatant by using Amicon® Ultra-15 units (*see* **Note 14**).
8. Aliquot the concentrated retroviruses and store at −80 °C (*see* **Note 14**).

3.4 Retroviral Transduction for Activated B Cells

1. After 48 h of stimulation, collect the activated B cells. Count the cell numbers and adjust to 2×10^5 cells/ml (*see* **Note 15**) (*see* Fig. 2).
2. Add Polybrene (*see* **Note 15**), anti-CD40 antibody and IL-21 cytokine to the cell suspension to a final concentration 8 μg/ml, 5 μg/ml, and 20 ng/ml, respectively. The concentrated retroviruses are diluted at 1:10 with complete B cell medium (*see* **Note 15**). 500 μl of the cell suspension is mixed with 500 μl of the diluted retroviral supernatant in a well of a 12-well plate.
3. Spin-transduce the cells in the plates by centrifuging at $1258 \times g$ for 90 min at 32 °C.
4. Put the plates into a 37 °C incubator. Six hours later, carefully remove 50% of the medium and replace it with fresh B cell medium (*see* **Note 15**).

3.5 Stimulation of Transduced B Cells on 40LB Feeder Cells

1. On the day of the spin transduction, prepare the 40LB feeder layer. Harvest the 40LB cells that were maintained in 10 cm dishes by trypsinization. Resuspend the cell pellets with 10 ml of complete 40LB medium and transfer to a new 10 cm dish. Irradiate the suspension with 12Gγ, count and plate at 5×10^4 cells/cm^2 per well in 12-well plates (*see* **Note 16**) (*see* Fig. 2).
2. One day after transduction, transfer the transduced B cells onto the 40LB feeder cell layer in the presence of 20 ng/ml of IL-21 (*see* **Note 16**).
3. Two days after transduction, add 1.25 μg/ml of puromycin to each well of cell culture plates (*see* **Note 16**) (*see* Fig. 2).
4. Split the transduced B cells every 2 days into fresh irradiated 40LB feeder layers. In addition, the transduced B cells can be analysed by flow cytometry or PCR/sequencing (*see* **Note 16**).

3.6 Measurement of Gene Knockout Efficiency

1. Centrifuge the transduced B cells and resuspend in FACS buffer.
2. Block the B cells with anti-FcγR antibody for 10 min (*see* **Note 3**) and stain with BV605-anti-CD19, BV785-anti-B220, and APC-anti-CD44 antibodies for 15 min at 4 °C.
3. Wash the stained cells with FACS buffer by centrifugation and resuspend with 200 μl FACS buffer. PI should be added at a 1:200 dilution (*see* **Note 17**).

4. Using the BD Fortessa, measure the knockout efficiency of B220 and CD44 genes in Cas9 (GFP^+)-expressing B cells by gating on BFP^+ cells (expressing sgRNA) (*see* Fig. 3).
5. Sort the $B220^-GFP^+$ or $CD44^-GFP^+$ B cells and extract the genomic DNA using the quick DNA extraction solution (*see* **Note 17**).
6. Amplify the targeted sequences by high-fidelity Herculase II DNA fusion Pol (*see* **Note 17**).
7. Purify the PCR products with the PCR clean purification kit (*see* **Note 17**).
8. Clone the PCR products into the pSTBlue-1 blunt vector.
9. Isolate the plasmids and sequence by Sanger method (*see* Fig. 4).

4 Notes

1. The human U6 promoter requires a G nucleotide at the start of the sequence to start transcription initiation; therefore a G nucleotide should be added at the 5′ end of the protospacer sequence.
2. We always use competent DH5α bacteria. However, other comparable competent bacterial strains can be used (TOP10, Stlb3, . . .).
3. To reduce unspecific binding of antibodies, Fcγ receptors are blocked with Anti-FcγR antibody (2.4G2).
4. Sterile erythrocyte lysis buffer should be autoclaved and stored at 4 °C.
5. Antibiotics should be added to prevent bacterial contamination. In particular, the use of gentamicin is recommended when primary B cells are isolated for in vitro cultures.
6. In cell line cultures, antibiotics should not be added to the culture medium. In the presence of antibiotics, mycoplasma can grow and the antibiotics also reduce the transfection efficiency.
7. To mimic activation by T cells, B cells are stimulated with anti-CD40 and IL4 or IL-21.
8. When FuGENE HD® reagent and DNA are mixed together, they form a complex that is able to enter Plat-E packaging cells. Complex formation is optimal in serum free medium such as Opti-MEM medium.
9. Amicon® Ultra-15 units are used to concentrate the retroviral particles. As sgRNA expressing MSCV vector contains a puromycin-resistance gene, puromycin can be used to eliminate non-transduced B cells.

10. For a quick isolation of the genomic DNA for PCR, DNA QuickExtract solution can be used (10 μl per 10^4 cells). To reduce error-prone PCR, we use high fidelity Fusion DNA polymerase. For sequencing, PCR products are end-converted and cloned into blunt end pSTBlue-1 vector with high efficiency.
11. After being annealed and phosphorylated, oligo duplexes are diluted 1:20 (500 nM) and checked on a 4% agarose gel. Oligo duplexes are further diluted 1:200 and used for cloning. After cloning into the MSCV vector, the inserts are sequenced and correct plasmids are expanded.
12. To avoid contamination, this step should be performed in the hood and researcher should wear the lab coat and the mask. The erythrocyte lysis buffer is toxic to cells, and therefore the incubation time should be minimized.
13. During the incubation with anti-mouse CD43 microbeads, the cells should be briefly vortexed every 5 min. The CD43 surface marker is expressed by T cells granulocytes/myeloid cells and activated B cells. Resting naïve B cells do not express CD43.
14. To reduce viral degradation, the transfected plates are cultured at 32 °C. Using this culture temperature, retroviruses show the highest multiplicity of infection. In addition, the retroviral supernatant collected at 48 h after transfection can be stored at 4 °C and combined with supernatant isolated at 72 h. To concentrate retroviral particles, 15 ml of retroviral supernatant is loaded into Amicon® Ultra-15 unit and centrifuged at 3220 × *g* for 10 min. The supernatant in the lower chamber is removed and the supernatant in upper chamber is transferred into a new 15-ml Falcon collection tube. A second round of concentration is performed reusing the same Amicon® Ultra-15 unit. The aliquots of the concentrated retroviruses should be stored at −80 °C for long time storage and multiple thawing–freezing should be avoided.
15. The infection rate of retroviruses is different from batch to batch and from scientist to scientist, and thus the infection rate should be optimized before each new experiment. Polybrene is added to increase the infection rate of retroviral transduction. Polybrene should be titrated for different cell types. We recommend titrating the infection rate of the concentrated retroviruses using the NIH3T3 cell line. Six hours after the spin transduction, new B cell medium supplied with cytokines should be added into the wells of transduced plates to reduce cell death and promote cell proliferation.
16. To get optimal stimulation of 40LB feeder cells, it is better to thaw new frozen 40LB cells. To stabilize the 40LB cells, they are selected with puromycin and neomycin for 3 days before using them as feeder layers. 40LB cells produce CD40L and

BAFF, and therefore anti-CD40 antibodies are not needed. To eliminate the non-transduced B cells, puromycin is added into B cell medium. The B cells expand very fast in the 40LB system, and therefore the cells should be split every 2 days.

17. To remove autofluorescent dying cells, PI should be included in the FACS buffer for acquisition. To determine the mutagenesis on B220 and CD44 loci, $B220^{-}GFP^{+}$ or $CD44^{-}GFP^{+}$ B cells are sorted directly into collection tubes containing quick DNA extraction solution. After denaturation, the DNA is ready for PCR. As CRISPR/Cas9-mediated mutagenesis caused large deletions in the targeted sequences, we isolated PCR products by using PCR clean kit, instead of the gel purification kit.

Acknowledgments

Supported by ERC Advanced Grant 268921 to K.R., and the Max-Delbrück-Center for Molecular Medicine.

Note Added in Proof

The experimental system described in this article including a new sgRNA design program has recently been published (Chu VT, Graf R, Wirtz T, Weber T, Favret J, Li X, Petsch K, Tran NT, Sieweke MH, Berek C, Kühn R, Rajewsky K. Efficient CRISPR-mediated mutagenesis in primary immune cells using CrispRGold and a C57BL/6 Cas9 transgenic mouse line. PNAS, 2016).

References

1. Cong L, Ran FA, Cox D, Lin S, Barretto R, Habib N, Hsu PD, Wu X, Jiang W, Marraffini LA, Zhang F (2013) Multiplex genome engineering using CRISPR/Cas systems. Science 339(6121):819–823. doi:10.1126/science.1231143
2. Hsu PD, Scott DA, Weinstein JA, Ran FA, Konermann S, Agarwala V, Li Y, Fine EJ, Wu X, Shalem O, Cradick TJ, Marraffini LA, Bao G, Zhang F (2013) DNA targeting specificity of RNA-guided Cas9 nucleases. Nat Biotechnol 31(9):827–832. doi:10.1038/nbt.2647
3. Wang T, Wei JJ, Sabatini DM, Lander ES (2014) Genetic screens in human cells using the CRISPR-Cas9 system. Science 343(6166):80–84. doi:10.1126/science.1246981
4. Mali P, Yang L, Esvelt KM, Aach J, Guell M, DiCarlo JE, Norville JE, Church GM (2013) RNA-guided human genome engineering via Cas9. Science 339(6121):823–826. doi:10.1126/science.1232033
5. Wang H, Yang H, Shivalila CS, Dawlaty MM, Cheng AW, Zhang F, Jaenisch R (2013) One-step generation of mice carrying mutations in multiple genes by CRISPR/Cas-mediated genome engineering. Cell 153(4):910–918. doi:10.1016/j.cell.2013.04.025
6. Yin H, Xue W, Chen S, Bogorad RL, Benedetti E, Grompe M, Koteliansky V, Sharp PA, Jacks T, Anderson DG (2014) Genome editing with Cas9 in adult mice corrects a disease mutation and phenotype. Nat Biotechnol 32(6):551–553. doi:10.1038/nbt.2884
7. Koike-Yusa H, Li Y, Tan EP, Velasco-Herrera Mdel C, Yusa K (2014) Genome-wide recessive genetic screening in mammalian cells with a

lentiviral CRISPR-guide RNA library. Nat Biotechnol 32(3):267–273. doi:10.1038/nbt.2800

8. Maruyama T, Dougan SK, Truttmann MC, Bilate AM, Ingram JR, Ploegh HL (2015) Increasing the efficiency of precise genome editing with CRISPR-Cas9 by inhibition of nonhomologous end joining. Nat Biotechnol 33(5):538–542. doi:10.1038/nbt.3190
9. Chu VT, Weber T, Wefers B, Wurst W, Sander S, Rajewsky K, Kuhn R (2015) Increasing the efficiency of homology-directed repair for CRISPR-Cas9-induced precise gene editing in mammalian cells. Nat Biotechnol 33(5):543–548. doi:10.1038/nbt.3198
10. Platt RJ, Chen S, Zhou Y, Yim MJ, Swiech L, Kempton HR, Dahlman JE, Parnas O, Eisenhaure TM, Jovanovic M, Graham DB, Jhunjhunwala S, Heidenreich M, Xavier RJ, Langer R, Anderson DG, Hacohen N, Regev A, Feng G, Sharp PA, Zhang F (2014) CRISPR-Cas9 knockin mice for genome editing and cancer modeling. Cell 159(2):440–455. doi:10.1016/j.cell.2014.09.014
11. Parnas O, Jovanovic M, Eisenhaure TM, Herbst RH, Dixit A, Ye CJ, Przybylski D, Platt RJ, Tirosh I, Sanjana NE, Shalem O, Satija R, Raychowdhury R, Mertins P, Carr SA, Zhang F, Hacohen N, Regev A (2015) A genome-wide CRISPR screen in primary immune cells to dissect regulatory networks. Cell 162(3):675–686. doi:10.1016/j.cell.2015.06.059
12. Radbruch A, Muehlinghaus G, Luger EO, Inamine A, Smith KG, Dorner T, Hiepe F (2006) Competence and competition: the challenge of becoming a long-lived plasma cell. Nat Rev Immunol 6(10):741–750. doi:10.1038/nri1886
13. Chu VT, Frohlich A, Steinhauser G, Scheel T, Roch T, Fillatreau S, Lee JJ, Lohning M, Berek C (2011) Eosinophils are required for the maintenance of plasma cells in the bone marrow. Nat Immunol 12(2):151–159. doi:10.1038/ni.1981
14. Rush JS, Hodgkin PD (2001) B cells activated via CD40 and IL-4 undergo a division burst but require continued stimulation to maintain division, survival and differentiation. Eur J Immunol 31(4):1150–1159. doi:10.1002/1521-4141(200104)31:4<1150::AID-IMMU1150>3.0.CO;2-V
15. Ozaki K, Spolski R, Ettinger R, Kim HP, Wang G, Qi CF, Hwu P, Shaffer DJ, Akilesh S, Roopenian DC, Morse HC 3rd, Lipsky PE, Leonard WJ (2004) Regulation of B cell differentiation and plasma cell generation by IL-21, a novel inducer of Blimp-1 and Bcl-6. J Immunol 173(9):5361–5371
16. Nojima T, Haniuda K, Moutai T, Matsudaira M, Mizokawa S, Shiratori I, Azuma T, Kitamura D (2011) In-vitro derived germinal centre B cells differentially generate memory B or plasma cells in vivo. Nat Commun 2:465. doi:10.1038/ncomms1475
17. Chu VT, Weber T, Graf R, Sommermann T, Petsch K, Sack U, Volchkov P, Rajewsky K, Kuhn R (2016) Efficient generation of Rosa26 knock-in mice using CRISPR/Cas9 in C57BL/6 zygotes. BMC Biotechnol 16(1):4. doi:10.1186/s12896-016-0234-4

Chapter 13

Germinal Center Formation with Retrovirally Transduced B Cells for Determining the Role of Specific Molecules In Vivo

Rinako Nakagawa

Abstract

Retrovirus-mediated gene transfer has become a powerful tool to investigate roles of specific molecules in B cells, due to its efficiency and expeditiousness. This technology is applicable to activated B cells in order to determine effects of a gene of interest during germinal center (GC) reactions in combination with adoptive transfer. To achieve this, B cells derived from SW_{HEL} mice expressing hen egg lysozyme (HEL)-specific B cell receptors (BCR) are stimulated with HEL antigen in vivo and then with anti-CD40 antibody ex vivo. These cells are then transduced with a retrovirus allowing bicistronic co-expression of a gene of interest and GFP, enabling differentiation of transgene positive cells. The retrovirally transduced cells are then adoptively transferred into immunized $CD45.1^{+}$ congenic recipient mice, to enable differentiation between donor and host cells. This chapter describes methods for (1) activation of HEL-specific mature B cells, (2) retroviral transduction of the activated B cells, (3) adoptive transfer of the cells into recipients, and (4) analysis of the resultant mice by flow cytometry.

Key words Antigen-specific B cells, GC-B cells, HEL, SW_{HEL}, Retroviral transduction, PEG-it, Adoptive transfer

1 Introduction

GCs develop upon extensive proliferation of antigen-activated mature B cells during the normal course of immune responses. Within GCs, affinity maturation of B cells occurs as a result of somatic hypermutation in the genes encoding BCR and clonal selection of higher affinity derivatives. During these processes, GC-B cells change their phenotype between proliferative centroblasts and nondividing centrocytes [1, 2]. GC-B cells hence experience a series of rapid changes in gene expression that places them under genetic stress. Unsurprisingly most types of lymphoma originate from GC-B cells or their descendants [3]. Gene expression must be stringently controlled in GC-B cells to prevent neoplastic transformation; however, the molecular mechanisms

Dinis Pedro Calado (ed.), *Germinal Centers: Methods and Protocols*, Methods in Molecular Biology, vol. 1623,
DOI 10.1007/978-1-4939-7095-7_13,

regulating this are not yet fully understood. Determining functions of specific genes in GC-B cells under the physiological environment is informative, and the analysis of genetically manipulated mice has had significant benefit. However, generation of such mice is too expensive and time-consuming for the purpose of simply testing effects of a gene of interest in vivo. In contrast, retroviral transduction is a rapid and inexpensive method, and thus is advantageous for studying the roles of specific proteins before progressing toward a mouse model.

Survival and proliferation of antigen-primed B cells is dependent on signals generated through BCRs and direct interaction with cognate T cells [1]. B cells interact with cognate T cells through a number of molecular pairings, and one of the key signals for GC formation and progression is mediated through CD40–CD40L interaction [4]. In vitro culture with agonistic anti-CD40 antibody (Ab) and a T cell secreted cytokine IL-4 induces expansion, class switching and plasmacytic differentiation of murine mature B cells. However, currently available in vitro culture systems cannot recapitulate in vivo GC environment and are thus unable to support the growth of GC-B cells appropriately. In efforts to analyze GC-B cells derived from retroviral transduced cells, the methods described here are a combination of B cell activation, retroviral transduction and adoptive transfer to mice. Firstly B cells bearing BCRs specific for a model antigen are primed by the soluble antigen in vivo, followed by treatment with anti-CD40 Ab ex vivo. This allows mature B cells to divide vigorously and prepares them for retroviral transduction [5, 6]. Next, retrovirally transduced B cells are adoptively transferred into congenic mice, enabling the development of GC-B cells in vivo upon further antigenic stimulation [5, 7]. The donor cells can be easily differentiated from host cells using a congenic marker by flow cytometry. Taking advantage of this system, we can study roles of specific molecules during GC reactions in vivo, in a relatively short time period and in a less labor-intensive manner. Moreover, RNAi can also be introduced via retroviral vectors for the reduction in expression of a gene of interest using this method [5, 7].

This chapter first describes the method for activating HEL-specific B cells derived from SW_{HEL} mouse strain [8] which carry the heavy and light chains of the HyHEL10 BCR that recognizes HEL. A protocol for retroviral production and transduction of splenic B cells follows. The retroviral vector used in this protocol produces bicistronic message containing the gene of interest and green fluorescent protein (GFP) allowing analysis of both retrovirally transduced GFP^+ cells and GFP^- internal control cells within the same mouse. Finally, flow cytometric analysis of the adoptively transferred HEL specific GC-B cells is described. An experimental scheme of the procedures is shown in Fig. 1.

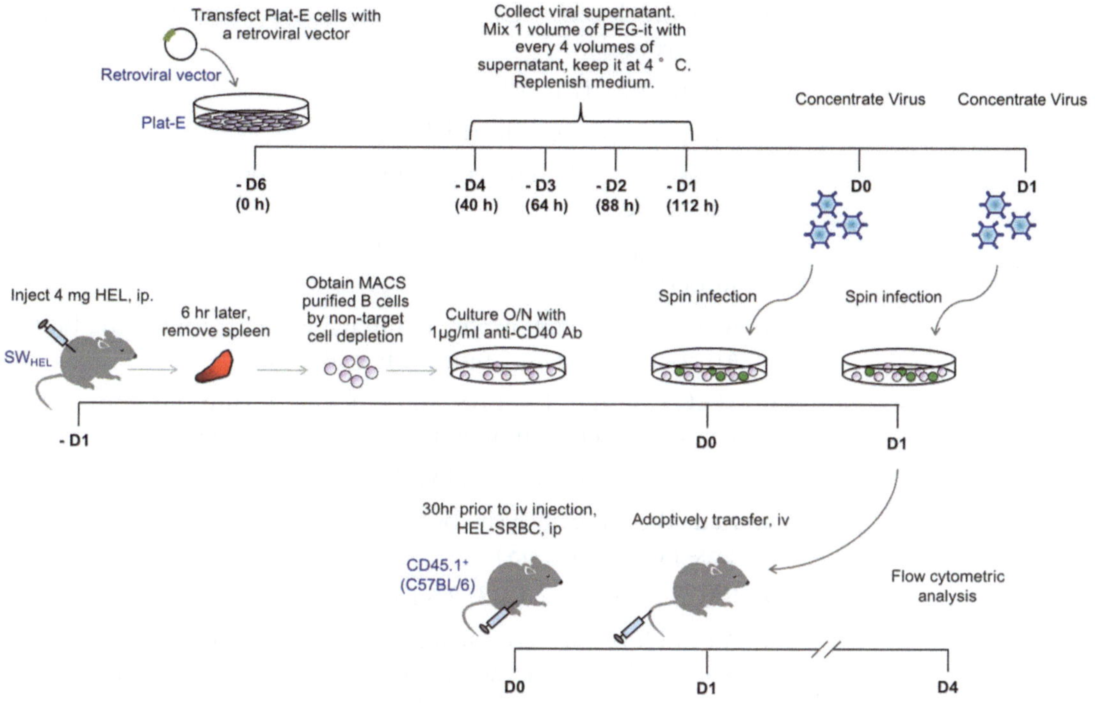

Fig. 1 An experimental scheme of the procedures

2 Materials

1. Platinum-E (PLAT-E) retroviral packaging line (obtained from Dr. Toshio Kitamura, The University of Tokyo, Japan).
2. MigRI retroviral vector (MSCV-based, Addgene).
3. FuGENE 6 Transfection Reagent (Promega).
4. B cell complete medium: RPMI 1640 medium containing 10% fetal bovine serum (FBS), 100 U/ml penicillin, 100 mg/ml streptomycin, 2 mM GlutaMAX, 50 μM 2-mercaptoethanol.
5. PLAT-E complete medium: DME medium containing 10% FBS, 100 U/ml penicillin, 100 mg/ml streptomycin, 2 mM GlutaMAX, 1 mM sodium pyruvate, 10 mM HEPES.
6. Serum-free PLAT-E transfection medium: DME medium containing 2 mM GlutaMAX, 1 mM sodium pyruvate, 10 mM HEPES.
7. 7 mg/ml hexadimethrine bromide (Polybrene) solution.
8. PEG-it viral preparation solution 5× (System Biosciences).
9. 6–12-week-old SW_{HEL} mice (C57BL/6 background; obtained from Dr. Robert Brink, Garvan Institute of Medical Research/University of New South Wales, Australia).

10. 8–16-week-old $CD45.1^+$ congenic mice (C57BL/6 background).
11. 20 mg/ml HEL/PBS solution.
12. Sheep red blood cells (SRBC).
13. 1-ethyl-3-(3-dimethylaminopropyl) carbodiimide hydrochloride (EDC).
14. Conjugation buffer: 0.35 M mannitol, 0.01 M NaCl, autoclaved.
15. B cell isolation kit, mouse (Miltenyi Biotec), containing biotin-Ab cocktail and anti-biotin MicroBeads.
16. MACS buffer: 1× PBS, 0.5% bovine serum albumin (BSA), 2 mM ethylenediaminetetraacetic acid (EDTA).
17. Anti-CD40 Ab (clone HM40-3; eBioscience).
18. Red Blood cell lysis buffer: 0.144 M NH_4Cl, 0.017 M Tris–HCl (pH 7.65).
19. FACS buffer: 1× PBS, 2% FBS.
20. Alexa Fluor 647-conjugated HyHEL9 Ab (obtained from Dr. Robert Brink, Garvan Institute of Medical Research/University of New South Wales, Australia).
21. Biotin-conjugated anti-CD45.1 Ab (clone A20; eBioscience).
22. Streptavidin MicroBeads (Miltenyi Biotec).
23. Fluorophore conjugated Abs: as listed in Subheading 3.5 (eBioscience unless otherwise stated).
24. Anti-mouse CD16/32 Ab (Clone 2.4G2) (BD Biosciences).
25. AutoMACS Pro Separator (Miltenyi Biotec).

3 Methods

3.1 Preparation of Retrovirus

3.1.1 Retroviral Vectors

Specific retroviral vectors are generated by subcloning a gene of interest into an MSCV-based MigRI retroviral backbone that contains sequence encoding GFP downstream of the internal ribosomal entry site. A bicistronic message allows expression of the gene of interest and GFP and retrovirally transduced cells are thus monitored by GFP fluorescence. An empty vector is used as a control.

3.1.2 Transfection and Collection of Retrovirus Containing Supernatant

A 293T cell line based ecotropic retroviral packaging cell line, PLAT-E line is used for its high potency for retroviral production [9] and the cell line is maintained in PLAT-E complete medium supplemented with 1 μg/ml puromycin, 10 μg/ml blasticidin (*see* **Note 1**). One day before transfection, seed 4–4.5 × 10^6 cells on a 10 cm dish per retroviral vector so that cells are ~80% confluent on the day of transfection. On the day of transfection, remove medium from the plated PLAT-E and replenish 8 ml PLAT-E complete medium (*see* **Note 2**). PLAT-E is transfected using the procedure described below.

1. Pipet 570 μl of serum-free PLAT-E transfection medium into a sterile 1.5 ml tube per retroviral vector.
2. Add 10 μg of retroviral plasmid to each of the tube containing medium and vortex.
3. Add FuGENE 6 Transfection reagent directly to medium (*see* **Note 3**) and mix immediately by vortex or flicking the tube.
4. Incubate the transfection reagent–DNA mixture for 10 min at room temperature (RT).
5. Add the mixture onto the PLAT-E cells in the dropwise manner.
6. 40 h after transfection, collect viral supernatant from each of the PLAT-E cultures, mix with 1/4 of original supernatant volume of PEG-it viral preparation solution and store at 4 °C until the day of infection (*see* **Note 4**).
7. Replenish 8 ml PLAT-E complete medium per dish and return the cells to the incubator.
8. 64 h and 88 h after transfection, repeat **step 6**.
9. 112 h after transfection, repeat **step 6**, then discard the transfected PLAT-E cells instead of replenishing the medium.

3.1.3 Concentration of Retroviral Supernatant

1. On the first day of retroviral infection, retrieve tubes containing supernatant–PEG-it mixture that were harvested 40 h and 64 h after transfection.
2. Centrifuge tubes at 1500 × *g* for 30 min at 4 °C. Thereafter, discard the supernatant ensuring that the pellet at the bottom of the tubes is not disturbed (*see* **Note 5**). Retro-vector particles usually appear as a beige pellet.
3. Resuspend the pellet in ~3 ml fresh PLAT-E complete medium per retroviral vector (*see* **Note 6**) and use it for retroviral transduction of activated B cells. Until then store the medium containing retrovirus at 4 °C. 1 ml viral containing medium is used for retroviral infection of B cells that are plated in a well of 6-well plate (*see* Subheading 3.3).
4. On the second day of retroviral infection, repeat the procedure using supernatant–PEG-it mixture that was harvested 88 h and 112 h after transfection.

3.2 Preparation and Stimulation of Splenic B Cells

3.2.1 Antigenic Stimulation of HEL-Specific B Cells In Vivo

A day before first retroviral transduction, inject 200 µl of 20 mg/ml HEL solution (4 mg) intraperitoneally into a 6–12-week-old SW_{HEL} mouse. Six hours later sacrifice the mouse and remove the spleen for B cell preparation.

3.2.2 B Cell Preparation

1. Crush the spleen to prepare a single cell suspension.
2. Centrifuge cell suspension at 300 × *g* for 5 min at 4 °C and remove supernatant.
3. Resuspend cell pellet in 120 µl MACS buffer and add 30 µl biotin-Ab cocktail per spleen.
4. Incubate the cells at 4 °C for 10 min and further add 90 µl MACS buffer and then 60 µl anti-biotin MicroBeads onto the cell suspension.
5. Incubate 10 min at 4 °C.
6. Wash cells in MACS buffer and centrifuge them at 300 × *g* for 5 min at 4 °C.
7. Resuspended the pellet in 1 ml MACS buffer and proceed with the AutoMACS depletion step. Run the sample through an AutoMACS column on the program "DepleteS."
8. Collect the negative cell fraction and pellet it by centrifugation at 300 × *g* for 5 min at 4 °C.
9. Additionally wash cells twice using B cell complete medium and centrifuge them at 300 × *g* for 5 min at 4 °C (*see* **Note 7**).

3.2.3 Splenic B Cell Culture

1. Resuspend the splenic B cell pellet at 0.8–1 × 10^7 cells/ml in B cell complete medium and add anti-CD40 Ab at the concentration of 1 µg/ml.
2. Seed 0.8–1 × 10^7 cells/well in a 6-well plate (1 ml of cell suspension) and then add a further 0.5 ml B cell complete medium supplemented with 1 µg/ml anti-CD40 Ab per well (total volume of 1.5 ml medium per well).
3. Place the plate(s) in an incubator overnight.

3.3 Retroviral Transduction of B Cells

1. On the day of first retroviral transduction, collect medium from the wells of 6-well plate containing splenic B cells that are designated for each type of virus into separate fresh tubes and centrifuge at 300 × *g* for 5 min at 4 °C. The 6-well plate that is used for overnight culture will be reused for retroviral spin infection (*see* **Note 8**).
2. Save the supernatant containing the B cell complete medium supplemented with anti-CD40 Ab in a tube per each type of retrovirus at 4 °C until retroviral infection is completed.

3. Take out the medium containing retrovirus (*see* Subheading 3.1.3) from 4 °C and add 7 mg/ml Polybrene solution in a 1:1000 dilution (final concentration: 7 μg/ml).
4. Resuspend the B cell pellet in the retrovirus containing medium supplemented with polybrene solution and use 1 ml of the cell suspension/well (*see* **Note 9**) in a 6-well plate.
5. Place the plates in a centrifuge and perform spin infection by centrifugation at 750 × *g* for 90 min at RT without brake.
6. Upon completion of spin infection process, collect the medium containing B cells that are infected with each retrovirus into a separate fresh tube and centrifuge at 300 × *g* for 5 min at 4 °C.
7. The plate used for spin infection is used for further B cell culture (*see* **Note 10**). In the mean time, add 0.5 ml/well fresh B cell complete medium supplemented with 1 μg/ml anti-CD40 Ab so that the wells do not dry out.
8. Discard supernatant from the centrifuged cells and resuspend the pellet in the saved B cell complete medium supplemented with anti-CD40 Ab.
9. Seed ~1.5 ml cell suspension/well in the 6-well plate that contains 0.5 ml fresh medium.
10. Return the plate into the incubator for further culture.
11. Next day (on the day of adoptive transfer), repeat the whole process with the retrovirus that was concentrated from the viral supernatant of 88 h and 112 h after transfection.
12. After secondary infection, culture the cells for 3 h and proceed with Subheading 3.4.3.

3.4 Adoptive Transfer of Retrovirally Transduced B Cells

3.4.1 HEL-SRBC Conjugation

1. One day before adoptive transfer, wash 1 ml SRBCs in 14 ml PBS twice and further wash in 14 ml conjugation buffer by centrifugation at 900 × *g* for 5 min at 4 °C.
2. Resuspend the SRBC pellet in 800 μl conjugation buffer and then add 100 μg HEL in 100 μl conjugation buffer.
3. Place the SRBC on a rocker for 10 min on ice.
4. Freshly prepare 100 μl of 100 mg/ml EDC solution in conjugation buffer and add to the tube containing SRBCs.
5. Rock the SRBCs for further 30 min on ice [10].
6. Wash the HEL conjugated SRBCs (HEL-SRBC) three times with 14 ml PBS and resuspend at a concentration of 1×10^9 cells/ml.
7. Confirm the conjugation by flow cytometry using Alexa Fluor 647-conjugated HyHEL9 Ab.

3.4.2 Immunization of CD45.1$^+$ Congenic Recipients

Thirty hours prior to adoptive transfer of retrovirally transduced cells, immunize CD45.1$^+$ congenic recipients intraperitoneally with 2×10^8 HEL-SRBC in 200 μl of PBS.

3.4.3 Adoptive Transfer of Retrovirally Transduced Cells

1. Harvest the B cells cultured for 3 h after secondary retroviral infection (*see* Subheading 3.3) from the 6-well plate and wash in 10 ml PBS once by centrifugation at $300 \times g$ for 5 min at 4 °C.
2. Resuspend the pellet in 1 ml PBS and use 30–40 μl of this cell suspension for flow cytometry to determine the frequency of GFP$^+$ B220$^+$ HEL$^+$ cells of total live cells.
3. For flow cytometric analysis, firstly stain cells with 100 μl 50 ng/ml HEL solution in FACS buffer for 10 min at 4 °C. Wash the cells in 1 ml FACS buffer once and then stain them with PE-conjugated B220 Ab (clone: RA3-6B2, 1/200 dilution) and Alexa Fluor 647-conjugated HyHEL9 Ab (1/1000 dilution). Typically, 10–20% cells are B220$^+$ HEL$^+$ and 5–20% of B220$^+$ HEL$^+$ cells are GFP$^+$ before adoptive transfer (Fig. 2).
4. Once the frequency of GFP$^+$ B220$^+$ HEL$^+$ B cells is determined, adjust the number of cells for injection so that each recipient mouse receives 1–2 × 10^5 GFP$^+$ B220$^+$ HEL$^+$ B cells.

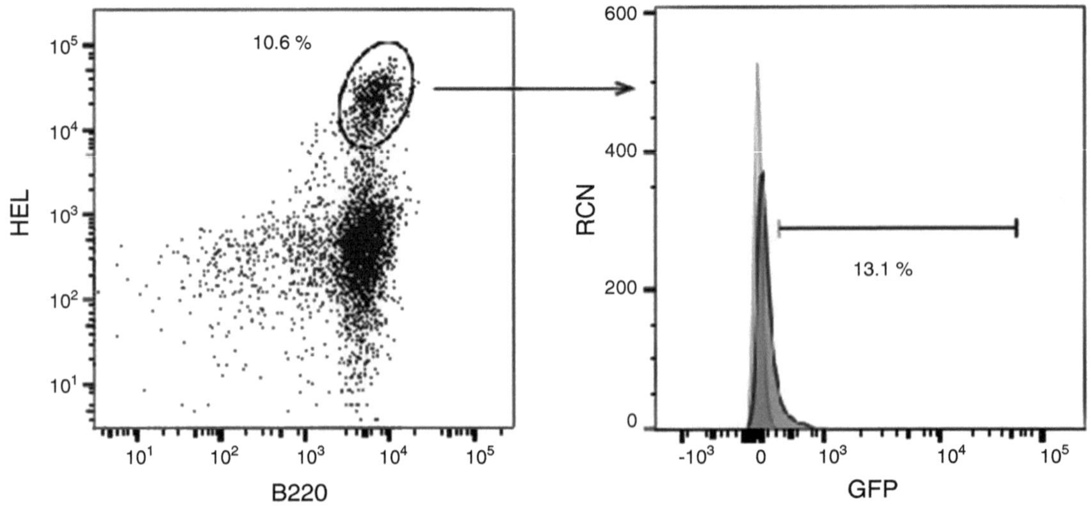

Fig. 2 Flow cytometric analysis of retrovirally transduced B cells before adoptive transfer. Activated B cells derived from SW_{HEL} were transduced with a retrovirus produced from pMigRI-IRES-GFP empty vector transfected PLAT-E. Flow cytometric gate for HEL-binding cells is shown (*left*). The right histogram indicates GFP expression level of the empty vector transduced cells (*dark grey filled*) and negative control (*light grey filled*). Negative control B cells were activated and cultured in the same method except retroviral transduction. Percentage of GFP$^+$ cells of B220$^+$ HEL-binders is shown (*right*). Cells were live and size gated

3.5 Flow Cytometric Analysis of Retrovirally Transduced GC-B Cells

1. Three days after adoptive transfer, sacrifice recipient mice and harvest their spleens.
2. Prepare a single suspension of splenocytes and add 500 μl of red blood cell lysis buffer per spleen.
3. Incubate for 2 min at RT and add immediately 200 μl of FBS.
4. Centrifuge cells at 300 × *g* for 5 min at 4 °C.
5. Resuspend the cell pellet; the color of which should be white; in 1 ml MACS buffer.
6. Add 1 μg of biotinylated anti-CD45.1 Ab (clone: A20) for 15 min at 4 °C.
7. Wash cells in FACS buffer once and resuspended in 200 μl FACS buffer containing 65 μl Streptavidin MicroBeads.
8. Incubate for 10 min at 4 °C, and subject the cell suspension to AutoMACS based negative selection by running the sample on the program "deplete."
9. Centrifuge the negatively selected cells at 300 × *g* for 5 min at 4 °C and remove the supernatant.
10. Block surface Fc receptors by incubating the cells with purified anti-mouse CD16/32 Ab (1/500 dilution).
11. Stain cells with fluorochrome-conjugated Abs: APCCy7 conjugated anti-CD45.2 Ab (clone: 104, 1/200 dilution), Alexa Fluor 700-conjugated anti-CD38 Ab (clone: 90, 1/300 dilution), PECy7 conjugated anti-CD95 Ab (clone: Jo2 obtained from BD Biosciences, 1/300 dilution), PE conjugated anti-CXCR4 Ab (clone: 2B11, 1/100 dilution), PerCPCy5.5 conjugated anti-CD86 Ab (clone: GL1 obtained from Biolegend, 1/200 dilution), eFluor 450 conjugated streptavidin (1/500 dilution) to delineate host-derived cells and Alexa Fluor 647-conjugated HyHEL9 Ab (1/1000 dilution), after initial HEL staining as described in Subheading 3.4.3.
12. Wash cells in 1 ml FACS buffer once and resuspended in 0.5–1 ml FACS buffer.
13. Add 7-AAD 5 min prior to flow cytometric analysis for dead cell discrimination.
14. Analyze by flow cytometry the donor derived GC-B cells ($CD45.1^-$ $CD45.2^+$ HEL^+ $CD38^{lo}$ $CD95^+$). A mixture of GFP^- and GFP^+ cells is injected into recipient mice, enabling a retroviral transgene negative internal control as GFP^- population within the same host mouse. In GFP^- and GFP^+ populations, centroblasts ($CXCR4^{hi}$ $CD86^{lo}$) and centrocytes ($CXCR4^{lo}$ $CD86^{hi}$) are further delineated (Fig. 3).

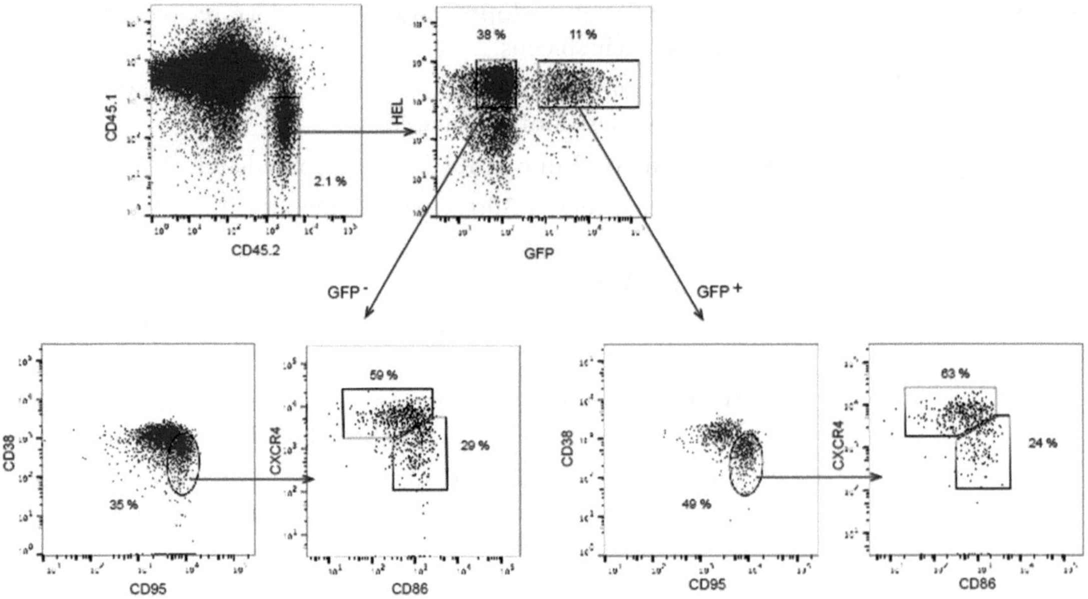

Fig. 3 Identification of retrovirally transduced GC-B cells. Splenocytes from a mouse that received SW_{HEL} derived activated B cells were analyzed by flow cytometry. Donor-derived cells were revealed by CD45.2 expression whereas host derived cells were identified by CD45.1 expression (*top left*). GFP^+ retrovirally transduced HEL-binders and GFP^- noninfected HEL-binders (*top right*) are further gated for GC-B cells ($CD38^{lo}$ $CD95^+$) and followed by centroblasts and centrocytes delineation using CXCR4 and CD86 (*bottom*). Gate frequency is shown in the plots. Cells were live and size gated

4 Notes

1. PLAT-E must be maintained in complete medium supplemented with 1 μg/ml puromycin, 10 μg/ml blasticidin to preserve retroviral packaging genes (gag, pol, and env). Under the drug selection pressure, PLAT-E keeps producing retrovirus efficiently up to 4 months [9]. Do not overgrow PLAT-E as the potency decreases significantly. Passage them when the culture reaches 70–90% confluency.
2. PLAT-E complete medium for FuGENE 6 transfection must not contain puromycin and blasticidin to maintain transfection efficiency.
3. Do not allow FuGENE 6 Transfection Reagent to contact with the sides of the tube that contains the serum-free PLAT-E transfection medium in order to avoid adversely influencing transfection efficiency.
4. Supernatant–PEG-it mixture should be refrigerated for at least 12 h before centrifugation for virus precipitation. The supernatant–PEG-it mixture is stable for 4–5 days at 4 °C.

5. The supernatant containing residual PEG-it solution can be transferred into a fresh tube instead of discarding it and recentrifuged at 1500 × *g* for 5 min at 4 °C. Remove the fluid completely by aspiration to avoid disturbing the precipitated viral particles in the pellet. Then combine pellets from the two centrifugations to yield more retroviral particles in total.
6. For a high titer virus preparation, resuspend 1/10 of original volume. The resuspension volume should be limited to 1/5 of original volume at maximum dilution. Concentrated virus can be stored at −80 °C; however, avoid freeze–thaw cycles for maximum virus recovery.
7. It is difficult to maintain sterility of an AutoMACS Pro Separator. Therefore washing the MACS-isolated cells in sterile complete medium at least twice is essential step to prevent contamination of the culture.
8. B cells can stick to the surface of the tissue culture plate. However, it is not necessary to pipette up and down to remove all of the cells if the same B cell culture plate is used for retroviral infection. This minimizes loss or damage of the activated B cells.
9. For retroviral infection, the well of 6-well plate should contain at least 1 ml of viral supernatant. However too much medium in a well may cause spillage during spin infection and 1–2 ml of viral supernatant is preferable.
10. For the same reasons as described in **Note 8**, most of the cells will adhere to the surface of the plate after spin infection. To minimize loss of cells keep using the same plate for the additional cultures.

References

1. MacLennan IC (1994) Germinal centers. Annu Rev Immunol 12:117–139
2. Victora GD, Nussenzweig MC (2012) Germinal centers. Annu Rev Immunol 30:429–457. doi:10.1146/annurev-immunol-020711-075032
3. Kuppers R (2005) Mechanisms of B-cell lymphoma pathogenesis. Nat Rev Cancer 5 (4):251–262. doi:10.1038/nrc1589. nrc1589 [pii]
4. Grewal IS, Flavell RA (1998) CD40 and CD154 in cell-mediated immunity. Annu Rev Immunol 16:111–135. doi:10.1146/annurev.immunol.16.1.111
5. Pereira JP, Kelly LM, Xu Y, Cyster JG (2009) EBI2 mediates B cell segregation between the outer and centre follicle. Nature 460(7259): 1122–1126. doi:10.1038/nature08226
6. Reif K, Ekland EH, Ohl L, Nakano H, Lipp M, Forster R, Cyster JG (2002) Balanced responsiveness to chemoattractants from adjacent zones determines B-cell position. Nature 416 (6876):94–99. doi:10.1038/416094a
7. Nakagawa R, Leyland R, Meyer-Hermann M, Lu D, Turner M, Arbore G, Phan TG, Brink R, Vigorito E (2016) MicroRNA-155 controls affinity-based selection by protecting c-MYC+ B cells from apoptosis. J Clin Invest 126 (1):377–388. doi:10.1172/JCI82914
8. Phan TG, Amesbury M, Gardam S, Crosbie J, Hasbold J, Hodgkin PD, Basten A, Brink R (2003) B cell receptor-independent stimuli trigger immunoglobulin (Ig) class switch recombination and production of IgG autoantibodies by anergic self-reactive B cells. J Exp Med 197 (7):845–860. doi:10.1084/jem.20022144. jem.20022144 [pii]

9. Morita S, Kojima T, Kitamura T (2000) Plat-E: an efficient and stable system for transient packaging of retroviruses. Gene Ther 7 (12):1063–1066. doi:10.1038/sj.gt.3301206

10. Goodnow CC, Crosbie J, Adelstein S, Lavoie TB, Smith-Gill SJ, Brink RA, Pritchard-Briscoe H, Wotherspoon JS, Loblay RH, Raphael K et al (1988) Altered immunoglobulin expression and functional silencing of self-reactive B lymphocytes in transgenic mice. Nature 334 (6184):676–682. doi:10.1038/334676a0

Chapter 14

Characterization of the B Cell Transcriptome Bound by RNA-Binding Proteins with iCLIP

Manuel D. Díaz-Muñoz, Elisa Monzón-Casanova, and Martin Turner

Abstract

Posttranscriptional regulation of gene expression shapes the B cell transcriptome and controls messenger RNA (mRNA) translation into protein. Recent reports have highlighted the importance of RNA binding proteins (RBPs) for mRNA splicing, subcellular location, stability, and translation during B lymphocyte development, activation, and differentiation. Here we describe individual-nucleotide resolution UV cross-linking and immunoprecipitation (iCLIP) in primary lymphocytes, a method that maps RNA–protein interactions in a genome-wide scale allowing mechanistic analysis of RBP function. We discuss the latest improvements in iCLIP technology and provide some examples of how integration of the RNA–protein interactome with other high-throughput mRNA sequencing methodologies uncovers the important role of RBP-mediated RNA regulation in key biological cell processes.

Key words Primary B cells, RNA binding proteins, Individual-nucleotide resolution UV cross-linking and immunoprecipitation (iCLIP), mRNA maturation, Intron, mRNA stability, mRNA translation, 3′ Untranslated regions (UTRs)

1 Introduction

The transcriptome of B lymphocytes undergoes qualitative and quantitative changes upon antigen-mediated B cell activation and during the germinal center (GC) reaction. Dynamic changes in transcription mediated by sequence specific DNA binding proteins regulate gene expression during B cell expansion, antibody affinity selection and B cell terminal differentiation in GCs [1, 2]. Although the importance of transcription factors (TFs) including NF-kB, Bcl6, IRF4, or Blimp1 is well established, less is known about the posttranscriptional regulatory mechanisms that control the transcriptome in GC B cells and how these influence protein abundance and function.

Recent reports have highlighted the importance of RNA-binding proteins (RBPs) and microRNAs (miRNAs) in B cell development, GC formation and antibody production. B cell

Dinis Pedro Calado (ed.), *Germinal Centers: Methods and Protocols*, Methods in Molecular Biology, vol. 1623,
DOI 10.1007/978-1-4939-7095-7_14, © Springer Science+Business Media LLC 2017

specific deletion of HuR, an RBP encoded by the *Elavl1* gene, impairs B cell activation and high-affinity antibody production [3, 4]. Similarly, the RNA-processing enzyme Dicer and individual miRNAs, including miR-155, are key regulators of the GC reaction and B cell terminal differentiation into antibody-secreting cells (ASC) [5–7]. RBP- and miRNA-dependent control of T follicular helper (Tfh) cell differentiation and function is also required for optimal GC responses [8–10], indicating that posttranscriptional regulation of gene expression is widely utilized by cell types involved in the GC reaction.

RBPs exert fundamental functions to guarantee the transfer of genetic information from DNA to protein. An atlas with more than 1500 curated RBPs has been compiled in the last few years [11]. This indicates that the number of RBPs is at least as high as the number of TFs. However, this number is likely to increase significantly as the unsuspected RNA binding properties of proteins are uncovered by a newly developed technique called interactome capture [12]. Messenger RNA (mRNA) associates with RBPs from the moment it is transcribed. In the nucleus RBPs control the mechanisms of RNA maturation such as splicing and polyadenylation. RBPs such as HuR and hnRNPLL alter the mRNA splicing profile of B cells and promote antibody production [3, 13]. The RBP Zfp318 has been shown to control transcription termination and thereby be essential for co-expression of IgM and IgD during B cell maturation [14, 15]. RBPs control mRNA translocation through the nuclear pore as well as sub-cellular location in the cytoplasm. RBPs regulate polyribosome assembly and the differential expression of the elongation initiation factors eIF4E1 and eIF4E3 has been linked to a switch in mRNA translation in diffuse large B cell lymphoma (DLBCL) [16]. Finally, RBPs modulate mRNA stability, a mechanism that can regulate mRNA abundance [17, 18].

Precise identification of direct RNA targets of RBPs and an understanding of the nature of RNA–protein complexes is necessary to elucidate the molecular processes regulated by RBPs. The first genome-wide approaches to identify RBP-specific RNA targets involved RNA immunoprecipitation and microarray analysis (RIP-ChIP). However, these approaches could not distinguish between direct and indirect interactions and did not provide any direct nucleotide sequence information related to the site of RNA–protein interaction. In 2003, Ule et al. developed a method called in vivo UV cross-linking and immunoprecipitation (CLIP) that enabled genome-wide characterization of direct RNA–protein interactions in intact cells [19]. Unlike other cross-linking agents, UV light is a "zero length" agent that generates covalent bonds between proteins and nucleic acids at their contact points [20, 21]. These covalent bonds are highly resistant to heat, detergent, and pH, allowing stringent washes during the protein purification steps.

The combination of high-throughput sequencing technologies with CLIP (called HITS-CLIP) has allowed the genome-wide identification of RBP RNA targets [22]. Originally HITS-CLIP cDNA library preparation required reverse transcription following the ligation of adaptor sequences at both the 5′ and 3′ ends of an RNA molecule. However, the retention of a short polypeptide at the site of RNA–protein cross-linking led to a failure of reverse transcription (RT) in 80% or more of molecules. This feature was later harnessed to develop an improved version of HITS-CLIP called individual-nucleotide resolution UV cross-linking and immunoprecipitation (iCLIP) [23, 24]. In this chapter, we describe how to apply iCLIP to uncover the molecular functions of RBPs in B cells. Unlike similar methods for studying RNA–protein interactions that use photoactivatable nucleotides such as 4-thioururidine, that must be taken up by cells and incorporated into nascent RNA, iCLIP can be applied to resting lymphocytes and to activated cells without introducing stress responses that might accompany the uptake of photoactivatable nucleotides. During iCLIP a single adaptor oligonucleotide is ligated to the 3′ end of each RNA molecule and used for reverse transcription of all RNA molecules targeted by a specific RBP. The principal advantage of iCLIP is that it takes advantage of the finding that RT stops at the site of UV cross-linking to map the cross-link site at a single nucleotide resolution as position −1 of any cDNA sequencing read aligned to the genome. During RT the 5′ adaptor sequence required for library preparation is added and ligated to the 5′ end of each cDNA upon molecule circularization (Fig. 1a).

When iCLIP is combined with other high-throughput sequencing methods used for to study the transcriptome (i.e., mRNAseq) and translatome (i.e., Ribosome footprinting profiling), it enables the generation of global RNA maps to uncover the distinctive functions of different RBPs in alternative splicing [25], polyadenylation [26], mRNA stability and translation [17, 18]. Recently reported improvements in iCLIP methodology (methods renamed as eCLIP [27] and irCLIP [28]) have increased the efficiency of library preparation by reducing both the amount of input material and handling times. Here we explain in detail the standard iCLIP protocol (summarized in Fig. 1a) and discuss possible ways to incorporate recent improvements into the protocol in order to gain a better understanding of how posttranscriptional regulation by RBPs controls B cell development and differentiation.

2 Materials

2.1 B Cell Isolation, Cell Activation, and UV Cross-Linking

1. Complete RPMI medium: RPMI 1640 Medium -Dutch modification-, 10% fetal calf serum (FCS), 100 U penicillin G/ml, 100 mg/ml streptomycin, 2 mM L-glutamine, 50 μM β-mercaptoethanol and 1 mM sodium pyruvate.

2. MACs buffer: PBS, 0.5% FCS, 2 mM EDTA.
3. Negative B cell isolation kit (e.g., Miltenyi Biotec)
4. Other materials: Cell strainers (40 μm), magnetic stand and LS separation columns (e.g., Miltenyi Biotec).
5. Mitogens (i.e., LPS, *E. coli* O127:B8) and cytokines (i.e., IL-4, IL-5, or IL-21) as required for in vitro B cell activation.
6. UV cross-linker with a 240 nm bulb (e.g., Stratalinker 2400).

2.2 iCLIP Buffers

1. Cell lysis buffer (RIPA buffer): 50 mM Tris–HCl (pH 7.4), 100 mM NaCl, 1% Igepal CA-630, 0.1% sodium dodecyl sulfate (SDS), 0.5% sodium deoxycholate, 1:200 protease inhibitor cocktail III (add fresh). Optional–If RNAse I is used for RNA digestion, 10 U of a ribonuclease inhibitor derived from human placenta can be used as it only inhibits ribonucleases A, B, and C (add fresh).
2. Cytoplasmic lysis buffer: 10 mM Tris–HCl (pH 7.4), 10 mM NaCl, 2.5 mM $MgCl_2$, 40 μg/ml digitonin, 1:200 protease inhibitor cocktail (add fresh), and 10 U of a ribonuclease inhibitor (if required, add fresh).
3. High-salt wash buffer: 50 mM Tris–HCl (pH 7.4), 1 M NaCl, 1 mM EDTA, 1% Igepal CA-630, 0.1% SDS, 0.3% sodium deoxycholate. Optional–Urea (up to 2 M) can be used for protein denaturalization, if required.
4. PNK wash buffer: 20 mM Tris–HCl (pH 7.4), 10 mM $MgCl_2$, 0.2% Tween 20.
5. 5× PNK buffer: 350 mM Tris–HCl (pH 6.5), 50 mM $MgCl_2$, 5 mM dithiothreitol (DTT).
6. 4× Ligation buffer: 200 mM Tris–HCl (pH 7.8), 40 mM $MgCl_2$, 4 mM DTT.
7. PK buffer: 100 mM Tris–HCl (pH 7.4), 50 mM NaCl, 10 mM EDTA.

2.3 RNA Digestion, RNA Labeling, and RNA–Protein Immunoprecipitation

1. RNase I (various concentrations, e.g., Thermo Fisher).
2. 2 U/μl DNAse (e.g., TURBO™ DNase, Thermo Fisher).
3. Sonicator (e.g., Misonix 3000) and thermomixer.
4. 20 U/μl RNase Inhibitor (e.g., SUPERase In™ or RNase-OUT, ThermoFisher).
5. 32γ-ATP (e.g., PerkinElmer).
6. Preadenylated adapter L3-App: rAppAGATCGGAAGAGCG GTTCAG/ddC/ (e.g., Integrated DNA Technologies).
7. T4 PNK including 10× PNK buffer (e.g., New England Biolabs).

8. T4 ssRNA ligase including 10× RNA ligase buffer, 10 mM ATP (e.g., New England Biolabs).
9. Polyethylene Glycol 8000 (PEG 8000).
10. Shrimp alkaline phosphatase including 10× shrimp alkaline phosphatase reaction buffer (e.g., Sigma-Aldrich).
11. Magnetic beads coupled to protein A or protein G for conventional IP (different sources). Protein A for human IgM. Protein G for mouse IgG1 and for rat, sheep, and goat IgG. Either protein A or protein G can be used for IP of human and rabbit IgG, and mouse IgG2a, IgG2b, and IgG3. Heavy chain antibodies from Camelidae species (VhH antibodies also called nanobodies) coupled with magnetic beads can be used alternatively for IP of tagged proteins (e.g., GFP-Trap®_M, ChromoTek).
12. Magnetic stand for 1.5 ml tubes (low binding).

2.4 RNA–Protein Complex Isolation and RNA Reverse Transcription

1. Pre-cast Bis–Tris–HCl polyacrylamide gels (e.g., NuPAGE® Novex®, Thermo Fisher). Polyacrylamide gel percentage varies depending on the molecular weight of the RNA–protein complexes.
2. Gel electrophoresis and transfer system.
3. LDS Sample Buffer (e.g., NuPAGE®, Thermo Fisher), MOPS SDS Running Buffer (e.g., NuPAGE®, Thermo Fisher) and Tris–Glycine Transfer Buffer with 10% methanol (e.g., Novex®, Thermo Fisher).
4. Nitrocellulose membranes (pore size 0.45 μm), Whatman paper and autoradiography films.
5. RNA isolation and precipitation: Proteinase K (DNAse and RNAse free), 10% SDS, phenol–chloroform–isoamyl alcohol 25:24:1 (pH 8), phase-lock gel heavy tube, glycogen, 3 M sodium acetate (pH 5.5), and absolute ethanol.
6. Thermocycler.
7. RNA reverse transcription: PCR tubes, dNTP mix (10 mM), RNAse inhibitors, 200 U/μl Reverse Transcriptase including 5× first-strand buffer (250 mM Tris–HCl; pH 8.3, 375 mM KCl, 15 mM $MgCl_2$), and 100 mM DTT (e.g., SuperScript® III, Thermo Fisher).
8. RtClip primers (from Huppertz et al. [24]):

Rt1clip - 5Phos/NNAACCNNNAGATCGGAAGAGCGTCG TGGATCCTGAACCGC.

Rt2clip - 5Phos/NNACAANNNAGATCGGAAGAGCGTCG TGGATCCTGAACCGC.

Rt3clip - 5Phos/NNATTGNNNAGATCGGAAGAGCGTCG TGGATCCTGAACCGC.

Rt4clip - 5Phos/NNAGGTNNNAGATCGGAAGAGCGTCGTGGATCCTGAACCGC.

Rt6clip - 5Phos/NNCCGGNNNAGATCGGAAGAGCGTCGTGGATCCTGAACCGC.

Rt7clip - 5Phos/NNCTAANNNAGATCGGAAGAGCGTCGTGGATCCTGAACCGC.

Rt8clip - 5Phos/NNCATTNNNAGATCGGAAGAGCGTCGTGGATCCTGAACCGC.

Rt9clip - 5Phos/NNGCCANNNAGATCGGAAGAGCGTCGTGGATCCTGAACCGC.

Rt11clip - 5Phos/NNGGTTNNNAGATCGGAAGAGCGTCGTGGATCCTGAACCGC.

Rt12clip - 5Phos/NNGTGGNNNAGATCGGAAGAGCGTCGTGGATCCTGAACCGC.

Rt13clip - 5Phos/NNTCCGNNNAGA TCGGAAGAGCGTCGTGGATCCTGAACCGC.

Rt14clip - 5Phos/NNTGCCNNNAGATCGGAAGAGCGTCGTGGATCCTGAACCGC.

Rt15clip - 5Phos/NNTATTNNNAGATCGGAAGAGCGTCGTGGATCCTGAACCGC.

Rt16clip - 5Phos/NNTTAANNNAGATCGGAAGAGCGTCGTGGATCCTGAACCGC.

2.5 Library Preparation

1. 6% Pre-cast denaturing polyacrylamide gels for purification of cDNA products from 150–250 bp (e.g., Novex® TBE-urea gels, Thermo Fisher).
2. TBE Running Buffer (e.g., Novex®, Thermo Fisher), TBE-Urea Sample Buffer (2×) (e.g., Novex®, Thermo Fisher) and SYBR (e.g., SYBR® Gold Nucleic Acid Gel Stain, ThermoFisher).
3. Dialyzer (e.g., D-Tube™ Dialyzer Maxi, Millipore).
4. Centrifuge tube filters/cellulose acetate membrane with pore size 0.22 μm (e.g., Corning® Costar® Spin-X®, Sigma) and GF/D Glass Circles, 10 mm, 100/pk (e.g., Sigma).
5. ssDNA Ligase including 10× reaction buffer and 50 mM $MnCl_2$ (e.g., CircLigase™ II, Epicentre Technologies).
6. Cut_oligo: GTTCAGGATCCACGACGCTCTTCAAAA.
7. BamHI with 10× digestion buffer (e.g., FastDigest, Thermo Fisher).
8. High-Fidelity PCR Master Mix (e.g., Phusion Flash, Thermo Fisher).
9. Solexa P5 adaptor primer: AATGATACGGCGACCACCGAGATCTACACTCTTTCCCTACACGACGCTCTTCCGATCT.

10. Solexa P3 adaptor primer: CAAGCAGAAGACGGCATACG AGATCGGTCTCGGCATTCCTGCTGAACCGCTCTTCC GATCT.
11. 8% TBE gels (e.g., Novex®, Thermo Fisher) and Nucleic Acid Gel Stain (e.g., SYBR® Gold Nucleic Acid Gel Stain, Thermo Fisher).
12. KAPA Library Quantification Kit from Kapa Biosystems.

3 Methods

3.1 Cell Isolation and Activation

1. Harvest spleen and lymph nodes in ice-cold complete RPMI medium
2. Smash tissues and pass cells through a 40 μm cell strainer.
3. Lyse red blood cells using any commercially available lysis buffer.
4. Generate a single cell suspension in ice-cold MACs buffer (*see* **Note 1**).
5. Isolate B cells by negative selection using a B cell isolation kit. Briefly, $CD43^+$, $CD4^+$ and Ter-119^+ cells are selectively labeled and depleted after sequential cell incubation for 15 min at 4 °C with specific biotinylated antibodies against CD43, CD4 and Ter-119 and with biotin-conjugated antibodies coupled to magnetic beads. Labeled cells are then passed through a LS column placed in a magnetic stand. After washing with abundant ice-cold MACs buffer, B cells contained in the buffer that flows through (90–95% purity) are pelleted and resuspended in complete RPMI medium (if B cells are further activated in vitro with mitogens) or in ice-cold PBS (if freshly isolated cells are used in iCLIP).
6. After in vitro B cell activation, count cells. 20–40 × 10^6 B cells (*see* **Note 2**) are centrifuged at 453 × *g* at 4 °C for 5 min and resuspended in 10 ml of ice-cold PBS.

3.2 In Vivo RNA–Protein Cross-Linking

1. For UV cross-linking of RNA–protein interactions, place the cell suspension in an ice-cold 10 cm diameter dish. With the lid off, irradiate cells with UV light (254 nm, 150 mJ, *see* **Note 2**) using a UV cross-linker. Place the dish back on ice.
2. Collect irradiated B cells into a 15 ml tube (use cell scrapers if necessary).
3. Centrifuge at 453 × *g* for 5 min at 4 °C.
4. Resuspend the cell pellet in 1 ml of ice-cold PBS and transfer it to a 1.5 ml tube.
5. For generation of total cell extracts, centrifuge sample tubes at 376 × *g* for 5 min at 4 °C, remove supernatant and snap-freeze

the cell pellet using dry ice and store samples at −80 °C for later use.

6. Alternatively, for cytoplasmic/nuclear protein extract isolation, B cells are centrifuged at 453 × *g* for 5 min at 4 °C. Resuspend cell pellets with 1 ml of ice-cold cytoplasmic lysis buffer. Incubate samples on ice for 10 min prior centrifugation at 845 × *g* for 5 min at 4 °C. Transfer the supernatant to a new 1.5 ml tube, snap-freeze on dry ice and store at −80 °C as the cytoplasmic B cell protein fraction.
7. Wash nuclear pellets with 1 ml of ice-cold cytoplasmic lysis buffer (3×). Then, add 1 ml of ice-cold total cell lysis buffer (RIPA buffer). Mix and keep samples mixing in the vortex for 30 min. at 4 °C. Centrifuge samples at 15,871 × *g* for 10 min at 4 °C. Collect supernatant in a new 1.5 ml tube, snap-freeze and store at −80 °C as the nuclear B cell protein fraction. Alternatively, nuclear samples can be processed as total cell extracts as described in the next section.

3.3 Immunoprecipitation and Labeling of RNA–Protein Complexes

1. Thaw samples for iCLIP by placing then on ice.
2. In the meantime prepare the beads. Wash 30 μl of protein A or protein G magnetic beads with 1 ml of cell lysis buffer. Add the antibody (*see* **Note 3**) to the beads resuspended in 0.5–1 ml of cell lysis buffer and incubate in rotation at room temperature for at least 30–60 min. Wash beads twice with cell lysis buffer after that.
3. Resuspend each cell pellet (20–40 × 10^6 B cells) with 1 ml of cell lysis buffer and keep on ice.
4. Sonicate samples with 10 s bursts at five decibels (three times each sample). Keep samples on ice during sonication. Place the probe in the middle of the tube without touching the walls and avoid foaming (*see* **Note 4**).
5. Centrifuge samples at 15,871 × *g* for 10 min at 4 °C.
6. Transfer the supernatant to a new 1.5 ml tube and add 10 U of DNase. Then, add a high amount (12 units per ml) or a low amount of RNase I (range concentration from 0.04 to 0.3 units per ml to obtain complete or partial RNA digestion. It needs testing (*see* **Note 5**)).
7. Incubate samples shaking in a thermomixer at 1000 rpm at 37 °C for 3 min precisely.
8. Add 40 units of RNase inhibitor (optional).
9. Transfer samples immediately into ice for at least 5 min.
10. Remove the buffer from the antibody-coupled magnetic beads and add the samples.

11. Immunoprecipitate the RNA–protein complexes by incubating samples at 4 °C for 1–18 h in rotation.
12. After incubation place the sample tubes in a magnet stand for 30 s.
13. Remove the supernatant (keep the supernatant for protein depletion check if needed), and wash samples with 1 ml high-salt wash buffer (*see* **Note 6**) and thoroughly mix samples for 30 s at 4 °C. Repeat **steps 12** and **13** for three times.
14. Wash samples with PNK wash buffer two times (*see* **Note 7**).
15. Remove PNK wash buffer completely and add 20 μl of 3′ RNA dephosphorylation PNK master mix: 17 μl of H_2O, 2 μl of 10× PNK buffer (pH 6.5) (*see* **Note 8**), 0.5 μl of RNase inhibitor, and 0.5 μl of PNK.
16. Incubate at 37° for 20 min in the thermomixer mixing at 1100 rpm.
17. Wash with 1 ml of high-salt wash buffer twice and once with PNK wash buffer.
18. Resuspend the beads in 1 ml of PNK wash buffer.
19. Transfer 100 μl of the mix (one tenth of beads in PNK wash buffer) to a new 1.5 ml tube for 5′ end RNA radioactive labeling. Keep it at 4 °C until next day to be used in **step 24**.
20. Remove completely the PNK wash buffer from the remaining 900 μl of the mix and add 20 μl of the 3′ end linker ligation mix: 9 μl H_2O, 2 μl of 10 mM ATP, 2 μl of 10x RNA ligase buffer (*see* **Note 8**), 4 μl of PEG 8000, 1 μl of T4 ssRNA ligase, 0.5 μl of RNase inhibitor, and 1.5 μl of preadenylated adapter L3-App (20 μM).
21. Incubate for 18 h at 16 °C in a thermomixer mixing at 1200 rpm.
22. Wash samples twice with high-salt wash buffer and once with PNK wash buffer.
23. Remove completely PNK wash buffer and add 12 μl of 1× LDS Sample Buffer. Reserve to be used in **step 28** after 5′ end RNA radioactive labeling is completed.
24. Take the sample fraction reserved in **step 19**. Wash once with PNK wash buffer and remove any supernatant completely before adding 20 μl of 5′ end dephosphorylation mix: 17 μl of H_2O, 2 μl of 10× shrimp alkaline phosphatase buffer, and 1 μl of shrimp alkaline phosphatase. Incubate at 37 °C for 15 min in a thermomixer mixing at 1200 rpm (*see* **Note 9**).
25. Wash samples twice with PNK wash buffer.
26. Remove the supernatant completely and resuspend the beads with 10 μl of 32γ-ATP labeling mix: 8.2 μl of H_2O, 1 μl of 10×

PNK buffer, 0.4 μl of PNK, and 0.4 μl of 32γ-ATP (0.148 MBq).

27. Incubate at 37 °C for 5 min in the thermomixer without mixing.
28. Place samples on the magnetic stand. Discard supernatant as radioactive waste.
29. Add in the cold-beads resuspended in 12 μl of 1× LDS Sample Buffer reserved in **step 23**.
30. Mix well and heat samples at 70 °C for 10 min in a thermomixer without mixing.

3.4 Detection of RNA–Protein Complexes and Isolation of RNA

1. Load samples in a 12-well pre-cast Bis–Tris–HCl polyacrylamide gel (*see* **Note 10**). Leave one or two empty wells between loaded lanes to avoid sample cross-contamination. Load samples and negative controls in the same gel.
2. Run at 180 V using 1× MOPS running buffer for 50 min or until the bromophenol blue dye is close to the bottom. Do not let the blue dye to run off the gel to avoid radioactive contamination of the electrophoresis apparatus.
3. Cut off the bottom of the gel containing the blue front dye and free 32γ-ATP and discard it as radioactive waste. Transfer the RNA–protein complexes in the upper part of the gel to a nitrocellulose membrane (pore size 0.45 μm) using Tris–glycine transfer buffer containing 10% methanol.
4. After transfer is completed, remove nitrocellulose membrane and rinse it with PBS. Then, wrap the membrane with plastic film (e.g., cling film) and attach it inside a radiographic cassette using cello tape. Include a luminescent sticker next to the membrane for positioning.
5. Expose to autoradiogram. With enough material and fresh 32γ-ATP, some signal should be detected after 30–60 min of exposure. Otherwise, leave overnight.
6. If the right amount of RNase I has been used for partial digestion of the RNA, RNA–protein complexes will appear as a smear above the molecular weight (MW) of the RBP of interest. This minimum MW is often marked by the weight of the small RNA–protein complexes generated in the sample treated with a high concentration of RNase I (the final MW is the weight of the RBP plus 6 KDa from the preadenylated adapter L3-App). Importantly, no signal should be detected in those lanes where the negative controls are loaded. If possible, we recommend to perform each iCLIP experiments using cell extracts from wild-type and knock out cells in parallel. Alternatively, isotype control antibodies can be used to assess antibody specificity and assay unspecific background.

7. Isolate those RNA–protein complexes running above 20–60 KDa from the expected molecular weight of the RBP with only the adapter (Fig. 1a and *see* **Notes 11** and **12**). With the help of the luminescent sticker, place the autoradiography film on the top of the nitrocellulose membrane and label the area of the nitrocellulose membrane containing the RNA–protein complexes to be isolated.
8. Cut each piece of the membrane marked in **step 7** out with a scalpel and transfer it to a 1.5 ml tube. Change the scalpel and your gloves between samples to avoid cross-contamination.
9. Add 200 μl of PK buffer and 15 μl of proteinase K.
10. Incubate at 37 °C for 20 min in a thermomixer mixing at 1200 rpm.
11. After that, add another 200 μl of PK buffer with 7 M urea and incubate samples further at 37 °C for another 20 min (*see* **Note 13**).
12. Transfer the sample supernatant to a phase-lock gel heavy tube.
13. Add 400 μl of phenol–chloroform–isoamyl alcohol 25:24:1 (pH 8).
14. Incubate for 5 min at 37 °C in a thermomixer mixing at 1200 rpm.
15. Centrifuge samples at 15,871 × *g* at room temperature for 5 min.
16. Transfer aqueous phase to a new 1.5 ml tube containing 2 μl of glycogen and 40 μl of 3 M sodium acetate (pH 5.5).
17. Vortex and add 1 ml of 100% ethanol. Vortex again and keep at −20 °C for at least 1 h.
18. Centrifuge samples at 15,871 × *g* at 4 °C for 30 min.
19. Wash RNA pellet with 1 ml 80% ethanol twice.
20. Let dry the RNA pellet and resuspend in 5 μl of H_2O.

3.5 Preparation of iCLIP cDNA Libraries for Sequencing

1. Add 1 μl of dNTP mix (10 mM) and 1 μl of Rtclip primer (0.5 μM) to the RNA.
2. Heat at 70 °C for 5 min using a thermocycler.
3. Cool sample down to 25 °C and add the following mix for RNA reverse transcription: 7 μl of H_2O, 4 μl of 5× first-strand buffer, 1 μl of 100 mM DTT, 0.5 μl of RNase inhibitor, and 0.5 μl of Reverse Transcriptase (*see* **Note 14**).
4. Run the following program in a thermocycler: 5 min at 25 °C, 20 min at 42 °C, 40 min at 50 °C, 5 min at 80 °C, hold at 4 °C.
5. Add 350 μl of Tris–EDTA buffer and 2 μl of glycogen. Vortex. Then, add 40 μl of 3 M sodium acetate (pH 5.5) and vortex again. Add 1 ml of 100% ethanol and precipitate the cDNA as

described for RNA precipitation in steps 17 to 19 under Subheading 3.3.

6. Resuspend cDNA in 6 μl of H_2O. Add 6 μl of TBE-Urea Sample Buffer (2×).
7. Heat the samples at 80 °C for 5 min.
8. Cool samples down on ice.
9. Load samples in a pre-cast 6% polyacrylamide–urea gel whose wells have been previously flushed with an excess of 1× TBE running buffer. Leave 1 or 2 empty lanes between samples. A DNA ladder (100 bp) could be loaded well far from the samples although it is not absolutely required.
10. Run gels at 180 V for 40 min.
11. (Optional) Add 50 ml of TBE running buffer in a small tray and add 3 μl of SYBR. Mix well. Transfer the gel into the tray and stain it by mixing for 2–3 min. Visualize the gel in a transilluminator.
12. Cut out a band from the gel corresponding to a DNA size of 80–200 nucleotides (*see* **Note 15**) and place it in a 0.5 ml tube with the bottom previously pierced with a 19–20G needle.
13. Place the 0.5 ml tube inside a 1.5 ml tube and centrifuge at 15,871 × *g* for 2 min. Recover all small gel pieces by using 400 μl of TE-buffer (or H_2O) + 2 μl 10% SDS + 67 μl 3 M NaOAc pH 5.5.
14. Incubate at 37 °C for 2 h in a thermomixer mixing at 1100 rpm.
15. Then, place the sample in dry ice for 2 min and incubate again in a thermomixer at 37 °C for another 2 h.
16. Transfer the supernatant and gel pieces into a Costar SpinX column to which two 1-cm glass pre-filters had been previously inserted and pushed to the bottom with a clean tip.
17. Centrifuge at 15,871 × *g* for 1 min.
18. Transfer the flow-through to a new 1.5 ml tube.
19. Add 2 μl of glycogen and 1 ml of 100% ethanol. Vortex and precipitate the cDNA as described above in **steps 17–19** under Subheading 3.3.
20. Resuspend the cDNA with 8 μl of ligation mix: 6.5 μl of H_2O, 0.8 μl of 10× Circligase Buffer II, 0.4 μl of 50 mM $MnCl_2$, 0.3 μl of CircLigase™ II ssDNA Ligase.
21. Incubate 1 h at 60 °C in a thermocycler.
22. Cool down at 22 °C and add 30 μl of Cut_oligo annealing mix: 26 μl of H_2O, 3 μl of BamHI FastDigest Buffer and 1 μl of 10 μM Cutc_oligo.

23. Run the following program in a thermocycler: 10 min at 95 °C, 70 cycles of 20 s starting at 95 °C and dropping the temperature by 1 °C in each cycle, hold at 25 °C.
24. Add 2 μl FastDigest BamHI. Incubate 30 min at 37 °C.
25. Heat-inactivate the enzyme at 80 °C for 5 min.
26. Add 350 μl of Tris–EDTA buffer and 2 μl of glycogen. Vortex. Then, add 40 μl of 3 M sodium acetate (pH 5.5) and vortex again. Add 1 ml of 100% ethanol and precipitate the cDNA as in **steps 17–19** under Subheading 3.3.
27. Resuspend cDNA in 20 μl of H_2O. Use 1 to 9.5 μl of cDNA template for PCR amplification and cDNA library generation (*see* **Note 16**).
28. Add the following PCR mix to the cDNA template: 0.25 μl of solexa P5 adaptor primer (10 μM), 0.25 μl of solexa P3 adaptor primer (10 μM), 10 μl of High-Fidelity PCR Master Mix (2×), and H_2O up to 20 μl.
29. Run the following program in a thermocycler: 5 min at 95 °C; then 20–30 cycles (*see* **Note 16**) of 30 s at 95 °C, 30 s at 65 °C, and 30 s at 72 °C. Keep at 72 °C for 3 min and hold at 25 °C.
30. Add 5 μl of TBE sample buffer (5×) and load iCLIP cDNA library in 2–3 lanes of a 8% TBE gel. Leave one or two empty lanes between samples. Load a DNA ladder (100 bp) in a lane far from samples. Run the gel at 150 V.
31. Stain gel with SYBR as described above in **step 11**. Visualize iCLIP cDNA library using a transilluminator and purify it by cutting the gel band and proceeding as described above in **steps 12–19**.
32. Resuspend iCLIP cDNA library in 20 μl of H_2O and quantify it by using a KAPA Library Quantification Kit.
33. Analyse cDNA library size, quantity and quality using an Agilent 2100 Bioanalyzer system.

3.6 Brief Description of Sequencing and Bioinformatics for iCLIP Data Analysis

1. Perform 50-bp single end Illumina sequencing as it is the most cost effective method for iCLIP library sequencing. iCLIP cDNA libraries are composed by short fragments of 160–185 bp (average length) of which 128 bp correspond to the Solexa adaptor primers and library barcodes. The use of unique barcodes enables cDNA library sample multiplexing (*see* **Note 17**) thus reducing sequencing cost. iCLIP data processing is summarized in Fig. 1b.
2. Once sequencing is completed, demultiplex samples based on the four known bases of the unique barcode. As an example, if we use the Rt1clip primer (barcode = NNAACCNNN) the first nine bases of each sequencing read will account for

barcode that will be read as NNNGGTTNN. Identification of bases in position 4–7 allows efficient demultiplexing.

3. Trim Solexa P3 adaptor sequences. RNA template used for iCLIP library preparation is very short in length and partial adaptor sequences are commonly found at the 3′end of each read.
4. Remove barcodes, align reads to the genome and remove duplicates based on mapping position in the genome and the five unknown bases of the barcode that allows identification of up to 1024 unique reads. We use TopHat2 [29] that uses Bowtie2 [30] for mapping reads against the transcriptome and the genome. The latest reference genome version can be downloaded from either Ensembl, Genecode or UCSC.
5. Upon read alignment, annotate position −1 as the unique RNA–protein cross-link site. Generate a BED file.
6. Visualize data using the UCSC Genome Browser or the Integrative Genomics Viewer (IGV) from the Broad Institute (*see* **Note 18**).
7. Use iCLIP data for RBP binding motif discovery and global annotation of RBP binding to distinctive genomic features (telomeres, ncRNAs, introns, 5′UTRs, 3′UTRs, open reading frames (ORF) or intergenomic regions—annotation from Ensembl) (*see* **Note 19**). Integrate cell transcriptome and translatome information (*see* **Note 20**) and incorporate new advances in iCLIP technology (*see* **Note 21**).

4 Notes

1. UV light can penetrate several layers of cells, and therefore UV irradiation can be performed prior to B cell isolation if desired.
2. The amount of cells and the energy used for RNA–protein UV cross-linking will vary depending on three factors: First the overall abundance of your RBP, second the specificity and affinity of the antibody for your RBP, and third the amino acid sequence of the RNA binding motif of your RBP. UV irradiation at 254 nm mediates the cross-linking of cysteine, methionine, tyrosine, leucine, histidine or phenylalanine with RNA at the point of contact. This is a highly inefficient process and the abundance of these amino acids in the RNA binding motif determines the energy requirements for efficient cross-linking and should be optimized for each RBP. As an example, UV cross-linking of RNA with the RNA recognition motifs of the RBP HuR is achieved with 150 mJ. HuR is an abundant protein in B cells. We aim to have 0.5–1 mg of total protein lysate from B cells to perform HuR iCLIP experiments.

3. iCLIP is dependent upon the quality and specificity of the antibodies used. The type and amount of antibody to use for RNA–protein immunoprecipitation needs to be defined empirically based on protein pull down efficiency and antibody coupling to protein A or G magnetic beads. We recommend testing these reagents first by using a traditional protocol for protein immunoprecipitation. Negative controls must be carried out in all iCLIP experiments to assess antibody specificity and experimental background. These controls include an isotype control antibody and cell extracts from knock out cells.
4. Sonication of total cell extracts and nuclear cell extracts clarifies samples by breaking genomic DNA. It is not required when iCLIP is performed with cytoplasmic cell extracts.
5. We use RNase I for RNA digestion as it shows no base preference and cleaves after all four bases. The RNase I concentration for digestion of RNA–protein complexes for library preparation must be tested thoroughly. Differences in factors such as the amount of input material; the nature of the RBP-RNA complexes; the batch of RNase I; and the time-length of the enzymatic reaction will have an impact on digestion efficiency. Large concentrations of RNase I (12 U/ml) digest the bound RNA almost completely allowing identification of RNA–protein complexes by SDS-PAGE after radioactive labeling of the small oligonucleotide remnant at the site of cross-linking. Titration of the RNase I concentration will result in generation of larger RNA–protein complexes that will be resolved as a smear of increasingly higher molecular weight in the acrylamide gels (*see* Fig. 1a).
6. Stringency of the wash buffer needs to be tested to reduce experimental background when preserving RNA–protein complexes. In our experience, initial concentration of 0.1% SDS can be increased to 0.3% in most immunoprecipitations without affecting the amount of material recovered. Urea (up to 2 M) can be used if there is a requirement for protein denaturalization, but this normally has an impact on RNA recovery.
7. PNK wash buffer is a minimum-salt containing buffer which will remove any excess of salt and detergent from the beads in preparation of the enzymatic reactions that will label the RNA 3′end with the L3-App adaptor oligonucleotide and the RNA 5′end with 32γ-ATP. The number of washes with PNK wash buffer can be increased as desired (a minimum of two is recommended) but PNK wash buffer needs to be completely removed prior to any enzymatic reaction.
8. Both commercial and homemade PNK buffer and Ligation buffer can be used for these reactions.

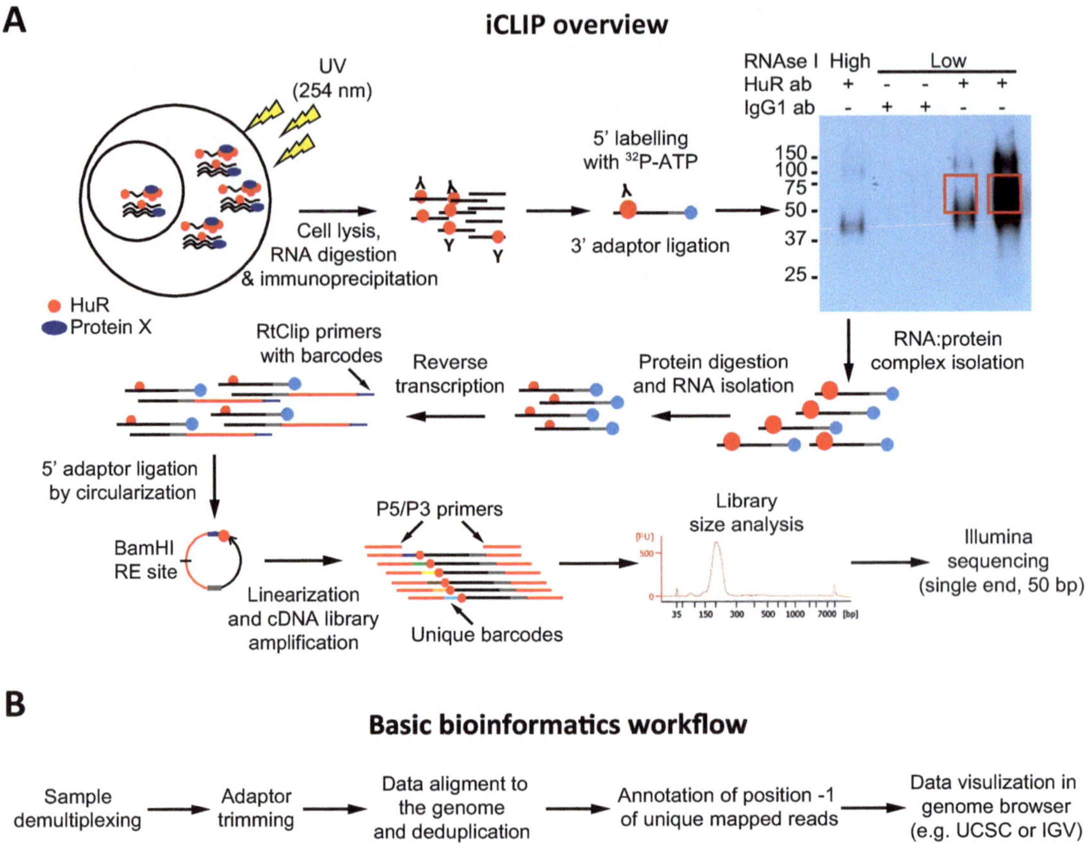

Fig. 1 iCLIP summary. (**a**) Overview of iCLIP methodology. (**b**) Basic bioinformatic data processing for identification of RNA–protein interactions

9. 5′end RNA dephosphorylation using shrimp alkaline phosphatase is optional as T4 PNK kinase has also phosphatase activity [24].
10. NuPAGE Novex 4–12% Bis–Tris polyacrylamide gels (from Thermo Fisher) are preferred as they maintain a pH close to 7 during electrophoresis preserving samples from RNA alkaline hydrolysis. The percentage of the pre-cast Bis–Tris–HCl polyacrylamide gels is chosen based on their capacity to efficiently separate proteins above the molecular weight of the RBP of interest. The number of wells is chosen based on the number of samples. Both negative iCLIP controls and iCLIP samples treated with a high or a low concentration of RNase I should be loaded in the same gel leaving at least one empty well in between samples to avoid sample cross-contamination.
11. RNA–protein complexes 20 KDa above the MW of the RBP of interest will be isolated in a standard iCLIP assay. These RNA–protein complexes will contain a RNA molecule that is at least 65 nt long and includes the preadenylated adapter L3-

App (20 nt. and MW = 6 KDa). The 45 nt remaining after removal of any adaptor sequence will enable the generation of iCLIP libraries with cDNA molecules that are long enough to be uniquely mapped to the genome.

12. Sometimes several discrete bands are seen in samples treated with a high concentration of RNase I. These are normally the result of ribonucleoprotein complexes in which two or more proteins are bound to the same RNA molecule. Although our personal experience is that including those will not change significantly the average length of the cDNA molecules in the iCLIP libraries, this needs testing for each RBP. If the different discrete bands are separated by more than 50–60 KDa in the bis–tris polyacrylamide gel, the RNA–protein complexes containing more than one protein can be excluded during complex isolation from the nitrocellulose membrane.
13. During protein digestion and RNA extraction from the nitrocellulose membrane, urea can be replaced with 0.2% SDS and samples can be incubated at 60 °C [28].
14. It has been shown that the use of group II intron reverse transcriptases (e.g., TGIRT-III from Ingex) has better processivity that SuperScript® III reverse transcriptase and their use can increase the efficiency during iCLIP library preparation [28].
15. Novex® TBE-Urea Sample Buffer (2×) contains bromophenol blue and xylene cyanol, two tracking dyes that separate differently during gel electrophoresis. In a 6% TBE-urea gel, bromophenol blue runs with the sample front and Rtclip primers (42 nt) run just behind. On the contrary, xylene cyanol runs at around 150 nucleotides which is the expected average size of cDNA after RT. Therefore, cutting 1 cm below and above the xylene cyanol dye enables cDNA isolation for iCLIP library preparation without the need of staining the 6% TBE-urea gel or the presence of a DNA ladder that may contaminate your sample.
16. The amount of cDNA template and the number of PCR cycles for iCLIP library preparation are defined experimentally (*see* Huppertz et al. [24]). The aim is to minimize cDNA read duplication by reducing as much as possible the number of cycles during iCLIP library PCR amplification.
17. During iCLIP cDNA library multiplexing we try to maintain a random distribution of bases in each position of the reads as this is important to avoid errors during Illumina base calling. The reads of iCLIP cDNA libraries are composed by the barcode of 9 bases followed by a unique read of 41 bases (if 50 bp single end Illumina sequencing is performed). Although the barcodes in the RtClip primers are selected to introduce random bases prior to the four known bases used for multiplexing,

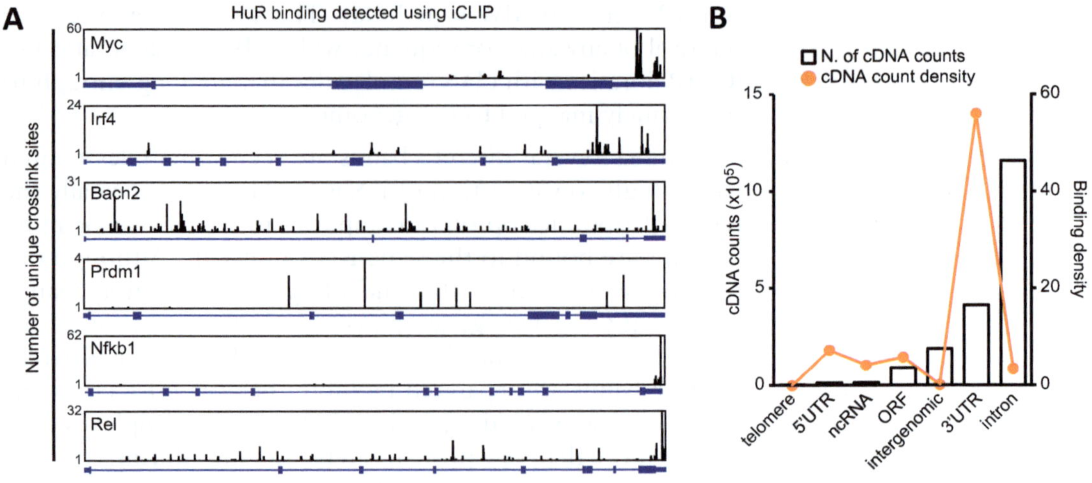

Fig. 2 Analysis of HuR binding to relevant mRNAs of germinal center B cells. (**a**) HuR iCLIP data visualized in the UCSC genome browser from LPS-activated B lymphocytes (GEO deposition number GSE62148 [3]). Binding of HuR to Myc, Irf4, Bach2 (transcript isoform 201), Prdm1, Nfkb1 (transcript isoform 005), and Rel mRNAs is shown as examples. (**b**) Global analysis of HuR binding associated to different genomic features (telomeres, ncRNAs, introns, ORFs, 5′ UTRs, 3′ UTRs, and intergenomic sequences). The latest genomic and transcript annotations can be downloaded from Genecode, UCSC, or BioMart

we recommended to multiplex libraries in such a way to account for 50% A/C and 50% G/T in each position of the barcode to avoid base miscalling.

18. iCLIP data visualization using UCSC or IGV genome browsers enables identification of RNA binding sites to study the RBP- dependent regulation of selected RNA targets. For example, the RBP HuR binds to important transcription factors that control the GC reaction and terminal differentiation into antibody-producing cells (Fig. 2a). The importance of these RNA–protein interactions in B cell development and differentiation can be then explored by conditional targeting and deletion of these binding sites as shown for Tnf and Bcl2 [31, 32].

19. In a global manner, iCLIP data can be used to characterize the RNA binding motifs of your RBP of interest (e.g., using HOMER motif analysis tool [33]) and their location in distinctive genomic features (ncRNA, introns, coding exons, UTRs, etc) as it might reflect its biological function (Fig. 2b).

20. Integration of iCLIP sequencing data with whole transcriptome and translatome analysis enables the study of the biological functions of RBPs in a genome-wide manner. For example, generation of iCLIP RNA maps combined with deep analysis of the cell transcriptome (by either RNAseq or microarray analysis) has revealed the global function of some

RBPs including hnRNPA1, hnRNPC, Tia1 and Tial1 in controlling mRNA splicing [3, 23, 34, 35] and mRNA stability [17, 18, 36]. Further integration of ribosome footprinting profiling data has been proven extremely useful to interrogate whether different mRNA transcripts generated due to alternative splicing encoded for different protein isoforms or are targeted for nonsense mediated decay [3].

21. Recent improvements of the iCLIP protocol have been reported. Both eCLIP [27] and irCLIP [28] introduce changes at key steps of the iCLIP protocol to increase sample preparation efficiency while decreasing the amount of total input, read duplication and handling time. eCLIP substitutes the circular ligation reaction performed with CircLigase™ II (**step 20** under Subheading 3.4 in iCLIP protocol) by a 3′ adaptor ligation reaction using T4 RNA ligase in the presence of PEG8000 and DMSO. By contrast, irCLIP changes are related to: 1—Use of an infrared dye instead of radioisotopes (**step 20** under Subheading 3.3). 2—RNA digestion on-beads rather than in-lysate (**steps 5–7** under subheading 3.3). 3—RNA–protein complex recovery by using SDS and high temperature instead of 7 M urea (**steps 9–11** under Subheading 3.3). 4—RT reaction efficiency that is increased when using a group II intron reverse transcriptase such as TGIRT-III (Ingex) instead of SuperScript® III reverse transcriptase (**step 3** under Subheading 3.4.). 5—The use of AMPure bead-isopropanol mixture for cDNA cleanup prior PCR amplification. Altogether, Zarnegar et al. [28] were able to reduce the starting cell number for irCLIP to just few thousands, a number which opens the possibility to the characterization of RNA–protein interaction in small populations of immune cells.

Acknowledgments

We thank Prof. Jernej Ule and Prof. Tomaz Curk for their valuable contribution in experimental setup and analysis. The authors were supported by a Biotechnology and Biological Sciences Research Council strategic Lola grant BB/J00152X/1 and program funding BBS/E/B/000C0409.

References

1. Basso K, Dalla-Favera R (2015) Germinal centres and B cell lymphomagenesis. Nat Rev Immunol 15(3):172–184. doi:10.1038/nri3814
2. De Silva NS, Klein U (2015) Dynamics of B cells in germinal centres. Nat Rev Immunol 15 (3):137–148. doi:10.1038/nri3804

3. Diaz-Munoz MD, Bell SE, Fairfax K, Monzon-Casanova E, Cunningham AF, Gonzalez-Porta M, Andrews SR, Bunik VI, Zarnack K, Curk T, Heggermont WA, Heymans S, Gibson GE, Kontoyiannis DL, Ule J, Turner M (2015) The RNA-binding protein HuR is essential for the B cell antibody response. Nat Immunol 16(4):415–425. doi:10.1038/ni.3115
4. DeMicco A, Naradikian MS, Sindhava VJ, Yoon JH, Gorospe M, Wertheim GB, Cancro MP, Bassing CH (2015) B cell-intrinsic expression of the HuR RNA-binding protein is required for the T cell-dependent immune response in vivo. J Immunol 195 (7):3449–3462. doi:10.4049/jimmunol.1500512
5. Thai TH, Calado DP, Casola S, Ansel KM, Xiao C, Xue Y, Murphy A, Frendewey D, Valenzuela D, Kutok JL, Schmidt-Supprian M, Rajewsky N, Yancopoulos G, Rao A, Rajewsky K (2007) Regulation of the germinal center response by microRNA-155. Science 316(5824):604–608. doi:10.1126/science.1141229
6. Vigorito E, Perks KL, Abreu-Goodger C, Bunting S, Xiang Z, Kohlhaas S, Das PP, Miska EA, Rodriguez A, Bradley A, Smith KG, Rada C, Enright AJ, Toellner KM, Maclennan IC, Turner M (2007) microRNA-155 regulates the generation of immunoglobulin class-switched plasma cells. Immunity 27 (6):847–859. doi:10.1016/j.immuni.2007.10.009
7. Xu S, Guo K, Zeng Q, Huo J, Lam KP (2012) The RNase III enzyme dicer is essential for germinal center B-cell formation. Blood 119 (3):767–776. doi:10.1182/blood-2011-05-355412
8. Bertossi A, Aichinger M, Sansonetti P, Lech M, Neff F, Pal M, Wunderlich FT, Anders HJ, Klein L, Schmidt-Supprian M (2011) Loss of Roquin induces early death and immune deregulation but not autoimmunity. J Exp Med 208(9):1749–1756. doi:10.1084/jem.20110578
9. Pratama A, Srivastava M, Williams NJ, Papa I, Lee SK, Dinh XT, Hutloff A, Jordan MA, Zhao JL, Casellas R, Athanasopoulos V, Vinuesa CG (2015) MicroRNA-146a regulates ICOS-ICOSL signalling to limit accumulation of T follicular helper cells and germinal centres. Nat Commun 6:6436. doi:10.1038/ncomms7436
10. Baumjohann D, Kageyama R, Clingan JM, Morar MM, Patel S, de Kouchkovsky D, Bannard O, Bluestone JA, Matloubian M, Ansel KM, Jeker LT (2013) The microRNA cluster miR-17 approximately 92 promotes TFH cell differentiation and represses subset-inappropriate gene expression. Nat Immunol 14(8):840–848. doi:10.1038/ni.2642
11. Gerstberger S, Hafner M, Tuschl T (2014) A census of human RNA-binding proteins. Nat Rev Genet 15(12):829–845. doi:10.1038/nrg3813
12. Castello A, Fischer B, Eichelbaum K, Horos R, Beckmann BM, Strein C, Davey NE, Humphreys DT, Preiss T, Steinmetz LM, Krijgsveld J, Hentze MW (2012) Insights into RNA biology from an atlas of mammalian mRNA-binding proteins. Cell 149(6):1393–1406. doi:10.1016/j.cell.2012.04.031
13. Chang X, Li B, Rao A (2015) RNA-binding protein hnRNPLL regulates mRNA splicing and stability during B-cell to plasma-cell differentiation. Proc Natl Acad Sci U S A 112(15): E1888–E1897. doi:10.1073/pnas.1422490112
14. Pioli PD, Debnath I, Weis JJ, Weis JH (2014) Zfp318 regulates IgD expression by abrogating transcription termination within the Ighm/Ighd locus. J Immunol 193 (5):2546–2553. doi:10.4049/jimmunol.1401275
15. Enders A, Short A, Miosge LA, Bergmann H, Sontani Y, Bertram EM, Whittle B, Balakishnan B, Yoshida K, Sjollema G, Field MA, Andrews TD, Hagiwara H, Goodnow CC (2014) Zinc-finger protein ZFP318 is essential for expression of IgD, the alternatively spliced Igh product made by mature B lymphocytes. Proc Natl Acad Sci U S A 111(12):4513–4518. doi:10.1073/pnas.1402739111
16. Landon AL, Muniandy PA, Shetty AC, Lehrmann E, Volpon L, Houng S, Zhang Y, Dai B, Peroutka R, Mazan-Mamczarz K, Steinhardt J, Mahurkar A, Becker KG, Borden KL, Gartenhaus RB (2014) MNKs act as a regulatory switch for eIF4E1 and eIF4E3 driven mRNA translation in DLBCL. Nat Commun 5:5413. doi:10.1038/ncomms6413
17. Galloway A, Saveliev A, Lukasiak S, Hodson DJ, Bolland D, Balmanno K, Ahlfors H, Monzon-Casanova E, Mannurita SC, Bell LS, Andrews S, Diaz-Munoz MD, Cook SJ, Corcoran A, Turner M (2016) RNA-binding proteins ZFP36L1 and ZFP36L2 promote cell quiescence. Science 352(6284):453–459. doi:10.1126/science.aad5978
18. Tiedje C, Diaz-Munoz MD, Trulley P, Ahlfors H, Laass K, Blackshear PJ, Turner M, Gaestel M (2016) The RNA-binding protein TTP is a global post-transcriptional regulator of feedback control in inflammation. Nucleic Acids Res 44:7418–7440. doi:10.1093/nar/gkw474

19. Ule J, Jensen KB, Ruggiu M, Mele A, Ule A, Darnell RB (2003) CLIP identifies Nova-regulated RNA networks in the brain. Science 302(5648):1212–1215. doi:10.1126/science.1090095
20. Pashev IG, Dimitrov SI, Angelov D (1991) Crosslinking proteins to nucleic acids by ultraviolet laser irradiation. Trends Biochem Sci 16 (9):323–326
21. Pelle R, Murphy NB (1993) In vivo UV-cross-linking hybridization: a powerful technique for isolating RNA binding proteins. Application to trypanosome mini-exon derived RNA. Nucleic Acids Res 21(10):2453–2458
22. Licatalosi DD, Mele A, Fak JJ, Ule J, Kayikci M, Chi SW, Clark TA, Schweitzer AC, Blume JE, Wang X, Darnell JC, Darnell RB (2008) HITS-CLIP yields genome-wide insights into brain alternative RNA processing. Nature 456 (7221):464–469. doi:10.1038/nature07488
23. Konig J, Zarnack K, Rot G, Curk T, Kayikci M, Zupan B, Turner DJ, Luscombe NM, Ule J (2010) iCLIP reveals the function of hnRNP particles in splicing at individual nucleotide resolution. Nat Struct Mol Biol 17 (7):909–915. doi:10.1038/nsmb.1838
24. Huppertz I, Attig J, D'Ambrogio A, Easton LE, Sibley CR, Sugimoto Y, Tajnik M, Konig J, Ule J (2014) iCLIP: protein-RNA interactions at nucleotide resolution. Methods 65 (3):274–287. doi:10.1016/j.ymeth.2013.10.011
25. Zarnack K, Konig J, Tajnik M, Martincorena I, Eustermann S, Stevant I, Reyes A, Anders S, Luscombe NM, Ule J (2013) Direct competition between hnRNP C and U2AF65 protects the transcriptome from the exonization of Alu elements. Cell 152(3):453–466. doi:10.1016/j.cell.2012.12.023
26. Yao C, Biesinger J, Wan J, Weng L, Xing Y, Xie X, Shi Y (2012) Transcriptome-wide analyses of CstF64-RNA interactions in global regulation of mRNA alternative polyadenylation. Proc Natl Acad Sci U S A 109 (46):18773–18778. doi:10.1073/pnas.1211101109
27. Van Nostrand EL, Pratt GA, Shishkin AA, Gelboin-Burkhart C, Fang MY, Sundararaman B, Blue SM, Nguyen TB, Surka C, Elkins K, Stanton R, Rigo F, Guttman M, Yeo GW (2016) Robust transcriptome-wide discovery of RNA-binding protein binding sites with enhanced CLIP (eCLIP). Nat Methods 13 (6):508–514. doi:10.1038/nmeth.3810
28. Zarnegar BJ, Flynn RA, Shen Y, Do BT, Chang HY, Khavari PA (2016) irCLIP platform for efficient characterization of protein-RNA interactions. Nat Methods 13(6):489–492. doi:10.1038/nmeth.3840
29. Kim D, Pertea G, Trapnell C, Pimentel H, Kelley R, Salzberg SL (2013) TopHat2: accurate alignment of transcriptomes in the presence of insertions, deletions and gene fusions. Genome Biol 14(4):R36. doi:10.1186/gb-2013-14-4-r36
30. Langmead B, Salzberg SL (2012) Fast gapped-read alignment with bowtie 2. Nat Methods 9 (4):357–359. doi:10.1038/nmeth.1923
31. Henry CJ, Casas-Selves M, Kim J, Zaberezhnyy V, Aghili L, Daniel AE, Jimenez L, Azam T, McNamee EN, Clambey ET, Klawitter J, Serkova NJ, Tan AC, Dinarello CA, DeGregori J (2015) Aging-associated inflammation promotes selection for adaptive oncogenic events in B cell progenitors. J Clin Invest 125 (12):4666–4680. doi:10.1172/JCI83024
32. Diaz-Munoz MD, Bell SE, Turner M (2015) Deletion of AU-rich elements within the Bcl2 3'UTR reduces protein expression and B cell survival in vivo. PLoS One 10(2):e0116899. doi:10.1371/journal.pone.0116899
33. Heinz S, Benner C, Spann N, Bertolino E, Lin YC, Laslo P, Cheng JX, Murre C, Singh H, Glass CK (2010) Simple combinations of lineage-determining transcription factors prime cis-regulatory elements required for macrophage and B cell identities. Mol Cell 38 (4):576–589. doi:10.1016/j.molcel.2010.05.004
34. Wang Z, Kayikci M, Briese M, Zarnack K, Luscombe NM, Rot G, Zupan B, Curk T, Ule J (2010) iCLIP predicts the dual splicing effects of TIA-RNA interactions. PLoS Biol 8(10): e1000530. doi:10.1371/journal.pbio.1000530
35. Huelga SC, AQ V, Arnold JD, Liang TY, Liu PP, Yan BY, Donohue JP, Shiue L, Hoon S, Brenner S, Ares M Jr, Yeo GW (2012) Integrative genome-wide analysis reveals cooperative regulation of alternative splicing by hnRNP proteins. Cell Rep 1(2):167–178. doi:10.1016/j.celrep.2012.02.001
36. Vogel KU, Bell LS, Galloway A, Ahlfors H, Turner M (2016) The RNA-binding proteins Zfp36l1 and Zfp36l2 enforce the thymic β-selection checkpoint by limiting DNA damage response signaling and cell cycle progression. J Immunol 197:2673–2685

Chapter 15

Flow-Cytometric Method Measuring B Cell Surface Immunoglobulin Avidity

Davide Angeletti, Gregory M. Frank, and Jonathan W. Yewdell

Abstract

The affinity of antibodies for their cognate antigens is a critical aspect of humoral immunity. The immune system has gone to great lengths to evolve a mechanism that enables real time increases in antibody affinity during the course of an immune response. This occurs in germinal centers (GC), which form in spleen and lymph nodes following immunization. GC B cell competition for limiting amount of antigen drives the selection of B cells expressing higher affinity Abs. Remarkably little is known about affinity maturation of B cells in immune responses to all but a handful of small model antigens. It has proven challenging to measure the avidity of specific Abs in polyclonal sera to more complex antigens, including viruses. In this chapter we present a simple, flow cytometry based, method that determines the average avidity of GC B cells for the influenza A virus hemagglutinin glycoprotein, the target antigen of traditional influenza vaccines. Flow cytometry using fluorescent hemagglutinin and B cell marker specific Abs enables high throughput qualitative and quantitative detection of individual B cells. By using a graded amount of antigen and gating on GC B cells we define the AC_{50} the amount of antigen required to stain 50% of hemagglutinin specific B cells. This number is in remarkable agreement with the avidity of the B cell population. This method can be generally employed to include antibody avidity measurements basic and clinical studies of immunity to viruses and other medically relevant immunogens.

Key words Avidity, Affinity, Immunoglobulin, B cell, Germinal center, Influenza, Hemagglutinin, Flow cytometry

1 Introduction

Germinal centers (GC) are structures that rapidly arise in secondary lymphoid organs upon immunization or infection with a foreign antigen. Naïve B cells recruited to GCs "affinity mature" by randomly diversifying their antibody (Ab) encoding genes positively selecting for cells that exhibit increased immunogen avidity [1]. Affinity maturation is critical for the development of a long-lived

Davide Angeletti and Gregory M. Frank have equally contributed to this work.
This work was supported by the Division of Intramural Research, NIAID.

Dinis Pedro Calado (ed.), *Germinal Centers: Methods and Protocols*, Methods in Molecular Biology, vol. 1623,
DOI 10.1007/978-1-4939-7095-7_15, © Springer Science+Business Media LLC 2017

immune response: with a maximum amount of Ig in human serum of 10–15 mg/ml, affinity maturation enables the immune system to maintain functional levels of Abs specific for thousands of potentially dangerous immunogens [2, 3]. It is difficult to measure the affinity of specific serum Abs due to technical limitations in measuring the concentrations of Abs specific for a given cognate antigen. Surface plasmon resonance (SPR) is able to measure serum Ab off-rates, but on-rate measurement also requires knowing the specific Ab concentration [4].

Here, we take advantage of the ability of surface-Ig expressing GC B cells to bind cognate antigen to measure the mean population avidity. By using a graded amount of antigen and staining a low number of cells we determine the AC_{50}, the concentration of antigen necessary to stain 50% of antigen specific cells. Previously, we reported that the AC_{50} correlates well with the affinity of hybridoma secreted Abs, and also changes during the course of an immune response in *wt* but not $AID^{-/-}$ mice, consistently with affinity maturation [5].

For the purpose of this protocol we use as an example the hemagglutinin (HA) specific GC B cell response after intranasal infection with mouse adapted influenza A virus (IAV). Mouse infection with IAV is a well-established model system with a robust GC response [5, 6]. Further, recombinant, natively folded HA are widely available [7, 8]. These proteins have a remarkable stability and conserve native secondary and tertiary structures as assessed by a large panel of monoclonal Abs [7]. Although HA naturally binds terminal sialic acids ubiquitously expressed on cell surfaces, a modified version of HA, abrogating its sialic acid binding capacity, has recently become available, greatly reducing background B cell binding and facilitating analysis [8].

The method presented in this chapter is discussed in the context of recombinant influenza HA modified to reduce sialic acid binding. It can be easily applied to measure avidity changes in GC B cells after immunization and upon appropriate validation, to a wide range of recombinant proteins.

2 Materials

2.1 Mouse Intranasal Infection

1. Mouse strain of interest (*see* **Note 1**).
2. Titrated IAV stock.
3. Induction chamber with supply oxygen gas and vaporizer.
4. Isoflurane.
5. 1.5-ml microtubes.

2.2 Removal of Mouse Lymphoid Organs and Single Cell Suspension Preparation for Analysis

1. Scissor and forceps for dissection.
2. 15-ml conical tubes.
3. 50-ml conical tubes.
4. RPMI 7%: RPMI 1640 supplemented with 7% fetal bovine serum (FBS).
5. Glass homogenizer.
6. 70-μm nylon mesh.
7. Refrigerated centrifuge capable of reaching 500 × *g*.
8. Hemocytometer.
9. Light microscope.
10. Ammonium-chloride potassium (ACK) lysis buffer.

2.3 Cell Staining with Fluorochrome-Conjugated Abs

1. 5-ml polystyrene FACS tubes.
2. FACS buffer: Phosphate buffer saline (PBS) pH 7.4, 0.1% bovine serum albumin (BSA).
3. Refrigerated centrifuge capable of reaching 500 × *g*.
4. Ethidium monoazide, bromide (EMA) live/dead fixable cell stain.
5. Unlabeled anti-Fcγ receptor clone 2.4G2.
6. Fluorochrome-labeled Abs appropriate to discriminate GC B cells. A possible panel includes anti-mouse CD3e, anti-mouse B220, anti-mouse CD38, and anti-mouse GL7.

2.4 Surface Ig Staining with the Antigen of Interest

1. FACS buffer: Phosphate buffer saline (PBS) pH 7.4, 0.1% bovine serum albumin (BSA).
2. Refrigerated centrifuge capable of reaching 500 × *g*.
3. Flow cytometer (e.g., BD LSRFortessa) capable of detecting at least five fluorochromes.
4. Purified tagged recombinant protein of interest (e.g., recombinant HA).
5. Fluorochrome-labeled Abs anti-mouse IgG and anti-mouse IgM.
6. Fluorochrome-labeled reagent specific for the protein tag (*see* **Note 2**).
7. 1.5% paraformaldehyde (PFA) for fixation: PBS supplemented with 1.5% PFA.

2.5 Analysis of Flow Cytometry Results and Calculation of B Cell Ig Avidity

1. Computer software for the analysis of flow cytometry data (e.g., FlowJo or BD FACSDiva).
2. Computer software for the analysis of the curve and avidity calculation (GraphPad Prism or similar).

3 Methods

3.1 Mouse Intranasal Infection

1. Dilute stock of virus to 2 Tissue Culture Infectious Dose $(TCID)_{50}$/μl in FACS buffer on ice.
2. Turn on gas to anesthesia chamber (3.5% isoflurane in O_2 at a flow of 1.5 L/min).
3. Place mouse in induction chamber and keep sealed until unconscious.
4. Remove mouse from the chamber, while supine instill into nostril 25 μl of virus solution diluted to $2TCID_{50}$/μl; final dose of $50TCID_{50}$/mouse (*see* **Note 3**).
5. Allow mice to recover in prone position (*see* **Note 4**).

3.2 Removal of Mouse Lymphoid Organs and Single Cell Suspension Preparation for Analysis

1. Prepare 1.5-ml microtubes containing RPMI 7%, 1 ml for spleen harvest, and 500 μl for lymph nodes and keep solutions on ice.
2. Euthanize mice according to the institutional guidelines and harvest lymphoid organs of interest (spleen or lymph nodes).
3. Place the spleen and/or lymph nodes in the tubes containing RPMI, 7%FBS. While spleens can be used individually, lymph nodes will have to be pooled in a single tube (five sets of nodes per experiment).
4. Use glass homogenizer to dissociate lymphoid organs into a single cell suspension.
5. Filter through 70-μm nylon cell strainer into a 50-ml conical tube and wash with 20 ml RPMI 7%.
6. Centrifuge cells at 500 × *g* for 5 min at 4 °C.
7. Aspirate or decant medium in excess.
8. Add 2 ml ACK lysis buffer, vortex, and incubate for 3 min at room temperature.
9. Wash with 20 ml RPMI 7%.
10. Centrifuge cells at 500 × *g* for 5 min at 4 °C.
11. Aspirate or decant medium in excess.
12. Resuspend cells in at least 10 ml RPMI 7% for spleen and 1 ml/node for lymph nodes.
13. Count cells on hemocytometer and store on ice or at 4 °C until further analysis.

3.3 Cell Staining with Fluorochrome-Conjugated Abs

1. Add 1 million cells per tube (*see* **Note 5**).
2. Wash with 1 ml FACS buffer.
3. Centrifuge cells at 500 × *g* for 5 min at 4 °C.
4. Aspirate or decant excess buffer.

5. Add 100 μl of EMA live/dead staining diluted 1:500 in FACS buffer (*see* **Note 6**).
6. Cover and incubate on ice or at 4 °C in the dark for 10 min.
7. Place on ice approximately 10 cm from bright fluorescent light for 10 min to photoactivate EMA
8. Wash with 2 ml FACS buffer.
9. Centrifuge cells at 500 × *g* for 5 min at 4 °C.
10. Add Fc block (anti-Fcγ receptor clone 2.4G2) 25 μl per tube at 10 μg/ml in FACS buffer. Incubate for 15 min at 4 °C.
11. Add titrated Abs specific for CD3e, B220, CD38, and GL-7 in 25 μl diluted in FACS buffer. Incubate for 30 min at 4 °C.
12. Wash with 1 ml FACS buffer.
13. Centrifuge cells at 500 × *g* for 5 min at 4 °C.

3.4 Surface Ig Staining with the Antigen of Interest

1. Prepare recombinant HA titration including a "no HA" control in 100 μl FACS buffer. Add to cells and incubate for 1 h at 4 °C (*see* **Notes 7–9**).
2. Wash with 2 ml FACS buffer.
3. Centrifuge cells at 500 × *g* for 5 min at 4 °C.
4. Add titrated secondary reagent in 50 μl diluted in FACS buffer. Incubate for 30 min at 4 °C (*see* **Notes 2** and **10**).
5. Wash with 1 ml FACS buffer.
6. Centrifuge cells at 500 × *g* for 5 min at 4 °C.
7. Resuspend cells in 250 μl of FACS buffer.
8. Cover cells with aluminum foil and store at 4 °C until ready to run on flow cytometer (*see* **Notes 11** and **12**).

3.5 Analysis of Flow Cytometry Results and Calculation of B Cell Ig Avidity

1. Analyze cells with a flow cytometer. Representative results are shown in Fig. 1 (*see* **Note 13**).
2. Set the gate of recombinant protein-binding GC B cells in the no-HA control so that frequency is less than 0.2%.
3. Calculate the frequency of HA-binding GC B cells in all the samples using the same gate.
4. Plot the frequency of HA-binding GC B cells vs molar protein concentration.
5. Fit the data using the one-site binding with Hill slope.
6. Avidity is determined by the concentration giving half of the maximum frequency of specific binding (Fig. 2a) (*see* **Note 7**).

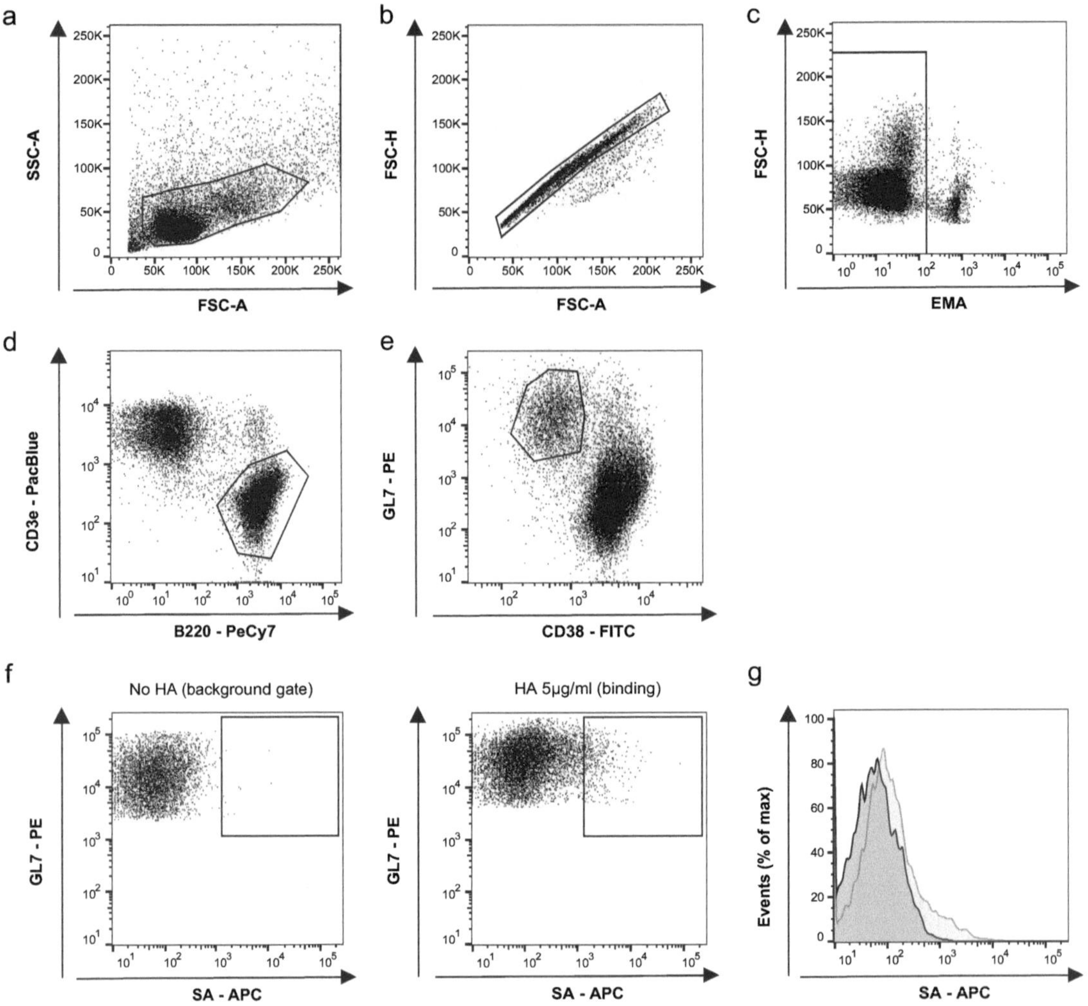

Fig. 1 Gating strategy to identify antigen specific GC B cells. Single cell suspension is stained with EMA live/dead cell staining, CD3e-PacificBlue, B220-PeCy7, CD38-FITC, GL7-PE, and Streptavidin (SA)-APC diluted 1:200 (**a**–**e**). (**f**) The gate to determine the antigen frequency is set by using the secondary reagent but no antigen and kept constant for all the antigen concentrations. (**e**) Same results as in (**f**) shown as histogram with no HA in *dark grey* and HA 5 µg/ml in *light grey*

4 Notes

1. The current protocol is optimized for B cells from C57/Bl6 mice. All steps, especially i.n. infection, should be validated and optimized in other mouse strains. The dose for each mouse strain and each virus strain must be determined empirically. Care must be taken to not infect mice with a lethal dose.
2. The secondary fluorochrome-labeled reagent will depend on the tag of the recombinant protein of interest. Streptavidin or avidin conjugated will allow detection of biotinylated proteins

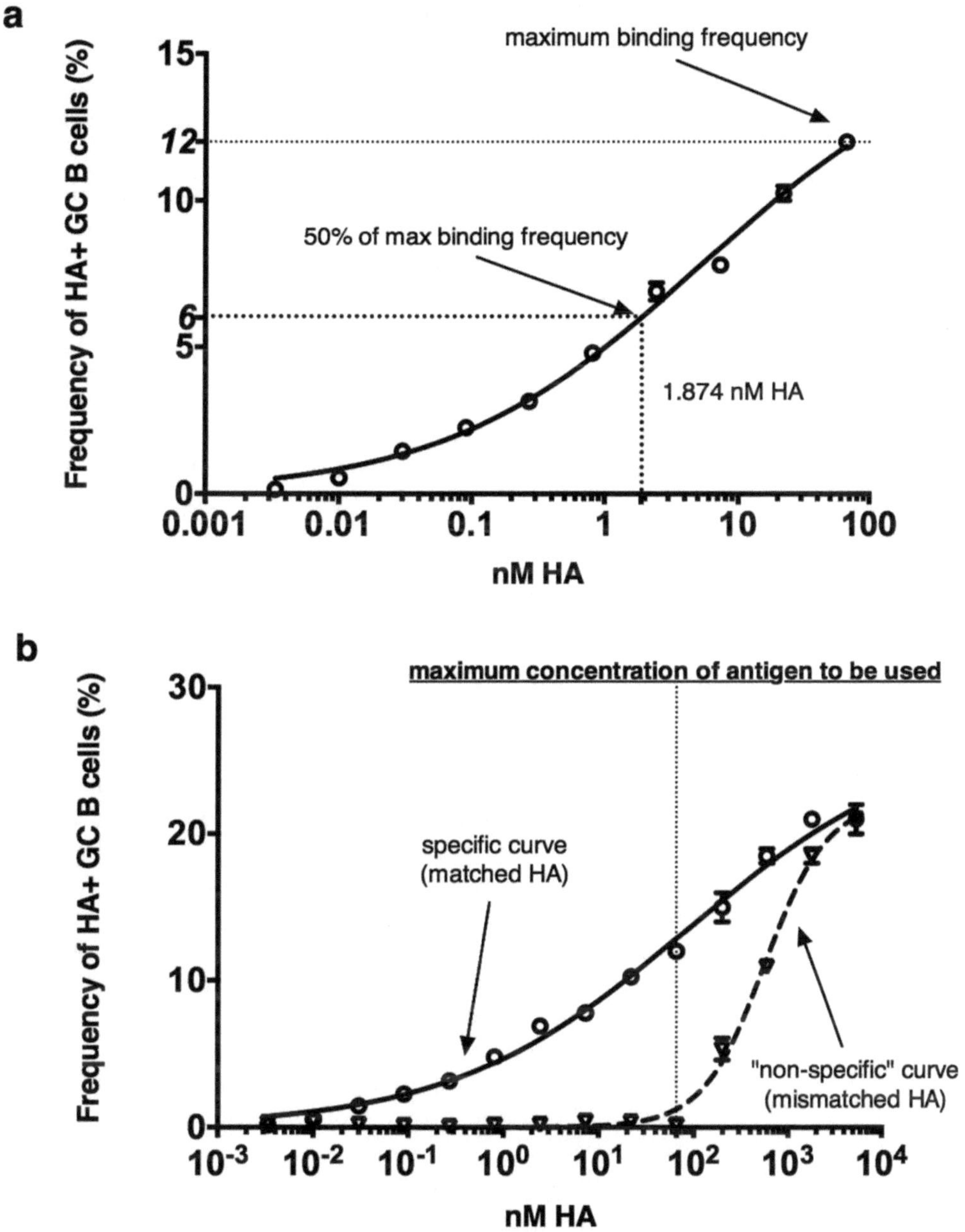

Fig. 2 Schematics representing a typical experiment. (**a**) Example of population avidity calculation defined as the concentration of HA giving 50% of the maximum binding frequency. In this case with a 12% maximum binding frequency, the population avidity is 1.874 nM, which reflects the amount of HA necessary to bind 6% of the cells. (**b**) Empirical determination of the maximum amount of HA to be used: this corresponds to the maximum amount of antigen staining exclusively B cells of mice infected with virus expressing a matched HA

while anti-his conjugated Abs will detect his-tagged proteins. Other tags can be detected with relevant reagents.

3. Check the breathing of the mice in the chamber. Once the breathing slows down, slowly pipet the virus solution on the nostrils taking care to avoid bubbling.
4. Germinal centers start to form 3–5 days after infection/immunization. The avidity measurement should be performed at early and late time points after challenge to visualize avidity differences and affinity maturation [5].
5. The cell number can be adjusted depending on the amount of recombinant protein used. It is crucial that the lowest concentration of antigen used should still be in molar excess compared to the amount of surface IgG on the B cells. The number of BCR per cell is approximately 50,000–150,000 per cell [9]. For example, if using 1,000,000 cells per tube and assuming 10% of lymph node cells are GC B cells this results in $0.5–1.5 \times 10^{10}$ bivalent BCRs per tube. Also critical is that once identified, the total number of cells stained remains consistent between experiments.
6. Other live/dead staining reagents can be used in place of EMA. Refer to the manufacture's instructions for directions.
7. The concentration range of the protein to be used needs to be determined empirically. There is no absolute need to reach saturation, as antigen binding is a continuum where high enough concentration of any protein will bind to germline BCR with low affinity. During the immune response Abs of high affinity will outcompete lower affinity Abs, reducing the relevance of the latter. Testing GC B cells for binding to the protein of interest, after infection or immunization with an antigenically unrelated virus or antigen prep, is a necessary step. This will give a population of cells that will allow to determine the maximum concentration of antigen that can be used to determine avidity above inevitable background binding (Fig. 2b).
8. Usually a threefold, ten points dilution curve will be sufficient to capture the relevant binding range. For the recombinant HA used herein we used protein concentration ranging between 6.66×10^{-12} M and 6.66×10^{-8} M.
9. If needed to quantitate GC B cell Ig isotype, labeled Abs specific for IgG or IgM can be added at this stage, together with the recombinant protein.
10. It is not advised to use directly labeled recombinant proteins as the fluorochrome could interfere with BCR binding.
11. It is advised to include appropriate flow controls, such as unstained cells, single stained and fluorescence minus one

(FMO) controls. This will help with compensation and gating strategies.

12. At this stage, the samples can be fixed and stored overnight at 4 °C wrapped in aluminum foil. Cells can be fixed with 100 μl 1.5% PFA for 30 min at 4 °C. It is advised to wash the 1.5% PFA before overnight storage if using tandem dyes. Stability of the recombinant protein staining post-fixation need to be determined empirically, i.e., HA staining is only stable for 24 h after fixation.
13. It is recommended to acquire a minimum of 200,000 events, which should be increased in case of low germinal center formation. Usually, a minimum frequency of 5% $GL7^+$ $CD38^-$ cells (among $B220^+$ B cells) is required to obtain reliable binding frequencies.

References

1. Victora GD, Nussenzweig MC (2012) Germinal centers. Annu Rev Immunol 30:429–457. doi:10.1146/annurev-immunol-020711-075032
2. Eisen HN, Siskind GW (1964) Variations in affinities of antibodies during the immune response. Biochemistry 3:996–1008
3. Stoop JW, Zegers BJ, Sander PC, Ballieux RE (1969) Serum immunoglobulin levels in healthy children and adults. Clin Exp Immunol 4 (1):101–112
4. Khurana S, Verma N, Yewdell JW, Hilbert AK, Castellino F, Lattanzi M, Del Giudice G, Rappuoli R, Golding H (2011) MF59 adjuvant enhances diversity and affinity of antibody-mediated immune response to pandemic influenza vaccines. Sci Transl Med 3(85):85ra48. doi:10.1126/scitranslmed.3002336
5. Frank GM, Angeletti D, Ince WL, Gibbs JS, Khurana S, Wheatley AK, Max EE, McDermott AB, Golding H, Stevens J, Bennink JR, Yewdell JW (2015) A simple flow-cytometric method measuring B cell surface immunoglobulin avidity enables characterization of affinity maturation to influenza A virus. MBio 6(4): e01156. doi:10.1128/mBio.01156-15
6. Rothaeusler K, Baumgarth N (2010) B-cell fate decisions following influenza virus infection. Eur J Immunol 40(2):366–377. doi:10.1002/eji.200939798
7. Magadán JG, Khurana S, Das SR, Frank GM, Stevens J, Golding H, Bennink JR, Yewdell JW (2013) Influenza A virus hemagglutinin trimerization completes monomer folding and antigenicity. J Virol 87(17):9742–9753. doi:10.1128/JVI.00471-13
8. Whittle JRR, Wheatley AK, Wu L, Lingwood D, Kanekiyo M, Ma SS, Narpala SR, Yassine HM, Frank GM, Yewdell JW, Ledgerwood JE, Wei C-J, McDermott AB, Graham BS, Koup RA, Nabel GJ (2014) Flow cytometry reveals that H5N1 vaccination elicits cross-reactive stem-directed antibodies from multiple Ig heavy chain lineages. J Virol. doi:10.1128/JVI.03422-13
9. Rabellino E, Colon S, Grey HM, Unanue ER (1971) Immunoglobulins on the surface of lymphocytes. I. Distribution and quantitation. J Exp Med 133(1):156–167

Chapter 16

Somatic Hypermutation and Affinity Maturation Analysis Using the 4-Hydroxy-3-Nitrophenyl-Acetyl (NP) System

Nicole Heise and Ulf Klein

Abstract

Somatic hypermutation of immunoglobulin variable region (IgV) genes and affinity maturation of the antibody response are the hallmarks of the germinal center (GC) reaction in T cell-dependent immune responses. Determining the consequences of the experimental manipulation of the GC response on somatic hypermutation and affinity maturation requires the availability of a system that allows measuring these parameters. Immunization of mice of the C57/Bl6 genetic background with the hapten 4-hydroxy-3-nitrophenyl-acetyl (NP) coupled to a carrier protein leads to the predominant usage of one particular IgV heavy chain gene segment, V186.2, among the responding B cells. Moreover, a specific somatic mutation in codon 33 of V186.2 that leads to a tryptophan to leucine amino acid exchange increases the affinity of the corresponding antibody by ~10-fold, thus representing a molecular marker for affinity maturation. In addition, due to the simplicity of the antigen and the virtual absence of NP-specific plasma cells prior to immunization, NP-based immunizations represent ideal tools to quantify the plasma cell response by measuring NP-specific antisera by ELISA and the generation of NP-specific plasma cells by ELISPOT analysis. We here describe approaches to (1) measure the anti-NP plasma cell response by ELISA and ELISPOT analysis, and to (2) amplify and sequence V186.2 rearrangements from GC B cells and plasma cells to determine the level of somatic hypermutation and the extent of affinity maturation in the anti-NP response.

Key words B cell, Plasma cell, Ig variable region gene, Germinal center, Somatic hypermutation, Affinity maturation, T cell-dependent, T cell-independent

1 Introduction

Antigen-specific memory B cells and plasma cells are generated during the germinal center (GC) reaction of T cell-dependent immune responses in secondary lymphoid tissues [1, 2]. Within the GC microenvironment, antigen-activated B cells undergo somatic hypermutation of the rearranged IgV genes with the aim to generate high-affinity antibodies that effectively bind to the invading pathogen, resulting in its elimination. In a defined area within the GC called dark zone, the rapidly proliferating "dark zone" cells hypermutate their IgV genes and then differentiate

Dinis Pedro Calado (ed.), *Germinal Centers: Methods and Protocols*, Methods in Molecular Biology, vol. 1623,
DOI 10.1007/978-1-4939-7095-7_16, © Springer Science+Business Media LLC 2017

into "light zone" cells that are positively selected for improved antigen binding in the GC light zone [2–5]. Selected light zone B cells recirculate to the dark zone to undergo additional rounds of mutation and selection to further improve antigen affinity before they eventually differentiate into memory B cells or plasma cells that exit the GC. The descendants of the GC reaction thus carry somatically mutated IgV genes and have often switched from IgM to other Ig classes which have different effector functions.

Understanding the molecular mechanism of the GC response is critical for the development of improved vaccines against microorganisms. Thus, for studies aimed at manipulating the GC response, which for example may include the deletion or overexpression of GC-associated genes or the use of particular adjuvants, it is imperative to have tools that allow the investigation of the processes that underlie the generation of antigen-specific memory B cells and plasma cells, i.e., somatic hypermutation and antibody affinity maturation. In C57/Bl6 mice, the immune response against the hapten 4-hydroxy-3-nitrophenyl-acetyl (NP) coupled to a carrier protein frequently results in a specific, affinity-enhancing hypermutation in codon 33 of the V186.2 gene segment that leads to an amino acid exchange resulting in a ~10-fold increase in affinity against NP [6–8], allowing for the generation of NP-specific GC B cells, memory B cells and plasma cells to be tracked at the molecular level. The resulting antibodies are mostly of the IgG1 class [6–8]. The characterization of the B-cell response against NP at the molecular level by several groups [9–11] provided the baseline for numerous studies on the extent of affinity maturation in situations where certain genes were either knocked out in the mouse germ line or conditionally deleted in GC B cells [12–16].

By immunizing mice with NP coupled to a carrier protein such as chicken gammaglobulin (CGG) or keyhole limpet hemocyanin (KLH), the characteristics and dynamics of the immune response against the hapten in the presence or absence of a certain signaling pathway or transcription factor, or a specific treatment, can be investigated by (1) determining the somatic hypermutation frequency in V186.2 gene rearrangements, (2) the fraction of B cells with the affinity-enhancing tryptophan to leucine amino acid exchange, (3) the quantity of NP-specific serum IgG1 secreted by GC-derived plasma cells by ELISA, and (4) the frequency of plasma cells in lymphoid tissues by ELISPOT analysis. The use of the NP system is not confined to T cell-dependent antibody responses, as the coupling of NP to lipopolysaccharide (LPS) or the polysaccharide Ficoll allows the study of the plasma cell response in T cell-independent responses type I and II, respectively, by ELISA.

2 Materials

2.1 Immunization

1. Sterile phosphate-buffered saline (PBS).
2. NP-KLH (e.g., NP_{28}-KLH, Biosearch Technologies; of note, the number of NP molecules conjugated to KLH varies among batches).
3. For T-independent immunizations (*see* **Note 1**), NP-LPS and NP-AECM-FICOLL (Biosearch Technologies).
4. Freund's adjuvant complete.
5. Freund's adjuvant incomplete.
6. 1-ml syringes.
7. Needles (18G1½ and $25G^5/_8$).
8. Vortex.
9. Sonicator.

2.2 Acquisition of Blood Samples and Serum Preparation

1. Razor blade or scalpel.
2. 1-ml or 2-ml syringes.
3. 21G1 needles.
4. 1.5-ml collection tubes.
5. Centrifuge.

2.3 ELISA

1. 96-well immune plates (Thermo Fisher Scientific).
2. 96-well culture plates.
3. Parafilm.
4. PBS.
5. 2% fetal bovine serum (FBS) in PBS.
6. *Wash buffer:* PBS containing 0.05% Tween 20.
7. *Capture antibody:* anti-mouse Ig(H+L) (Southern Biotec).
8. NP-bovine serum albumin (BSA) with low and high hapten coating (e.g., NP_9-BSA and NP_{25}-BSA).
9. Mouse Ig (e.g., IgM, IgG1).
10. *Detection antibody:* alkaline phosphatase (AP)-conjugated anti-mouse Ig (e.g., IgM, IgG1).
11. PNPP (*p*-nitrophenylphosphate; Southern Biotec).
12. *Substrate buffer* (500 ml): ddH_2O + 24.5 mg $MgCl_2 \times 6H_2O$ + 48 ml diethanolamine; adjust pH to 9.8 with 5 N HCl.
13. Microplate reader.

2.4 Sample Generation from Mouse Tissues Following NP Immunization

1. PBS–0.5% BSA
2. Glass slides; alternatively 40-μm cell strainer, 50-ml Falcon tube, and plunger of a 5-ml syringe.
3. 10-ml syringes.
4. 22G1 needles.
5. RBC lysis buffer (e.g., Red Blood Cell Lysis HybriMax; Sigma-Aldrich); alternatively, prepare lysis buffer: ammonium chloride (NH_4Cl) 0.15 M, 8.29 g per l; potassium bicarbonate ($KHCO_3$) 10 mM, 1 g per l; EDTA 0.1 mM, 0.037 g per l; H_2O, 1 l; filter solution with 0.45-μm filter, store at 4 °C.
6. 15-ml Falcon tubes.
7. Refrigerated centrifuge.
8. Optional: cell counter (e.g., Countess, Invitrogen).

2.5 ELISPOT

1. 96-well filtration plates (Millipore, Cat# MSIPS4510).
2. 96-well culture plates.
3. Plastic wrap.
4. 0.2 μm filter.
5. 35% ethanol (diluted in molecular grade H_2O).
6. Sterile PBS.
7. RPMI culture medium supplemented with 10% fetal bovine serum (FBS), 1% penicillin/streptomycin, and 0.1% β-mercaptoethanol.
8. *Wash buffer:* PBS containing 0.05% Tween 20.
9. PBS–2% BSA.
10. NP-BSA with high hapten coating (e.g., NP_{25}-BSA).
11. *Detection antibody:* AP-conjugated anti-mouse IgG1.
12. Nitro blue tetrazolium chloride-5-bromo-4-chloro-3-indolyl phosphate (NBT/BCIP; Roche).
13. *NBT/BCIP substrate buffer:* 0.1 M Tris + 0.1 M NaCl + 0.05 M $MgCl_2$; pH 9.5.
14. Optional: cell counter.
15. Microplate reader (alternatively, colonies can be counted using a magnifying glass and pictures of wells can be taken with a camera that has a macro function).

2.6 Isolation of B-Cell Subsets from Immunized Mice by FACS

1. RBC lysis buffer (*see* above).
2. PBS–0.5% BSA.
3. Glass slides; alternatively 40-μm cell strainer, 50-ml Falcon tube, and plunger of a 5-ml syringe.
4. 15-ml Falcon tubes.

5. MidiMACS Separator (Miltenyi Biotec).
6. MidiMACS Separation Units (columns; Miltenyi Biotec).
7. "Untouched" B-cell isolation kit (Miltenyi Biotec).
8. Antibodies: B220-PercP, CD95-PE, CD138-PE (conjugates available from various companies), PNA-biotin (Vector Laboratories), streptavidin-FITC (or streptavidin-APC if mouse cells are GFP^+ or YFP^+).
9. Fluorescent-activated cell sorter (FACS; e.g., provided by the institution's staff-operated Flow Cytometry Core Facility).
10. 15-ml, 5-ml, or 1.5-ml collection tubes.
11. *Collection buffer:* PBS–BSA with 20% fetal bovine serum (FBS).
12. Refrigerated centrifuge.

2.7 Sequencing of V186.2 Gene Rearrangements Amplified from RNA

1. TRIzol Reagent (Invitrogen).
2. 26G½ needles and 1 ml syringes.
3. Chloroform.
4. Isopropanol.
5. Glycogen.
6. DEPC-treated water.
7. UV spectrometer (e.g., NanoDrop ND-8000 Spectrophotometer; NanoDrop Technologies).
8. cDNA synthesis kit (e.g., from Invitrogen or an equivalent kit).
9. PCR thermocycler.
10. PCR tubes.
11. dNTPs.
12. Gene-specific primers (*see* Table 1).
13. Taq polymerase and reagents for PCR amplification.
14. Gel electrophoresis unit with appropriate buffer.
15. Agarose and agarose plates.

Table 1
Primers for cDNA synthesis and PCR amplification

Cγ1-cDNA	5′-CATGGAGTTAGTTTGGGCAG-3′
Cμ-cDNA	5′-CCACCAGATTCTTATCAGAC-3′
V186.2-leader	5′-AGCTGTATCATGCTCTTCTTGGCA-3′
V186.2-nested	5′-CATGCTCTTCTTGGCAGCAACAG-3′
Cγ1-PCR	5′-ATCCAGGGGCCAGTGGATAGAC-3′
Cμ-PCR	5′-CGCAGGAGACGAGGGGGAAGAC-3′

16. Ethidium bromide.
17. Gel imaging system.
18. TA-cloning vector; pGEM-TEasy (Promega).
19. Competent bacteria (DH5α).
20. SOC medium.
21. Ampicillin.

3 Methods

3.1 Primary and Secondary Immunization with NP

The general approach is summarized in Fig. 1.

3.1.1 Preparation of a NP-KLH Emulsion

1. Dissolve NP-KLH in sterile PBS at a concentration of 1 mg/ml.
2. For primary immunizations, NP-KLH is mixed with complete Freund's adjuvant at a 1:1 ratio (for the use of an alternative adjuvant, alum, *see* **Note 2**). Prepare 2.5–3× of the volume needed to ensure sufficient immunization emulsion for all experimental mice.

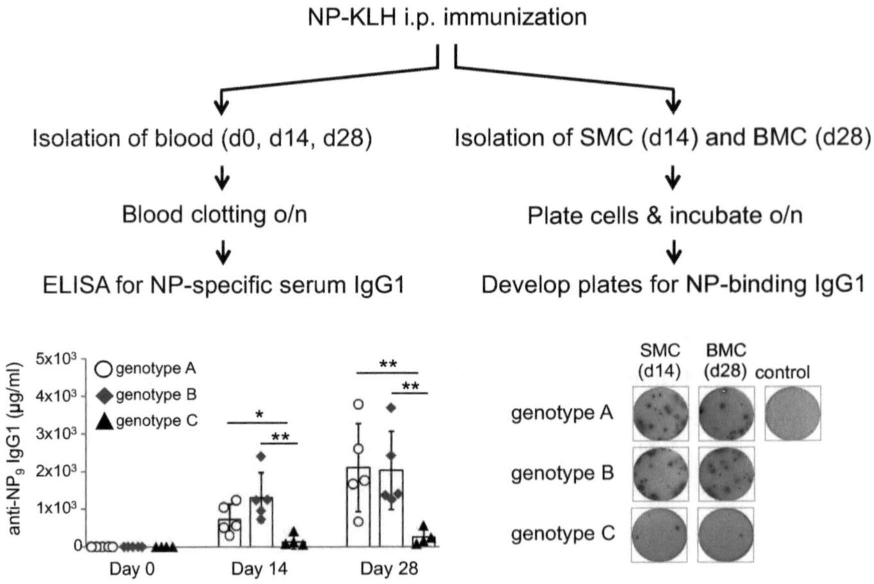

Fig. 1 Experimental flow: analysis of NP-specific serum Ig by ELISA and ELISPOT for anti-NP-specific IgG1 plasma cells. The major experimental steps are indicated and representative examples for ELISA and ELISPOT results are shown. *SMC* splenic mononuclear cells, *BMC* bone marrow cells

3. Vortex the mixture for 10 min and subsequently sonicate 3× for 5 min at 30 s intervals.
4. For secondary challenges, mix NP-KLH at a 1:5 ratio with sterile PBS. The diluted NP-KLH solution is then mixed with incomplete Freund's adjuvant at a 1:1 ratio and processed as described.
5. For T-independent immunizations (*see* **Note 1**).

3.1.2 Immunization

1. Fill a 1 ml syringe with the NP-KLH emulsion by using an 18G1½ needle. Use the same syringe for all mice. If immunizing many mice, fill as few syringes as possible.
2. Remove the needle and compress the mixture by drawing the punch up and down several times to remove the air.
3. For injection, attach a 25G$^5/_8$ needle and intraperitoneally inject 200 μl per mouse.

3.2 Detection of NP-Specific Ig by ELISA

3.2.1 Collection and Preparation of Serum from Immunized Mice

1. Prior to and at different time-points following NP immunization (7 days before immunization, day 14 and day 28 after immunization), take a few drops of blood from the tail vein of the mice using a razor blade or scalpel or by an alternative IACUC-approved method from live mice and collect into 1.5 ml collection tubes (important: no addition of heparin). At the final time point, withdraw peripheral blood (any volume from a drop to several hundred microlitres is fine) with a 21G1½ needle and a 1 or 2 ml syringe from the heart of sacrificed mice and transfer to a 1.5 ml reaction tube.
2. Keep reaction tube at RT for the remainder of the day and at 4 °C o/n to allow clotting.
3. Centrifuge at 12000 × *g* for 5 min and carefully transfer the clear supernatant (serum) to a new 1.5 ml reaction tube.
4. Repeat the process to remove any clotted material and store serum at −20 °C until use.

3.2.2 Detection of NP-Specific Ig

1. Coat one or more 96-well immune plates with 50 μl per well of 2.5 μl/ml NP_9-BSA, and the same amount of plates with NP_{25}-BSA diluted in PBS (the latter is necessary only if one plans to determine also the specificity of the secreted antibodies for an affinity maturation analysis; *see* Subheading 3.2.2, **step 12** below). In doing so, skip the first row of each plate and coat instead with 2.5 μl/ml capture antibody diluted in PBS to later generate a standard curve. Seal the plates with Parafilm and incubate overnight at 4 °C. Discard the solution and wash the wells 3× with 200 μl wash buffer.
2. Apply 200 μl of PBS–2% FBS per well, seal the plates with Parafilm and incubate at room temperature for at least 1 h.

3. Discard the solution and wash the wells 3× with 200 μl wash buffer.
4. In the first row of each plate, generate a standard curve (of for example IgM, IgG1) with each 50 μl of 400—200—100—50—25—12.5—6.25—0 ng/ml.
5. In a separate 96-well plate, generate 1:2 serial dilutions in PBS–2% FBS of each serum sample, starting with a 1:8000 dilution and ending with a 1:64,000 dilution. Do every condition in duplicate. Add 50 μl of each sample dilution to one test plate coated with NP_9-BSA, and one test plate coated with NP_{25}-BSA (use the same order of samples and dilutions on both plates!). Seal the plates with Parafilm and incubate for 2 h at room temperature.
6. Discard the samples and wash the wells 3× with 200 μl wash buffer.
7. Add 50 μl of detection antibody diluted 1:2000 in PBS–2% FBS to each well. Seal the plates and incubate at room temperature for 1 h.
8. Discard the solution and wash the wells 5× with 200 μl wash buffer.
9. Dissolve one PNPP 5 mg tablet in 5 ml substrate buffer and apply 50 μl to each well. Incubate at RT for 10–30 min, depending on the intensity of color development.
10. Read the plates at 405 nm OD. Save the "raw data" as well as the "blanked data" and "calculated concentrations" in an Excel data sheet.
11. For analysis of the amount of secreted NP-specific antibodies, plot the standard curve and include all sample values of the NP_9-BSA-coated plate that lie within the linear range of the standard curve, indicated by similar Ig concentrations at different sample dilutions. An average of at least two different dilutions in duplicate is recommended for analysis.
12. To estimate the specificity of the secreted antibodies, analyze the NP_{25}-BSA-coated plate in the same way as described above, and calculate the ratio of high-affinity (NP_9-BSA):total NP-specific antibodies (NP_{25}-BSA) for each sample and dilution in the linear range of the standard curve. If for example in a hypothetical knockout genotype the NP_9/NP_{25} ratio was lower compared to the proficient controls, this would indicate that affinity maturation in this genotype is impaired.
13. For T-independent immunization (*see* **Note 1**).

3.3 Detection of Antibody Secretion by ELISPOT

All steps on day 1 and day 2 have to be performed under the hood to keep the plates sterile. Do every condition in duplicate. The general approach is summarized in Fig. 1.

3.3.1 Preparation of the Plates (Day 1)

1. Pre-wet each well of a filtration plate with 15 μl of 35% ethanol for a maximum of 1 min.
2. Wash each well of the plate 3× with 200 μl sterile PBS.
3. Coat each well with 100 μl of 5 mg/ml NP-BSA (e.g., NP_{25}-BSA) diluted in PBS. Wrap the plate tightly in plastic wrap and incubate at 4 °C o/n.

3.3.2 Sample Preparation (Day 2)

1. At least 2 h before addition of the samples, discard the coating solution and wash the plate 3× with sterile PBS.
2. Block the plate with 100 μl per well with culture medium (RPMI) for at least 2 h at 37 °C.
3. In the meantime, collect the spleen and/or bone marrow samples from NP-immunized mice at day 14 or day 28, respectively. Isolate splenic mononuclear cells (SMC) in PBS–0.5% BSA by mincing the spleen between glass slides (rough surfaces) or using a 40-μm Falcon cell strainer placed on top of a 50 ml Falcon tube and a plunger from a 5-ml syringe to mince the spleen with occasionally flushing with ice-cold PBS–0.5% BSA. Bone marrow cells are isolated by flushing the femurs with PBS–0.5% BSA.
4. Spin down at 4 °C in a 15-ml Falcon and resuspend in RBC lysis buffer for 5 min.
5. Add 10 ml PBS–BSA and spin; after resuspending in 10 ml, let the "ghosts" settle and transfer supernatant into a new 15 ml tube.
6. Count the cells (e.g., using a cell counter) and resuspend at the desired cell density (e.g., 1×10^6 cells/ml for SMCs, 8×10^6 cells/ml for bone marrow) in prewarmed culture medium (RPMI).
7. Generate 1:2 dilution series for each sample in a separate 96-well culture plate, starting with 250,000 cells for SMCs, and 2×10^6 cells for bone marrow cells.
8. Decant the blocking medium of the filtration plate and add 100 μl of each sample dilution, resulting in a starting cell number of 100,000 for SMCs and 800,000 cells for bone marrow samples.
9. Incubate the samples overnight at 37 °C and 5% CO_2. During this time, do not open the incubator door, as it may result in smeared spots difficult to read.

3.3.3 Spot Development (Day 3)

The following steps can now be performed on the benchtop.

1. Discard the cell solution and wash the plate 6× with wash buffer.
2. To each well, add 100 μl of AP-conjugated anti-mouse IgG1 diluted 1:2000 in PBS–2% BSA. Filter the detection antibody solution prior to use (0.2 μm filter). Incubate for 1 h at RT.
3. Discard the detection antibody and wash the plate 3× with wash buffer, followed by three washes with PBS-only to avoid interference of Tween with spot development.
4. Take 200 μl NBT/BCIP stock solution per 10 ml NBT/BCIP substrate buffer, pass the solution through a 0.2-μm filter and apply 100 μl per well. Incubate for 5–10 min.
5. Stop spot development under running water and wash the plate extensively. While washing, remove the underdrain and continue rinsing.
6. Blot the plate on paper towels and leave it to dry under the hood in the dark.
7. Read the plate in a microplate reader or determine number of spots by alternative methods.

3.4 Isolation of B-Cell Subsets from Immunized Mice by FACS

3.4.1 Isolation of Mononuclear Cells and B Cells

1. Isolate splenic mononuclear cells in PBS–0.5% BSA at day 14 after immunization (*see* **Note 3**) by mincing the spleen between glass slides (rough surfaces) or using a 40-μm Falcon cell strainer placed on top of a 50-ml Falcon tube and a plunger from a 5-ml syringe to mince the spleen with occasional flushing with PBS–0.5% BSA.
2. Spin down in a 15 ml Falcon and resuspend in RBC lysis buffer for 5 min, then add 10 ml PBS–0.5% BSA and spin; after resuspending in 10 ml, let the "ghosts" settle and transfer supernatant into a new 15-ml tube, count cells. This fraction should be used when flow-cytometrically isolating plasma cells ($CD138^{hi}B220^{low}$).
3. Isolate mouse B-cells by magnetic cell separation using the MidiMACS system with the "untouched B-cell isolation kit" (Miltenyi Biotec) according to the manufacturer's instructions or an equivalent system from another manufacturer; count cells.

3.4.2 Isolation of B-Cell Subsets by FACS

The cell count of B cells after magnetic cell separation is approx. 10^7 cells, although it can vary. Staining for flow cytometric isolation should be performed in a 15-ml Falcon tube. Antibody concentrations need to be titrated which can be done on an analyzer (e.g., FACSCalibur or equivalent instrument) using less cells and antibody, which is then scaled up for sorting. As a rule of thumb, staining of 10^7 B cells can be performed in 100 μl total staining volume with each 5 μl of antibodies purchased from Pharmingen;

antibodies from other companies including Biolegend or Bioscience usually require titration. For the isolation of splenic GC B cells, stain B cells isolated by magnetic cell separation with antibodies recognizing B220 and CD95 and PNA (Fig. 2; for alternative strategies, *see* **Notes 4** and **5**). For the isolation of splenic plasma cells, stain mononuclear cells with antibodies recognizing B220 and CD138 (Fig. 2). Staining is for 10 min on ice, followed by washing with PBS–0.5% BSA and resuspending the cells in approx. 2 ml PBS–0.5% BSA (inquire the appropriate cell density of the fraction to be sorted from the cell sorter operator).

1. Prepare collection tubes according to the following criteria: Depending on the number of cells that are expected to be recovered, either 15-ml Falcon tubes or 5-ml tubes or 1.5-ml reaction tubes should be used (inquire with the cell sorter operator about the best strategy). As a rule of thumb, for the isolation of GC B cells from immunized normal mice, 15-ml Falcon tubes are recommended; if the genotype of the mice is associated with the occurrence of reduced numbers of GC B cells, or if a subset of GC B cells (e.g., $eGFP^+$ GC B cells) are sorted, 5 ml or 1.5 ml tubes represent the best choices. Generally, plasma cells represent a minute population and should be sorted into 1.5 ml tubes. Regardless of the size of the tubes, fill the tubes completely with PBS–0.5% BSA and incubate for 10 min to coat the surface; this ensures that the sorted cells slide to the bottom early into the sort. After discarding the PBS–0.5% BSA, add 20% FBS in PBS–0.5% BSA (and up to 50% FBS if the expected cell number is very low) at approx. 1/5th of collection tube's volume (e.g., 3 ml for a 15 ml tube).
2. For the isolation of GC B cells, set a lymphocyte scatter gate to exclude dead cells and small debris, a second gate to identify $B220^+$ cells, and a third gate to identify $CD95^{hi}PNA^{hi}$ GC B cells (Fig. 2). For the isolation of plasma cells, set a lymphocyte scatter gate (make certain to include large cells, as plasma cells are larger than B cells), and a second gate to identify $CD138^{hi}B220^{low}$ cells (Fig. 2). Sort cells into collection tubes. Optional: As control for unmutated V186.2 rearrangements, $CD95^{-}PNA^{lo}$ B cells can be isolated in parallel, although for this purpose we suggest the use of an alternative cell population as control (*see* **Note 6**).
3. Spin down sorted cells, carefully remove the supernatant and lyse the cell pellet in 800 μl TRIzol, incubate for 10 min at RT and continue with Subheading 3.5.1, or freeze at −20 °C if not continuing with the RNA isolation.
4. For an alternative approach that does not involve the isolation of B-cell subsets by FACS, (*see* **Note 7**). For studies where the specific analysis of peripheral blood memory B cells is of interest, please refer to ref. [12].

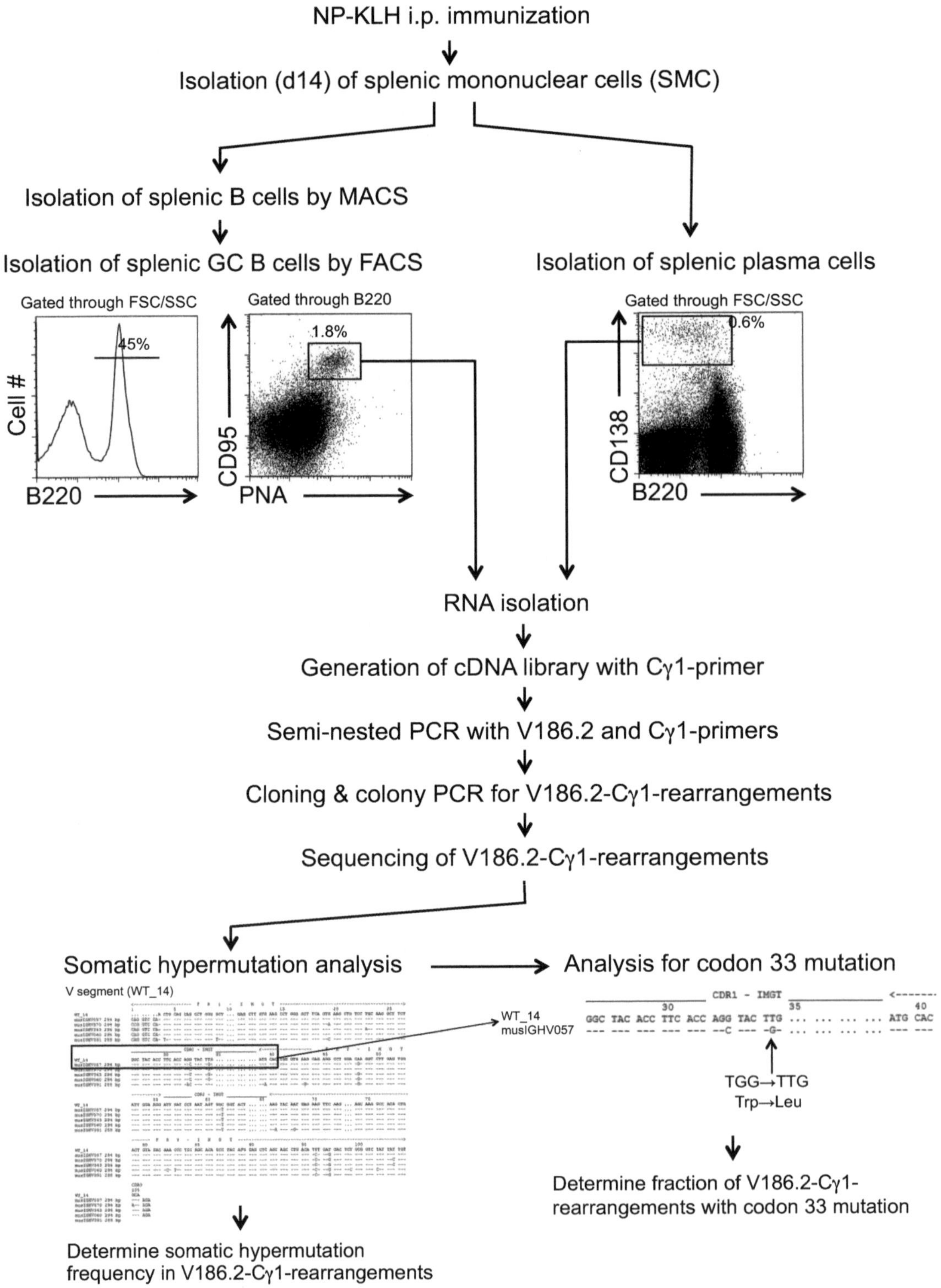

Fig. 2 Experimental flow: analysis of V186.2-Cγ1 rearrangements for somatic hypermutation and affinity maturation. The major experimental steps are indicated. Representative examples for the isolation of

3.5 Sequencing of V186.2 Gene Rearrangements Amplified from RNA

3.5.1 RNA Isolation and cDNA Synthesis

1. Following incubation for 10 min at RT in 800 μl TRIzol, pass 2× through a G26½ needle using a 1 ml syringe to shear the DNA.
2. Add 160 μl chloroform and shake vigorously with hand for 15 s. Incubate for 3 min at RT. Centrifuge at 4 °C for 15 min at 12000 × *g*.
3. Carefully remove the upper layer of the interface and transfer to a new tube without touching the interphase. Add 2 μl glycogen to supernatant. Add 400 μl isopropanol, mix and incubate for 10 min at RT. Centrifuge at 4 °C for 15 min at 12000 × *g*.
4. Remove the isopropanol (leave ~100 μl); centrifuge again briefly and carefully remove remaining supernatant (the RNA pellet is transparent and thus difficult to see in isopropanol). Wash pellet with 1 ml 75% ethanol in DEPC-treated water (now pellet becomes visible).
5. Remove ethanol (the remaining drop can be removed using a G26½ needle placed on a pipette) and dry the pellet on the bench for 15 min, until ethanol has evaporated. Dissolve RNA in 7 μl DEPC-treated water. RNA can be stored at −80 °C if not continuing with the cDNA synthesis.
6. Make 2 mM dilutions of the Cγ1-cDNA (and if applicable the Cμ-cDNA) primers (Table 1) using DEPC-treated water.
7. Follow the cDNA synthesis protocol according to the manufacturer's instructions using 7 μl RNA (in the case that the total reaction volume is 20 μl) or less depending on the number of sorted cells, and 1 μl of the Cγ1-cDNA primer and, if amplifying μ-transcripts in parallel (*see* **Note 8**), 1 μl of the Cμ-cDNA primer. Perform the cDNA reaction as instructed up to the incubation with RNase H. cDNA can be stored at −20 °C if not continuing with the PCR reaction.

3.5.2 Seminested PCR Amplification

1. For the 1st round of PCR, prepare a master mix consisting of 8 μl dNTP solution (1.25 mM), 5 μl 10× PCR reaction buffer, 5 μl $MgCl_2$ (25 mM), 1 μl V186.2-leader primer, 1 μl Cγ1-PCR primer (for PCR primers, *see* Table 1), and 27.5 μl H_2O. Add 2 μl of cDNA reaction mixture to master mix and then 0.5 μl of a mix of Taq polymerase with 1/60 Pfx polymerase (alternatively, a high fidelity polymerase that allows the amplification of low amounts of DNA can be used). The PCR program consists of 95 °C 2 min, 20× (95 °C 30 s, 70 °C 30 s,

Fig. 2 (Continued) B220^{+}PNA^{hi}CD95^{hi} GC B cells and B220^{lo}CD138^{hi} plasma cells are shown. A representative example for the analysis of somatic hypermutation and the analysis for codon 33 mutation using VBASE2 [17] are indicated; in VBASE2, codon 33 of V186.2 is found at position 34. FSC, forward scatter; SSC, side scatter

72 °C 90 s), 72 °C 5 min, and 10 °C pause. There is no agarose gel analysis at this step.

2. For the 2nd round of PCR, prepare a master mix consisting of 8 μl dNTP solution (1.25 mM), 5 μl 10× PCR reaction buffer, 3 μl $MgCl_2$ (25 mM), 1 μl V186.2-nested primer, 1 μl Cγ1-PCR primer (for PCR primers, *see* Table 1), and 28.5 μl H_2O. Add 3 μl of 1st round PCR product of each of the experimental samples and the water control to master mix and then 0.5 μl of a mix of Taq polymerase with 1/60 Pfx polymerase (alternatively, a high fidelity polymerase that allows the amplification of low amounts of DNA can be used). The PCR program consists of 95 °C 2 min, 30× (95 °C 30 s, 70 °C 30 s, 72 °C 90 s), 72 °C 5 min, and 10 °C pause.
3. Run 5 μl of the PCR product on an agarose gel. Of note: there should be two water controls run along with the experimental samples, one which went through the nested PCR to control for potential contamination in the 1st round of amplification, and one that controls the 2nd round of amplification.
4. For a genomic DNA-based strategy to amplify V186.2 rearrangements (*see* **Note 9**).

3.5.3 Cloning of V186.2 Segments

1. *Gel isolation:* Prepare a 1.5% agarose gel with wells large enough for loading of 50 μl per well. Add 5 μl loading buffer to the remaining 45 μl PCR reaction and run on the agarose gel; load marker to identify correct length of PCR product. Cut out band with a scalpel using a UV lamp to visualize band and transfer into a 1.5 μl reaction tube. Isolated band can be stored at 4 °C if not continuing with the gel purification.
2. Isolate PCR product using a suitable gel extraction method such as the QIAquick Gel Extraction kit (Quiagen) or an equivalent kit using the manufacturer's instructions. Elute PCR product with 30 μl H_2O. Measure the DNA concentration, e.g., using NanoDrop methodology or agarose gel-based quantification methods. Optional: 20 ng of DNA can be loaded onto an agarose gel to verify that all bands appear at the same intensity.
3. *A-tailing to allow ligation of PCR product into pGEM-TEasy vector:* Dilute sample to a concentration of 62.5 ng DNA in 6.4 μl. If the DNA concentration is too low, the volume can be reduced using a SpeedVac centrifuge. Add 1 μl 10× PCR reaction buffer, 0.8 μl $MgCl_2$ (25 mM), 1.6 μl dNTPs (1.25 mM), and 0.2 μl Taq polymerase. Mix and incubate for 20 min at 70 °C; place on ice.
4. Set up a ligation reaction using the pGEM-TEasy vector kit with 5 μl ligation buffer, 0.5 μl cloning vector, 2 μl PCR product, 1.5 μl H_2O, and 1 μl T4 ligase, which should be added last. Mix and incubate for 1 h at RT.

5. Using a standard transformation protocol with DH5α competent bacteria, add 2 μl of the ligation reaction to 50 μl DH5α. Following heat shock, add 500 μl SOC medium, of which after incubation for 1 h at 37 °C, 100 μl are spread onto an agar plate that contains ampicillin. The remaining bacteria in SOC medium should be kept at 4 °C and can be spread onto agarose plates to obtain additional V186.2-containing clones.
6. *Colony PCR:* Carefully lift colony from plate and transfer into a 0.5 ml reaction tube containing 40 μl H_2O, vortex vigorously. Prepare a master mix consisting of 28.5 μl H_2O, 8 μl dNTP solution (1.25 mM), 5 μl 10× PCR reaction buffer, 3 μl $MgCl_2$ (25 mM), 1 μl V186.2-nested primer, 1 μl Cγ1-PCR primer (for PCR primers, *see* Table 1), and 0.4 μl Taq polymerase. Add 3 μl of the sample to the master mix. The PCR program consists of 95 °C 2 min, 32× (95 °C 30 s, 70 °C 30 s, 72 °C 90 s), 72 °C 5 min, and 10 °C pause. Determine DNA concentration in PCR product on agarose gel with suitable quantification marker. Send samples for sequencing together with an appropriate primer (e.g., T7 primer) using the sequencing company's guidelines; e.g., some companies require 4 ng/μl for direct sequencing of PCR products; usually, the company will purify the PCR product in house before sequencing, so that removal of excess PCR primers is not required.

3.5.4 Sequence Analysis

The VBASE2 database (http://www.vbase2.org/) [17] is a convenient tool for the sequence analysis of V186.2 sequences. Following editing the sequence in the spherogram, which includes removing any primer sequences from the 5′ and 3′ ends and verifying the nucleotide sequence versus the A, G, T and C peaks, save the sequence in FASTA format and run it against the VBASE2 database. V186.2 corresponds to musIGHV057. If the sequence is more homologous to other V genes, it should be excluded from further analysis. The codon encoding the affinity enhancing tryptophan to leucine amino acid exchange is at position 34 in the VBASE2 representation. Most somatic mutations lead to a G to T nucleotide exchange (TGG to TTG), but are not restricted to this particular exchange. To determine the number of somatic mutations in the V186.2 gene segment, all nucleotide exchanges relative to musIGHV057 should be counted with the exception to those that are directly at the junction to the diversity (D) gene segment, as those nucleotide differences could represent non-germ line encoded (N) nucleotides added by terminal deoxynucleotidyl transferase (TdT). The V186.2 mutation frequency is determined by dividing the number of somatic mutations through the number of V186.2 nucleotides sequenced, excluding N, D, and joining (J) sequences.

4 Notes

1. For the study of T-independent type-I and type-II responses, mice are immunized i.p. with 20 μg NP-LPS or 30 μg of NP-AECM-FICOLL, respectively, dissolved in sterile PBS respectively. Serum is isolated from blood as described in Subheading 3.2.1 7 days following immunization. ELISA is performed as described in Subheading 3.2.2, except that the anti-NP serum response of IgM and IgG3 antibodies is measured.
2. The adjuvant alum is an alternative for Freund's adjuvant and can be used interchangeably in the study of the T-dependent response to NP. Before deciding on an adjuvant, it is recommended to inquire with the institution's IACUC which adjuvant is acceptable.
3. The analysis for somatic hypermutation in V186.2 rearrangements following NP-immunization is commonly performed around days 12 to 14 after immunization, which is considered to be at the height of the GC reaction.
4. Different antibody combinations can be used to identify GC B cells, including the commonly used anti-CD95/anti-CD38 combination to identify $CD95^{hi}CD38^{lo}$ GC B cells [13], and anti-CD19 is a suitable alternative for anti-B220. GL7, which is highly expressed on GC B cells, can be used as an alternative for CD95, e.g., in the combination anti-GL7/anti-CD38.
5. When using an antibody-conjugate with PerCP, it need be ascertained that this fluorescence is compatible with the lasers and filters of the FACS sorter. As fluorescent colors, we suggest to use FITC, PE and APC. If the cells are GFP-positive, consult with the operator of the flow cytometer which fluorescence can be used as an alternative to FITC.
6. The most suitable control for unmutated V186.2 rearrangements from antigen-inexperienced naïve B cells are $B220^{+}IgM^{+}IgD^{+}$ splenic B cells (CD19 can be used as an alternative for B220). A suitable fluorochrome combination is B220-FITC, IgM-APC and IgD-PE. cDNA libraries generated with the Cμ cDNA primer as described in Subheading 3.5.1 are amplified with the V186.2-leader and Cμ-PCR primer in the 1st round and the V186.2-nested and Cμ-PCR primer in the 2nd round (for PCR primers, *see* Table 1).
7. In principle, it is not necessary to flow-sort GC B cells or plasma cells for somatic hypermutation and affinity maturation analysis of rearranged V186.2 genes upon NP-immunization. cDNA generated from whole splenic mononuclear cells or B cells isolated by magnetic cell separation can be used for the amplification of IgG1-expressing V186.2 gene segments [18].

There is one caveat, however. Since plasma cells have much larger amounts of antibody-encoding Ig transcripts than memory B cells or GC B cells, the cDNA-based V186.2 gene analysis is heavily skewed towards transcripts derived from plasma cells. To avoid this problem, V186.2 Cγ1-membrane transcripts can be selectively amplified using the following strategy [18]: RNA is reverse-transcribed with an oligonucleotide hybridizing in the Cγ1-membrane-encoding exon (mCγ1-cDNA oligo: 5′-TGACAGCAGCGCTGTAGCAC-3′) and amplified for 30 cycles with the amplification conditions described in Subheading 3.5.2, **step 1** using the V186.2-leader primer (Table 1) and a reverse primer hybridizing in the Cγ1-membrane-encoding exon (mCγ1-PCR oligo: 5′-CAGCACAGGTCTCGTCCAGTTG-3′). Generally, no nested PCR strategy is required, but can be implemented using the mCγ1-PCR and the V186.2-nested primer described in Subheading 3.5.2, **step 2**. All other cloning steps can be performed as described in Subheadings 3.5.2–3.5.4.

8. Somatic hypermutation analyses for V186.2 rearrangements are usually performed for IgG1 antibodies. Since somatic hypermutation occurs in both switched and non-switched GC B cells, V186.2 μ-transcripts may be analyzed as well. cDNA libraries generated with the Cμ-cDNA primer as described in Subheading 3.5.1 are amplified with the V186.2-leader and Cμ-PCR primer in the 1st round and the V186.2-nested and Cμ-PCR primer in the 2nd round (for PCR primers, *see* Table 1).
9. A strategy has been published that allows the amplification of V186.2 rearrangements from genomic DNA isolated from B-cell subsets [14]. In our own experience, this strategy leads to the amplification of a large percentage of non-V186.2 gene segments (~40%), and unless there is a particular requirement to amplify V186.2 rearrangements from genomic DNA, in our opinion the cDNA-based approach is more cost-effective.

Acknowledgments

We thank Nilushi De Silva, Kathryn Silva, and Qiong Shen for their involvement in the establishment of the protocols described here. This work was supported by the NCI/NIH and a fellowship of the German Research Council (DFG) to NH.

References

1. Rajewsky K (1996) Clonal selection and learning in the antibody system. Nature 381 (6585):751–758. doi:10.1038/381751a0
2. MacLennan IC (1994) Germinal centers. Annu Rev Immunol 12:117–139. doi:10.1146/annurev.iy.12.040194.001001
3. Allen CD, Okada T, Cyster JG (2007) Germinal-center organization and cellular dynamics. Immunity 27(2):190–202. doi:10.1016/j.immuni.2007.07.009
4. Victora GD, Nussenzweig MC (2012) Germinal centers. Annu Rev Immunol 30:429–457. doi:10.1146/annurev-immunol-020711-075032
5. De Silva NS, Klein U (2015) Dynamics of B cells in germinal centres. Nat Rev Immunol 15 (3):137–148. doi:10.1038/nri3804
6. Cumano A, Rajewsky K (1986) Clonal recruitment and somatic mutation in the generation of immunological memory to the hapten NP. EMBO J 5(10):2459–2468
7. Blier PR, Bothwell A (1987) A limited number of B cell lineages generates the heterogeneity of a secondary immune response. J Immunol 139 (12):3996–4006
8. Allen D, Simon T, Sablitzky F, Rajewsky K, Cumano A (1988) Antibody engineering for the analysis of affinity maturation of an anti-hapten response. EMBO J 7(7):1995–2001
9. Weiss U, Rajewsky K (1990) The repertoire of somatic antibody mutants accumulating in the memory compartment after primary immunization is restricted through affinity maturation and mirrors that expressed in the secondary response. J Exp Med 172(6):1681–1689
10. McHeyzer-Williams MG, McLean MJ, Lalor PA, Nossal GJ (1993) Antigen-driven B cell differentiation in vivo. J Exp Med 178 (1):295–307
11. Smith KG, Light A, Nossal GJ, Tarlinton DM (1997) The extent of affinity maturation differs between the memory and antibody-forming cell compartments in the primary immune response. EMBO J 16(11):2996–3006. doi:10.1093/emboj/16.11.2996
12. Blink EJ, Light A, Kallies A, Nutt SL, Hodgkin PD, Tarlinton DM (2005) Early appearance of germinal center-derived memory B cells and plasma cells in blood after primary immunization. J Exp Med 201(4):545–554. doi:10.1084/jem.20042060
13. Victora GD, Schwickert TA, Fooksman DR, Kamphorst AO, Meyer-Hermann M, Dustin ML, Nussenzweig MC (2010) Germinal center dynamics revealed by multiphoton microscopy with a photoactivatable fluorescent reporter. Cell 143(4):592–605. doi:10.1016/j.cell.2010.10.032
14. Dominguez-Sola D, Victora GD, Ying CY, Phan RT, Saito M, Nussenzweig MC, Dalla-Favera R (2012) The proto-oncogene MYC is required for selection in the germinal center and cyclic reentry. Nat Immunol 13 (11):1083–1091. doi:10.1038/ni.2428
15. Heise N, De Silva NS, Silva K, Carette A, Simonetti G, Pasparakis M, Klein U (2014) Germinal center B cell maintenance and differentiation are controlled by distinct NF-kappaB transcription factor subunits. J Exp Med 211 (10):2103–2118. doi:10.1084/jem.20132613
16. McHeyzer-Williams LJ, Milpied PJ, Okitsu SL, McHeyzer-Williams MG (2015) Class-switched memory B cells remodel BCRs within secondary germinal centers. Nat Immunol 16 (3):296–305. doi:10.1038/ni.3095
17. Retter I, Althaus HH, Munch R, Muller W (2005) VBASE2, an integrative V gene database. Nucleic Acids Res 33(Database issue): D671–D674. doi:10.1093/nar/gki088
18. Klein U, Casola S, Cattoretti G, Shen Q, Lia M, Mo T, Ludwig T, Rajewsky K, Dalla-Favera R (2006) Transcription factor IRF4 controls plasma cell differentiation and class-switch recombination. Nat Immunol 7(7):773–782. doi:10.1038/ni1357

Chapter 17

Targeting Gene Function in Germinal Center B Cells: A Practical Approach

Valentina Petrocelli and Stefano Casola

Abstract

The germinal center (GC) reaction represents an essential phase of an adaptive immune response. Dysfunction of GC B cells can lead to life-threatening diseases including autoimmune disorders, lymphomas, and opportunistic infections. Defining the molecular circuitries controlling GC B cell physiology is crucial to understand the pathogenesis of GC B cell disorders, as well as to develop improved vaccines against foreign pathogens. Conditional gene targeting based on the Cre/loxP recombination system has substantially accelerated our comprehension of the genetic networks controlling GC B cell function. Several independent studies in the past 10 years have highlighted the many advantages and the few limitations and pitfalls associated to conditional gene manipulation in GC B cells using the Cre/loxP recombination system. Here, we describe the basic features of GC B cell-specific gene targeting experiments. We provide indications on the type of Cre transgene and controls to be chosen, way-out strategies to overcome leakiness of the Cre/loxP system, and approaches to minimize the number of experimental animals and to speed up analyses on conditional mutant GC B cells.

Key words B cells, Germinal center, Conditional gene targeting, Cre/loxP, Cγ1-cre, Aicda-cre

1 Introduction

The germinal center (GC) reaction represents a crucial phase of an adaptive immune response. During the GC reaction, antigen-specific B cells undergo intense proliferation ensuring rapid clonal expansion of rare responders. GC B cell proliferation is coupled to the accumulation of non-templated point mutations within immunoglobulin (Ig) heavy (H) and light (L) chain variable (V) region genes through a process called Ig somatic hypermutation (SHM), leading to the rapid diversification of the Ig repertoire expressed by these cells. Cells expressing mutated B cell antigen receptors (BCR) undergo competition for antigen binding and presentation to T follicular helper cells. This step is crucial for few, highly selected, GC B cells to receive the survival and mitogenic signals needed for a subsequent round of clonal expansion and Ig V gene diversification

Dinis Pedro Calado (ed.), *Germinal Centers: Methods and Protocols*, Methods in Molecular Biology, vol. 1623, DOI 10.1007/978-1-4939-7095-7_17, © Springer Science+Business Media LLC 2017

[1]. Repeated cycles of antibody gene mutation and selection ensure the ultimate generation of rare B cells expressing high affinity BCRs, which exit the GC after differentiating into long-lived memory B cells or antibody-secreting plasma cells [2, 3].

Understanding the complex set of molecular and cellular events guiding the transition of B cells through the GC reaction represents a major focus of investigation. This knowledge can help the design of better vaccines against life-threatening pathogens [4] as well as to define the causes of GC derived diseases including autoimmune disorders and B cell malignancies (reviewed in [5]).

Conditional gene targeting offers the unique opportunity to study the role of candidate genes in a cell type- and stage-specific fashion within a living organism that is otherwise fully competent for the same gene function. Since its first application in the mouse model in 1993, the Cre/loxP recombination system has become the method of choice to manipulate gene function in vivo in a cell type- and stage-specific, and time-controlled fashion.

Conditional gene targeting in vivo requires, at first, the generation of mice engineered to express phage P1-derived Cre recombinase in a cell-type and stage-specific fashion. Transgenic mice, selectively expressing Cre in the target cells, are bred to knockin animals carrying loxP sites flanking critical exon(s) of candidate genes (also called "floxed" genes). Recognition of loxP sites by dimers of Cre recombinase triggers recombination between them, leading to deletion (loxP in the same orientation) or to inversion (loxP in the opposite orientation) of the floxed gene segment, resepctively.

The Cγ1-cre knockin strain has been the first reported transgenic line expressing Cre recombinase in GC B cells [6]. To induce Cre expression, Cγ1-cre mice rely on the property of the GC B cells to undergo Ig class switch recombination (CSR) [6]. Specifically, the insertion of the Cre coding sequence within the 3′ region of the IgG1 constant region genomic locus, preceded by an internal ribosome entry site (IRES), allows for expression of the recombinase in GC B cells undergoing IgG1 CSR [6]. Upon immunization with T cell dependent antigens that recruit type-2 helper T cells secreting IL-4, early induction of Ig heavy chain Cγ1 germ line transcripts [7] promotes Cre recombinase expression in GC B cell founders, leading to Cre-mediated recombination in most GC B cells [6]. Cγ1-cre mice have been successfully used to either inactivate [6, 8–15] or induce [9, 16] expression of genes of interest in GC B cells and their progeny of long-lived IgG1-class switched memory B cells and antibody-secreting plasma cells [6]. Other Cre mouse strains allowing inducible gene targeting in GC B cells have recently been described [17–20]. Cre transgenic lines alternative to the Cγ1-cre strain were generated placing the Cre coding sequence under the control of *cis* regulatory sequences of the activation induced cytidine deaminase (Aicda or AID) gene. AID is the enzyme responsible for Ig SHM and CSR, which is primarily, yet not exclusively, induced in GC B cells [21, 22].

$C\gamma1$-cre mice have been instrumental to elucidate the importance in GC B cells of transcription factors [12, 13], histone modifiers [8, 14, 23], signaling molecules [6, 11] and apoptotic determinants [10]. This has led to an acceleration of our understanding of the critical molecular circuitries controlling GC B cell proliferation, selection and terminal differentiation. In addition, through conditional gene targeting experiments, mechanisms leading to malignant transformation of GC B cells, or to their aberrant selection promoting autoimmunity, have started to be unveiled [9, 10, 16].

The success of in vivo conditional gene targeting studies relies on accurate knowledge of the experimental system. The following paragraphs intend to help the future users of this methodology to optimize the design of a conditional gene targeting experiment to study gene function in GC B cells.

1.1 Choosing the Most Appropriate GC B Cell-Specific Cre Strain

The choice of the Cre transgenic line represents a crucial step in the design of a conditional gene targeting experiment. The $C\gamma1$-cre strain allows tight, GC B cell-specific, expression of the Cre recombinase following activation of Ig CSR. In the first 1–4 days following immunization with T cell-dependent antigens, antigen-specific B cells induce the expression of $C\gamma1$ germ line transcripts as a result of the interaction with cognate T cells. This leads to the production in GC B cell founders of $C\gamma1$-IRES-cre bicistronic transcripts, which ensures sufficient Cre expression to carry out site-directed recombination [6]. Biallelic expression of $C\gamma1$ germ line transcripts ensures $C\gamma1$-cre transgene expression in all GC B cells of $C\gamma1$-cre heterozygous mice, including those expressing the functional IgH V region gene rearrangement from the second IgH locus [6]. Expression of $C\gamma1$-cre transcripts anticipates IgG1 CSR and continues after its completion in IgG1-class switched GC B cells and in their long-lived progeny of $IgG1^+$ memory B cells and plasma cells [6]. The efficiency of Cre-mediated recombination is partial in GC B cells of $C\gamma1$-cre mice, which undergo switching to other Ig isotypes, especially in gut-associated lymphoid tissues where these cells mainly switch to IgA [6]. This limitation is overcome in Aicda-cre transgenic animals. In the latter lines the expression of Cre recombinase is transient and follows that of the endogenous *Aicda* gene. Aicda-cre lines ensure gene targeting in a broader pool of activated B cells than $C\gamma1$-cre mice, both within and outside the GC. Cre-mediated recombination by the Aicda-cre transgene targets all Ig isotype-switched B cells participating to both T cell-dependent and -independent immune responses. In Aicda-cre animals, post GC memory B cells and plasma cells, which lack AID expression, are not chronically exposed to Cre expression, in contrast to $C\gamma1$-cre transgenic mice. The possibility that Aicda is

Table 1
List of mouse Cre lines for conditional gene targeting in GC B cells

Mouse strain	Type	Reference
Ighg1$^{tm1(IRES\text{-}cre)Cgn}$	Knockin	[6]
Tg(Aicda-cre)1Rcas	BAC transgene	[17]
Aicda$^{tm1(cre)Mnz}$	Knockin	[20]
Tg (Aicda-cre) 9Mbu	BAC transgene	[19]
Aicda$^{tm1.1(cre/ERT2)Crey}$	Knockin	[18]

induced in a fraction of immature B cells requires a careful evaluation of the utility of Aicda-cre lines for experiments in which gene targeting needs to be strictly restricted to GC B cells. The choice to use Aicda-cre BAC transgenic mice rather than knockin animals deserves further considerations. Aicda-cre knockin lines offer a tighter control on Cre expression than BAC transgenic lines, given the insertion of the Cre minigene as a single-copy into the endogenous *Aicda* locus. At the same time, the disruption of the endogenous *Aicda* locus through the insertion of the Cre transgene is expected to reduce by half the overall amount of AID protein expressed by Cre- knockin GC B cells. Aicda-cre BAC transgenic mice carry one or more copy of the transgene inserted randomly in the genome rendering difficult the distinction between Cre heterozygous and homozygous compound mutants. A side-by-side comparison between Aicda-cre knockin and Aicda-cre BAC transgenic lines in terms of efficiency of Cre-mediated recombination and cell- and stage-restricted expression of the recombinase has yet to be reported.

Noteworthy is the generation by the Reynaud and Weill laboratories of the Aicda-creERT$_2$ mouse knockin line [18]. In these animals, the Cre transgene is expressed as a fusion protein with a minimal human estrogen receptor ligand-binding domain (ERT$_2$). Aicda-creERT$_2$ animals allow for temporal control of Cre-mediated recombination in AID-expressing B cells through the feeding of transgenic animals with the estrogen analogue tamoxifen (*see* Table 1).

1.2 Monitoring Cre-Mediated Recombination in GC B Cells

The possibility to track in vivo, at single cell resolution, GC B cells undergoing conditional gene disruption offers the advantage to follow the fate of (and possibly to isolate) the mutant cells within a population of cells still proficient for the candidate gene. To achieve this purpose, alternative approaches can be pursued (Fig. 1): (1) *Use of Reporters for Cre-mediated recombination*: this represents a popular approach to monitor Cre-mediated recombination in GC B cells. Specifically, mice are engineered to carry extra

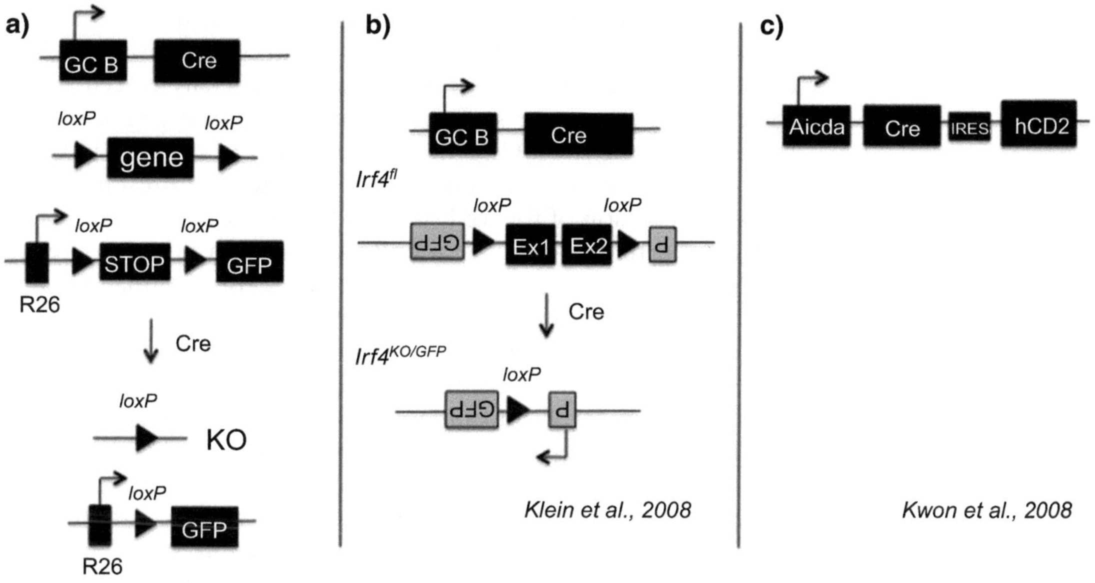

Fig. 1 Monitoring Cre-mediated recombination in vivo. Schematic view of alternative approaches to monitor Cre activity/expression in GC B cells. (**a**) Compound mutants carry a loxP-flanked candidate gene together with a reporter gene, which is induced by Cre-mediated recombination; (**b**) Cre-mediated deletion of the floxed allele triggers the expression from the same locus of a reporter gene; (**c**) Cre activity in GC B cells is detected through the expression of a reporter gene inserted in the same trangenic construct

Table 2
Table summarizing examples of conditional reporter alleles currently available

Mouse strain	Reference
Gt(ROSA)26Sor$^{tm9(CAG\text{-}tdTomato)Hze}$	[24]
Gt(ROSA)26Sor$^{tm6(CAG\text{-}ZsGreen1)Hze}$	[24]
Gt(ROSA)26Sor$^{tm4(ACTB\text{-}tdTomato,\text{-}EGFP)Luo}$	[25]
Gt(ROSA)26Sor$^{tm1.1(CAG\text{-}tdTomato,\text{-}EGFP)Pjen}$	[26]
Tg(CAG-Bgeo/GFP)21Lbe	[27]
Gt(ROSA)26Sor$^{tm1(EYFP)Cos}$	[28]
Gt(ROSA)26Sor$^{tm13(CAG\text{-}MYC,\text{-}CD2^{*})Rsky}$	[15]
Gt(ROSA)26Sor$^{tm1(CAG\text{-}Brainbow2.1)Cle}$	[29]

copies of genes coding for fluorescent or functionally inert surface markers that are preceded by a loxP-flanked transcriptional and translational termination sequence (also defined as STOP cassette; a list of the most popular conditional reporter genes is shown in Table 2; *see* **Note 1**). In compound mutants, conditional reporter alleles are combined to the GC B cell Cre transgene and to loxP-flanked alleles of the candidate gene through consecutive rounds of breeding. Conditional reporter genes are generally placed under the control of promoters of ubiquitously expressed genes, such as the *Rosa26* gene [30]. Cells that express Cre recombinase in GC B cells activate the expression of the reporter gene as a result of Cre-mediated excision of the loxP-flanked STOP sequence. Expression of the reporter gene can be monitored through flow cytometric, immunohistochemical, immunofluorescence, or live imaging analyses. This approach ensures the monitoring of B cells undergoing Cre-mediated recombination in the GC, and, at the same time the tracking of their progeny within and outside the GC reaction, including memory B and plasma cells [6]. Another approach to monitor Cre-mediated recombination is based on the tracking of Cre-recombined alleles through de novo expression from the targeted locus of a reporter gene. This approach was successfully used to monitor *Irf4* mutant GC B cells and their post-GC progeny [12]. (2) *Use of reporter of Cre transgene expression*: this approach was employed by Busslinger and colleagues to construct an Aicda-cre BAC transgenic line in which the expression of the recombinase is combined to that of a signaling-incompetent human CD2 receptor through an IRES sequence [19].

1.3 Leakiness of GC B Cell Cre Mouse Lines

Expression of Cre recombinase alleles from cell-type and stage-specific promoters does not exempt knockin lines from showing in some instances unexpected patterns of Cre-mediated recombination. Indeed, one common feature of several Cre transgenic lines is the transient expression of the recombinase in male and/or female germ cells. This condition may lead to Cre-mediated recombination of co-segregating loxP-flanked genes in germ cells, leading to the transmission of mutant alleles through the germ line. This phenomenon is thought to occur as a result of transient transcriptional activation of tissue specific genes in germ cells during meiosis. Cγ1-cre belongs to the transgenes that are occasionally expressed in oocytes and less frequently in sperm cells [6]. Hence, to avoid unwanted recombination events, it is recommended to ensure transmission of the Cγ1-cre allele together with the loxP-flanked gene through the male germ line. Germ cell leakiness of Cre expression (including parental bias) has yet to be reported for Aicda-cre knockin and BAC transgenic lines. Based on these considerations, genotyping for germ line Cre-recombined alleles is recommended before analysing compound experimental mice (e.g., carrying two floxed alleles in combination with the Cre transgene).

Finally, to determine the most appropriate Cre line to use, the chromosome location of the transgene should be also carefully considered. Indeed, mapping of the Cre transgene and the floxed allele to the same chromosome may render the analysis of compound mutants extremely time consuming, given the high numbers of littermates to screen to obtain experimental mice (e.g., gene$^{flox/flox}$; Cre).

1.4 Choosing Controls for Conditional GC B Cell Studies

A particular attention should be given to the choice of control animals for conditional gene targeting experiments. The selection for the best type of controls benefits from possible preexisting information on the function of the gene of interest. (1) Expression of Cre recombinase has been reported to exert toxic effects in different cell types including B cells. Hence, control mice will strictly require the presence of the Cre transgene. To the best of our knowledge, heterozygous Cγ1-cre mice do not suffer from defects in the formation of normal sized GCs and in the number of memory B and plasma cells. (2) Pre-existing experimental evidence pointing to possible biological effects resulting from haploinsufficiency of the gene of interest should prevent the use as controls of mice heterozygous for the loxP-flanked allele and carrying the Cre transgene. In these instances, mice inheriting only the Cre transgene represent the best control. Mice carrying only loxP-flanked alleles should be omitted from the control group, as they are uninformative. (3) The genetic background of experimental and control mice should be matched. Cγ1-cre mice are available through The Jackson laboratory on both the BALB/c and C57BL/6J genetic backgrounds. If experiments are carried out on a mixed genetic background, Cγ1-cre-only and/or Cγ1-cre; gene$^{flox/+}$ controls should be generated from breeding pairs that are on a similar mixed genetic background than those giving rise to experimental animals.

1.5 Studying Candidate Cancer Genes in GC B Cells

GC B cells represent a common target of malignant cell transformation. This has led to a strong interest in developing mouse models of human GC-derived B cell lymphomas to understand the mechanisms and identify the determinants of lymphoma initiation and/or progression, as well as to test in vivo novel therapeutics. Conditional gene targeting offers the opportunity to modulate expression/activity of candidate proto-oncogenes and/or tumor suppressor genes selectively in GC B cells mimicking somatic genetic alterations that are acquired in the course of lymphomagenesis. For instance, Cγ1-cre mice have been used to support a role for *Blimp-1* and *Kmt2d* as GC B cell tumor suppressors [9, 14]. At the same time, constitutive activation in GC B cells of the *c-MYC* proto-oncogene together with an active form of PI3KCA from the *Rosa26* locus has led to the development of the first GC-derived human Burkitt lymphoma model [16].

The use of Cγ1-cre or any of the Aicda-cre transgenes to modulate the expression/activity of cancer genes has potential pitfalls to be carefully considered. Leakiness of Cre expression in rare cells belonging to other lineages may lead to dysregulation of candidate cancer genes in unwanted cell types, tissues, and organs. This condition may drive the development of cancers targeting non GC B cells before the possible appearance of lymphomas of GC origin [16, 31]. This problem was recently overcome reconstituting lethally irradiated animals with bone marrow of animals carrying the Cγ1-cre allele in combination with Cre-inducible proto-oncogenes [16].

1.6 Accelerating GC B Cell Gene Targeting Experiments In Vivo

Conditional gene targeting experiments are time consuming and very costly. A substantial breeding effort is required to generate experimental mice (6–8 months are required to generate a sizeable cohort of animals homozygous for the loxP-flanked gene and carrying the Cre transgene), especially if additional reporter alleles are included in the experimental design. The size of the cohort of control and experimental mice to reach statistically sound results represents a further factor that can considerably extend the time interval before experiments are started. Finally, mice have to reach full maturation of their immune system before GC B cell responses can be thoroughly analyzed. Analysis of GC B cells from compound mutants can be substantially accelerated through bone marrow reconstitution analyses. Briefly a small cohort (4–6 mice/group) of experimental and control animals, is used as donor of bone marrow-derived hematopoietic stem cells to reconstitute a large cohort (at least 10 animals/donor) of lethally irradiated syngenic animals. Reconstituted animals are allowed to recover from the transplantation and ultimately used for immunization experiments to study GC responses and/or to determine the occurrence of GC-derived diseases, such as autoimmune disorders or lymphomas. To reduce possible long-term toxic effects of lethal irradiation, sublethal irradiation of recipient animals may be considered. In the latter case, B cell deficient mice, including μMT and JHT mice, should be considered as bone marrow recipients.

This chapter outlines protocols to study the effects of conditional gene manipulation on GC B cells numbers and function.

2 Materials

2.1 Buffers and Media

1. Tail lysis buffer: 100 mM Tris–HCl pH 8.5, 5 mM EDTA, 200 mM NaCl, and 0.2% SDS.
2. B cell medium: DMEM, 10% fetal calf serum, 2 mM glutamine, 1 mM nonessential amino acids, 50 μM β-mercaptoethanol, and 1 mM sodium pyruvate.
3. Erythrocytes lysis buffer: mix solutions A (0.17 M Tris pH 7.6) and B (0.83% NH_4Cl) at 1:9 ratio (v/v).
4. FACS buffer: 1× PBS, 1% BSA, and 0.05% NaN_3.
5. MACS buffer: 1× PBS, 0.5% BSA, and 2 mM EDTA.
6. SDS buffer: 0.5% SDS, 50 mM Tris–HCl pH 8.0, 5 mM EDTA pH 8.0, 100 mM NaCl, and 0.02% NaN_3.
7. Triton buffer: 5% Triton X-100, 100 mM Tris–HCl pH 8.0, 5 mM EDTA pH 8.0, 100 mM NaCl, and 0.02% NaN_3.

8. Immunoprecipitation (IP) buffer: Add SDS buffer to Triton buffer at 2:1 ratio (v/v).
9. Protease Inhibitor cocktail set III, EDTA-free (Calbiochem).
10. 150 mM Wash Buffer: 1% Triton-X 100; 150 mM NaCl; 20 mM Tris–HCl pH 8; 0.1% SDS; 2 mM EDTA pH 8.
11. 500 mM Wash Buffer: 1% Triton X-100; 500 mM NaCl; 20 mM Tris–HCl pH 8; 0.1% SDS; 2 mM EDTA pH 8.
12. De-cross-linking buffer: 1% SDS and 0.1 M $NaHCO_3$.
13. 10× Phosphate-buffered saline (PBS).
14. Formaldehyde.
15. Ethylenediaminetetraacetic acid (EDTA).
16. NaCl.
17. Tris–HCl pH 6.5.
18. 50 × Tris-acetate EDTA (TAE).
19. Seakem® GTG® Agarose

2.2 Immunogens and Adjuvants

1. Defribinated sheep red blood cells.
2. 4-Hydroxy-3-nitrophenylacetyl (NP) coupled to chicken gamma globulin (CGG) at a 25:1 ratio (Biosearch Technologies).
3. Alumn (Imject® Alum, aqueous solution of aluminum hydroxide (40 mg/mL) and magnesium hydroxide (40 mg/mL), Pierce) mixed with $NP_{25}CGG$ (4-hydroxy-3-nitrophenylacetyl hapten conjugated to Chicken gamma globulin, Biosearch Technologies).

2.3 Molecular Biology and Immunoprecipitation

1. Gapdh primers for genomic DNA analysis: Gapdh_Fw: 5′-AGCGCTGACCTTGAGGTCTCCTTG-3′; Gapdh_Rv: 5′-GTTGCCTACGCAGGTCTTGCTGAC-3′.
2. SHM analysis: V_H186.2Fw_Ex: 5′-TCTTTACAGTTACT-GAGCACACAGGAC-3′; V_H186.2Fw_In: 5′-CAGTAGCAG GCTTGAGGTCTGGAC-3′; J_{H2}Rv: 5′-GGGTCTAGAGG TGTCCCTAGTCCTTCATGACC-3′.
3. pGEM®-T easy kit (Promega).
4. Real-time PCR machine (LightCycler480, Roche).
5. Protein-G bound magnetic beads (ThermoFisher).
6. Magnetic support for 1.5-mL polypropylene tubes.
7. Proteinase K.
8. RNase A.

2.4 Cell Staining, Separation, and Analysis

1. Peanut agglutinin (PNA), Vector Laboratories.
2. Anti-mouse FAS/CD95 (Clone Jo2).
3. Anti-mouse CD19 (clone eBio1D3).
4. Anti-mouse CD38 (clone 90).
5. Biotin conjugated anti-mouse CD43 (clone S7).
6. Biotin conjugated anti-mouse CD11b (clone M1/70).
7. Biotin conjugated anti-mouse CD11c (clone N418).
8. Biotin conjugated anti-mouse Ter-119 (clone Ter-119).
9. Biotin conjugated anti-mouse CD8 (clone 53-6.7).
10. Biotin conjugated anti-mouse CD4 (clone GK1.5).
11. Biotin conjugated anti-mouse CD3e (clone 145-2c11).
12. FITC-conjugated Streptavidin.
13. Anti-biotin coupled magnetic microbeads.
14. Cell magnetic separation system.
15. Reagents to detect active caspase-3, e.g., CASPGLOW.
16. 5 mL polystyrene round-bottom FACS sorting tube.
17. Frosted glass slides.
18. 70 μm, 40 μm nylon mesh and filters.
19. FACS instrument equipped with blue argon (488 nm) and red diode lasers (635 nm).

3 Methods

3.1 Genotyping of Conditional Mutant Mice

1. Lyse 1 cm of caudal biopsy in 400 μL of tail lysis buffer containing Proteinase K (100 μg/mL) overnight at 56 °C, shaking in a thermomixer.
2. Centrifuge lysed material for 1 min at top speed in a tabletop centrifuge and transfer it into a fresh 1.5 mL tube.
3. Precipitate genomic DNA adding 1 mL of isopropanol, mix by inversion and centrifuge at top speed for 1 min in a tabletop centrifuge.
4. Air-dry DNA pellet to remove residual traces of isopropanol and resuspend in 300 μL of milliQ water.
5. Use 1 μL of genomic DNA solution to perform polymerase chain reaction (PCR) analysis using primer pairs to discriminate wild-type from floxed and Cre-recombined alleles (*see* **Note 2**).

3.2 T Cell Dependent Immunization with Sheep Red Blood Cells

1. To immunize animals with Sheep Red Blood Cells (SRBC) take 2–5 mL of SRBC from the original stock depending on the number of mice that will be immunized.

2. Dilute blood in 10 volumes of sterile 1× PBS. Centrifuge at 300 × *g* for 10 min at 4 °C.
3. Repeat wash and centrifuge three times more.
4. After the third wash take an aliquot and count cells. Centrifuge at 300 × *g* for 10 min at 4 °C.
5. Resuspend SRBC in sterile 1× PBS at a concentration of 2×10^9/mL.
6. Inject intraperitoneally 100 μL/mouse (2×10^8 cells).
7. Collect the spleens of immunized animals 8 days after immunization.

3.3 T Cell-Dependent Immunization with NP-CGG

1. To immunize animals with NP-CGG mix alum-based Imject® (*see* **Note 3**) and NP_{25}CGG at a 1:1 ratio (v/v). For primary immune responses use 50–100 μg NP_{25}CGG per mouse (*see* **Note 4**).
2. Incubate for at least 1 h at 37 °C, shaking in a thermomixer.
3. Inject intraperitoneally 100 μL of 1 mg/mL NP_{25}CGG/alum.
4. To analyze the peak of the GC response, sacrifice animals 12 days after immunization.

3.4 Preparation of Single Cell Suspensions from Secondary Lymphoid Organs

In immunized mice, GC B cells can be isolated from gut-associated lymphoid organs (GALT, including Peyer's patches (PP) and mesenteric lymph nodes (MLN)). Upon active immunization, multiple lymphoid organs including spleen and lymph nodes can be isolated to analyze GC responses. Active immunization allows for synchronization of GC responses in secondary lymphoid organs. Instead, GALT GCs are formed chronically as a result of the continuous stimulation of local B cells by the intestinal microflora.

1. Sacrifice immunized animals and isolate the spleen and, possibly, other lymphoid organs (*see* **Notes 5** and **6**).
2. Collect organs in DMEM, 10% FBS
3. Smash organs between two frosted glass slides.
4. Filter the cell suspension through 70 μm nylon filters.
5. Pellet cellular suspensions by centrifugation at 300 × *g* for 10 min.
6. Resuspend cells in 1 mL/spleen of erythrocyte lysis buffer and incubate the samples for 3 min on ice (*see* **Note 7**).
7. Stop the lysis by adding 10 volumes of B cell medium.
8. Count cells.
9. Resuspend cell pellet at 10^7 cells/mL in 1× PBS.

3.5 Flow Cytometric Analysis of GC B Cells

1. Transfer 100 μL of cell suspension (10^6 cells) into the well of a 96 well plate.
2. Centrifuge for 5 min at 300 × *g*, at 4 °C.
3. Aspirate carefully the supernatant and resuspend cells in 50 μL of FACS buffer containing a mixture of antibodies against PNA, FAS/CD95, CD19, and CD38 (*see* **Note 8**).
4. Mix well and incubate for 20 min at 4 °C in a light-protected environment.
5. Stop the staining reaction diluting the sample with 150 μL of FACS buffer.
6. Spin the 96-well plate for 30 s at 1610 × *g* at 4 °C
7. Repeat **step 5** for two more times.
8. Resuspend cell pellet in 180 μL FACS buffer.
9. Transfer cells into 1.2-mL FACS collection polystyrene tubes.
10. Dilute the sample further with 220 μL of FACS buffer.
11. Acquire samples on an FACS instrument equipped with blue argon (488 nm) and red diode lasers (635 nm).

3.6 Purification of GC B Cells by Cell Sorting (See Note 9)

1. Stain single cell suspensions resuspended in 1× PBS with 100 μL of antibody staining solution/10^7 cells containing a combination of biotin-conjugated monoclonal antibodies against CD43, CD11b, CD11c, Ter-119, CD8, CD4, and CD3e (*see* **Note 8**).
2. Mix well and incubate for 20 min at 4 °C.
3. Add 10 volumes of degassed MACS buffer to stop the staining reaction.
4. Centrifuge for 5 min at 300 × *g*, at 4 °C.
5. Remove supernatant and resuspend cell pellet in 100 μL/10^7 cells of MACS buffer containing anti-biotin coupled microbeads at a 1:10 ratio (v/v).
6. Mix well and incubate for 20 min at 4 °C.
7. Stop the reaction by diluting the cell suspension with 10 mL of ice-cold degassed MACS buffer
8. Centrifuge sample for 10 min at 300 × *g*, at 4 °C.
9. Resuspend cell pellet up to 10^8 cells/500 μL of degassed MACS buffer.
10. Load LS MACS column with 500 μL of cell suspension.
11. Wash LS MACS column with 3 mL ice-cold degassed MACS buffer
12. Repeat **step 11** three times.
13. Collect the flow-through, containing the B cell fraction.
14. Count the cells.

15. Stain 5×10^5 cells in 50 μL FACS buffer containing antibodies against the pan-B cell marker CD19 for 20 min at 4 °C, in a light-protected ice-bucket.
16. Stop the reaction by adding 10 volumes of FACS buffer, centrifuge cells at 300 × *g*, for 5 min at 4 °C.
17. Resuspend cells in 200 μL FACS buffer.
18. Acquire samples on a FACS instrument to determine if purity of the MACS-purified B cell population is higher than 90% in order to proceed with the staining of the remaining population of purified B cells for cell sorting.
19. Stain purified B cells (10^8 cells/mL) in MACS buffer containing a mixture of antibodies against, respectively, FAS/CD95, CD19, and CD38 (*see* **Notes 10** and **11**).
20. Mix well and incubate for 20 min at 4 °C in a light-protected environment.
21. Stop the staining by diluting the sample with 10 mL of ice-cold MACS buffer.
22. Centrifuge for 10 min at 300 × *g* at 4 °C.
23. Resuspend the cell pellet in 3–5 mL of ice-cold MACS buffer.
24. Filter sample through a 40 μm nylon mesh.
25. Collect cells in a 5 mL polystyrene round-bottom FACS sorting tube.

3.7 Assessing the Efficiency of Cre-Mediated Recombination in GC B Cells by Detection of a Fluorescent Cre Reporter Gene

1. Resuspend single cell suspensions obtained from lymphoid organs at 10^7 cells/mL.
2. Transfer 100 μL of cell suspension (10^6 cells) into the well of a 96 well plate.
3. Centrifuge for 5 min at 300 × *g*, at 4 °C.
4. Aspirate carefully the supernatant and resuspend cells in 20 μL of FACS staining solution containing antibodies against FAS/CD95, CD19, and CD38 (*see* **Note 12**).
5. Mix well and incubate for 20 min at 4 °C in a light-protected environment.
6. Stop staining reaction diluting the sample with 170 μL of FACS buffer.
7. Centrifuge for 30 s at 1610 × *g* at 4 °C; in a Beckman centrifuge (5810R) equipped with rotor accommodating 96-well plate adapters.
8. Repeat **steps 5** and **6** for two more times.
9. Resuspend cell pellet in 200 μL FACS buffer, transfer cells into FACS collection tubes (5-mL Polystyrene round-bottom tube, BD) and dilute the sample further with 200 μL of FACS buffer.
10. Acquire sample on a FACSCalibur instrument.

3.8 Assessing the Efficiency of Cre-Mediated Recombination in GC B Cells by Quantification of loxP-Flanked Alleles in GC B Cells

1. Genomic DNA from freshly sorted GC B cells is extracted by resuspending the cell pellet in 50 μL/10^5 cells of 10 mM Tris–HCl complemented with freshly added proteinase K (0.5 μg/mL).
2. Incubate for 2.5 h at 50 °C.
3. Heat-inactivate proteinase K incubating the sample at 95 °C for 10 min in an thermomixer [32] (*see* **Note 13**).
4. Use 1 μL of the DNA solution as template to perform quantitative PCR, using primers annealing both within the genomic region flanked by loxP sites.
5. Normalize DNA input performing a second qPCR analysis on a reference gene.
6. Set up qPCR reactions using primers at a final concentration of 250 nM.
7. Add to the mixture 10 μL of LightCycler® 480 SYBR Green I Master mix containing Hot-start polymerase (Roche), genomic DNA (1–5 μL), and water to a final volume of 20 μL.
8. The following PCR conditions represent a starting point for setting up the best condition of amplification for the candidate gene: initial denaturation at 95 °C for 10 min followed by 45 cycles at 95 °C for 10 s, 60 °C for 10 s, and 72 °C for 10 s. Final extension is carried out at 72 °C for 10 min. Similar conditions are used to amplify a reference gene, such as *Gapdh*, using primers listed in materials.
9. The efficiency of Cre-mediated recombination is calculated quantifying the amount of residual floxed gene present in mutant GC B cells. To normalize for DNA input, the same DNA samples are subjected to qPCR for a reference gene (*Gapdh*).

3.9 Assessment of Expression of the Floxed Gene in Mutant GC B Cells

1. Extract total RNA from purified GC B cells (from 5×10^4 to 1×10^5 cells), using the AllPrep DNA/RNA micro-extraction kit (Qiagen), following the manufacturer's instructions.
2. The entire amount of total RNA extracted from 10^5 cell equivalent is reverse transcribed using the Reverse Transcriptase SuperScript II kit (Life Technologies) according to the manufacturer's instructions.
3. Perform Real Time PCR (qRT-PCR) adding primers at a final concentration of 250 nM to 10 μL of LightCycler 480 SYBR Green I Master mix containing Hot polymerase (Roche) and 10–15 ng of cDNA in a final volume of 20 μL per reaction, in a 96-well plate (Roche);
4. PCR conditions are described in Subheading 3.8.

3.10 Analysis of GC B Cell Apoptosis in Conditional Mice

1. Prepare single cell suspensions from lymphoid organs of immunized mice as described in Subheading 3.4.
2. Stain 1×10^6 spleen cells using antibodies against GC B cell markers as described in Subheading 3.5.
3. Stained cells are washed and centrifuged as described in Subheading 3.6.
4. Resuspend cell pellet in 100 μL/10^6 cells of B cell medium containing 1 μL of the cell-permeable pan caspase inhibitor FITC-VAD-FMK (CaspGLOW™ Fluorescein Active Caspase Staining Kit, BioVision) according to the manufacturer's instructions.
5. Incubate samples for 1 h at 37 °C/5% CO_2 in a cell culture incubator.
6. Stop the reaction adding 500 μL of CaspGLOW™ wash buffer and centrifuge the cells at 300 × *g* for 5 min.
7. Resuspend cells in 400 μL FACS buffer and acquire sample on a FACS instrument.

3.11 Assessing In Vivo Proliferation of GC B Cells

1. Inject immunized mice intraperitoneally with the BrdU analog EdU (200 μg in 100 μL 1× PBS) 1.5–2 h before their sacrifice.
2. Prepare single cell suspension from spleen and possibly other lymphoid organs, as described in Subheading 3.4.
3. Perform first surface staining for GC B cells using antibodies against FAS/CD95, CD19, and biotin CD38, in combination with FITC-conjugated Streptavidin.
4. Centrifuge the cells at 300 × *g* for 5 min.
5. Resuspend in 50 μL of Click-iT EdU reaction buffer (Click-iT™ EdU Alexa Fluor® 647, Life Technologies).
6. Incubate for 15 min at 4 °C, in a light-protected environment.
7. Wash the cells with 100 μL of FACS buffer.
8. Centrifuge for 30 s at 1610 × *g* at 4 °C; in a Beckman centrifuge (5810R) equipped with rotor accommodating 96-well plate adapters.
9. Resuspend the cells in 100 μL of 1× Click-iT® permeabilization and wash buffer.
10. Incubate for 15 min at 4 °C, light-protected.
11. Pellet the cells and incubate them for 30 min at RT, light-protected, with Click-iT® reaction mix, following the manufacturer's instructions.
12. Stop the reaction with 100 μL of 1× Click-iT® permeabilization and wash buffer.
13. Perform two final washes in FACS buffer prior to the acquisition of samples on a FACS instrument.

3.12 Analysis of Ig Somatic Hypermutation

This protocol can be applied to the analysis of conditional mutant mice immunized with the hapten–carrier conjugate NP-CGG, in alum.

1. Purify GC B cells from day 12–14 NP_{25-32}-CGG immunized mice and extract genomic DNA as described in Subheading 3.8.
2. Genomic DNA from 10^5 cell equivalents is used as template for first round genomic PCR to amplify the V_H186.2 gene using the V_H186.2Fw_Ex and J_{H2}Rv and Phusion® High-Fidelity Taq Polymerase (New England Biolabs).
3. Five microliters of the first round PCR reaction are used as template for a second-round seminested PCR reaction (final volume 50 μL) using a V_H186.2Fw_In primer and the same J_H2Rv [33].
4. PCR product tailing reaction is performed incubating the second-round PCR product in the presence of fresh Taq polymerase (1 unit/sample; Go Taq, Promega), $MgCl_2$ (25 mM) and dNTP 10 mM for 30 min at 72 °C.
5. PCR product is run on a 1× TAE; 1% agarose gel, excised and purified using the PCR Gel Extraction Kit.
6. Purified PCR products are cloned in the pGEM®-T Easy Vector (Promega).
7. Mini-preparation of plasmid DNA extracted from 20–40 individual colonies is subjected to Sanger sequencing using primers adjacent to the cloning site (*see* **Note 14**).
8. Identify unique V_H186.2 rearrangements analyzing the CDR3 sequence.
9. Quantify the number of mutations collected in unique V_H186.2 rearrangements (excluding CDR3 region) and divide it by the total number of nucleotides sequenced (contributing to the unique rearrangements) to obtain the mutation frequency. For clonal rearrangements (e.g., sharing the same CDR3 sequence) each mutation is counted only once.
10. To assess antibody affinity maturation, the frequency of replacement mutations leading to a tryptophan to leucine substitution at codon 33 of V_H186.2 is determined (*see* **Note 15**).

3.13 Analysis of Genome Wide Distribution of Histone Marks by Chromatin Immunoprecipitation in GC B Cells

1. Collect spleens and prepare single cell suspensions from mice immunized 8 days earlier with SRBC as described in Subheadings 3.2 and 3.4 (*see* **Note 16**).
2. Purify GC B cells by cell sorting as described in Subheading 3.6.
3. Pool sorted GC B cells from independent mice of the same genotype.

4. Centrifuge cells in a tabletop centrifuge for 5 min at 3300 × *g*, at 4 °C.
5. Transfer cells into a 15 mL falcon tube and add 10 mL/10^7 cells of freshly prepared 1% formaldehyde in 1× PBS.
6. Fix cells at RT for 10 min while gently rotating (30 rpm).
7. Stop fixation by adding glycine to a final concentration of 0.125 M (1 mL of 1.37 M/10^7 cells).
8. Incubate the samples for 5 min, at RT, on rotation (30 rpm). *Perform the following steps on ice.*
9. Centrifuge samples for 5 min at 300 × *g*, at 4 °C (*see* **Note 17**).
10. Wash once with 1 mL of ice-cold 1× PBS.
11. Transfer cells onto a new 1.5 mL tube.
12. Centrifuge for 5 min at 3300 × *g* in a tabletop centrifuge, at 4 °C.
13. Resuspend samples in 1 mL SDS buffer plus 1 mM PMSF and 5 mM of protease inhibitors/10^7 cells.
14. Store fixed cells for at least one overnight at −80 °C.
15. Gradually thaw the sample in a water bath.
16. Centrifuge for 10 min at 1500 × *g*, at 15 °C.
17. Resuspend in 1 mL immunoprecipitation buffer (IP)/2 × 10^7 cells, containing 2 volumes of SDS buffer and 1 volume of Triton buffer with final 1 mM PMSF and 5 mM protease inhibitors.
18. Sonicate multiple 130 μL aliquots of each sample using Covaris microTUBEs in the S220 ultrasonicator (Covaris).
19. Set machine parameters to a duty factor of 10%, peak incidence power of 175 W and 200 cycles per burst for 150 s according to manufacturer's guidelines, to generate average size fragments of 200 bp.
20. Centrifuge sonicated chromatin at maximum speed (16,100 × *g*) at 4 °C for 10 min.
21. Transfer to a new 1.5 mL tube to remove cellular debris.
22. Dilute sonicated chromatin in 1 mL of IP buffer.
23. Remove 1% of the volume of sonicated material chromatin before immunoprecipitation (Input).
24. Immunoprecipitate chromatin overnight at 4 °C, on rotation by adding antibodies against candidate histone marks (*see* **Note 18**) to 1 mL of IP buffer containing tube.
25. Add 20 μL of protein-G bound magnetic beads/sample (Thermo Fisher), previously equilibrated with 1 mL of IP buffer, to tubes containing the chromatin/antibody mixture and incubate for 2–3 h at 4 °C on rotation.

26. Place tubes into an ice-cold magnetic support (Thermo Fisher).
27. Remove supernatant.
28. Tubes are taken out from the magnetic field and beads are resuspended in 1 mL ice-cold 150 mM Wash Buffer.
29. Tubes are placed back into the magnetic support and supernatant is removed.
30. Repeat twice the wash step described in **steps 25** and **26**.
31. Tubes are taken out from the magnetic field and histone-bound chromatin attached to the beads are resuspended in 1 mL ice-cold 500 mM Wash Buffer.
32. Repeat **step 26**.
33. Tubes are taken out from the magnetic field and beads are resuspended in 1 mL ice-cold 1× TE (10 mM Tris–HCl; 1 mM EDTA) buffer.
34. Repeat **step 26**.
35. Remove tubes from the magnet and resuspend beads in 100 μL of de-cross-linking buffer. De-cross-link also the input sample described in **step 20** adding 90 μL of de-cross-linking buffer.
36. Incubate for 1 h at 30 °C.
37. Immunoprecipitated samples are placed onto the magnetic support and supernatant transferred to a fresh 1.5-mL tube.
38. Add NaCl (to a final concentration of 0.2 M) and RNase A (to a final concentration of 25 μg/μL) to the de-cross-linked material and incubate overnight at 65 °C.
39. Stop the reaction adding EDTA (final10 mM); Tris–HCl pH 6.5 (final 40 mM) and Proteinase K (final 0.5 μg/μL).
40. Incubate for 1 h at 45 °C.
41. Purify DNA adding 600 μL of PB buffer (Qiagen) to the de-cross-linked material.
42. Load the mix onto Qiagen columns provided with the gel extraction kit and complete the purification following the manufacturer's protocol.
43. Elute DNA in 25 μL of milliQ water.
44. Quantify the DNA using Qubit Fluorometric quantitation (Thermo Fisher Scientific).
45. Perform gene-specific PCRs or send samples for deep sequencing.

4 Notes

1. A comprehensive list of Cre reporter lines can be obtained visiting the Jackson Laboratory website: https://www.jax.org/mouse-search/?

linkBuilder=1&stocktype=&diseaseterm=&straingroup=&researcharea=Research%20Tools&researcharea2=Cre-lox%20System&researcharea3=%26lt%3Bi%26gt%3BloxP%26lt%3B%2Fi%26gt%3B-flanked%20Sequences%3A%20Test%2FReporter&strainstatus=&genesymbol=&allelesymbol=&geneticbackground=

2. Detection by genomic PCR of the Cre-recombined allele in the DNA extracted from the tail biopsy of conditional mutant mice carrying the Cγ1-cre allele is highly suggestive of leaky Cre expression in germ cells or in the first stages of embryonic development. To detect Cre-recombined alleles, PCR primers are designed to anneal to genomic sequences outside the loxP-flanked region. Such strategy will give rise to (possibly) two PCR products of different size, with the smaller fragment representing the mutant allele.
3. Alternative alum formulations are available. A comparison of their property to boost T cell-dependent immunization is described in [34].
4. It is recommended to prepare an amount of Alum/NP_{25}CGG that slightly exceeds that necessary to immunize the predicted number of experimental mice. Recall responses are performed using 10–25 μg of NP_{25}CGG resuspended in 1× PBS and intravenously injected. The day of the analysis depends on the purpose of the experiment and on the type of immunogen used. GC nucleation is analyzed within the first 72–96 h after immunization. Molecular assessment of antibody affinity maturation is commonly studied at the peak of the GC response. For example, for NP_{25}-CGG immunization protocols, it is recommended to sacrifice 12–14 days after immunization.
5. Peyer's patches are isolated from the antimesenteric side of the small intestine. MLNs are embedded within the intestinal fat pad between the layers of the mesentery.
6. If numbers of GC B cells in PP is needed, collecting the same number of PPs is recommended.
7. Preparation of single cell suspensions from lymph nodes and PP does not require the step of erythrocytes lysis.
8. Optimal antibody concentration should be determined prior to the analysis in a pilot experiment.
9. GC B cells are highly susceptible to undergo apoptosis. Hence, if purification of GC B cells is needed to perform downstream molecular and/or biochemical analyses, it is recommended to minimize the time required to isolate such cells. Purification of GC B cells is commonly achieved by FACS-assisted cell sorting. To reduce the sorting time, single cell suspensions isolated from lymphoid organs of immunized mice can be subjected

to a pre-sorting B cell purification step. This can be achieved through magnetic cell sorting. A depletion protocol is recommended as it ensures the isolation of untouched B cells avoiding possible influences of staining antibodies with the activation/survival of the cells.

10. By binding to the Fas receptor, the anti-CD95/Fas Jo2 mAb can trigger apoptosis of Fas-expressing cells. Hence, it is recommended that cells stained with the Jo2 antibody are strictly kept on ice and quickly processed. If GC B cells require purification by cell sorting, preparing aliquots of cells that are freshly stained with Jo2 should be considered in order to minimize the sorting time. Alternatives to Fas detection to isolate GC B cells include peanut agglutinin (PNA) and GL-7.
11. The addition of the anti-CD23 antibody to the staining cocktail allows for concomitant sorting of, respectively, GC ($CD19^{+}CD38^{lo}CD95^{hi}$), Follicular ($CD19^{+}CD38^{+}CD23^{+}$), and MZ ($CD19^{+}CD23^{lo}CD38^{hi}$) B cells. In addition, the staining of B cells with anti-CXCR4 (2B11, eBioscience) and -CD86 (B7-2, eBioscience) antibodies ensures the discrimination of centroblasts from centrocytes [35].
12. If a fluorescent reporter allele has been included in the experimental design to monitor Cre-mediated recombination in GC B cells, the staining mixture to detect the latter cells should not contain antibodies coupled to fluorochromes that emit light at the same wave length as the fluorescent reporter.
13. Extraction of genomic DNA from sorted GC B cells can also be pursued with commercially available kits such as the AllPrep DNA/RNA microkit (Qiagen), which has the advantage to allow concomitant isolation of both genomic DNA and total RNA.
14. Sanger sequencing can be replaced by Next Generation Sequencing technology, which ensures the analysis of a much higher number of rearrangements at a substantially lower cost.
15. At the peak of the GC response (12–14 days post immunization) the expected frequency of rearrangements carrying the tryptophan to leucine substitution is around 50% [36].
16. To maximize the yield of GC B cells to be employed for chromatin immunoprecipitation studies five or more mice should be considered for immunization. Given their ability to trigger robust GC B cell responses, it is recommended to use sheep red blood cells as immunogen. GC B cells purified from individual mice are pooled before proceeding with ChIP experiments.
17. To minimize cell loss, fixed cells may be aliquoted into microfuge tubes and spun in an Eppendorf® tabletop centrifuge.

18. Due to its low sensitivity, the Bio-Rad DC™ protein quantification method is not recommended to accurately quantify chromatin extracted from limited numbers of GC B cells ($<5 \times 10^6$ cells). For 5×10^6 (or less) GC B cells, the entire sample is used for chromatin immunoprecipitation. For every 2.5×10^6 GC B cells, we use respectively 2.5 μg of anti-H3K4me3 (Ab8580, Abcam), 10 μg of anti-H3K27me3 (C36B11, Cell Signaling), and 2.5 μg of anti-H3K4me1 (Ab8895, Abcam) antibodies.

Acknowledgments

We are grateful to the many collaborators who have shared over the past 10 years their results generated with the Cγ1-cre strain. S. Casola wishes to thank the 1997–2001 members of the K. Rajewsky "Cre club" at the Institute of Genetics in Cologne for suggestions and help in the development of the Cγ1-cre strain. S. Casola and V. Petrocelli wish to thank members of the Casola group for helpful discussions. S. Casola is supported by the Italian Association for Cancer Research (AIRC), the CARIPLO Foundation and by the Telethon Foundation. V. Petrocelli is supported by the U. Veronesi Foundation.

References

1. Victora GD, Nussenzweig MC (2012) Germinal centers. Annu Rev Immunol 30:429–457. doi:10.1146/annurev-immunol-020711-075032
2. Rajewsky K (1993) Immunology. The power of clonal selection. Nature 363(6426):208. doi:10.1038/363208a0
3. Weisel FJ, Zuccarino-Catania GV, Chikina M, Shlomchik MJ (2016) A temporal switch in the germinal center determines differential output of memory B and plasma cells. Immunity 44 (1):116–130. doi:10.1016/j.immuni.2015.12.004
4. Victora GD, Wilson PC (2015) Germinal center selection and the antibody response to influenza. Cell 163(3):545–548. doi:10.1016/j.cell.2015.10.004
5. Basso K, Dalla-Favera R (2015) Germinal centres and B cell lymphomagenesis. Nat Rev Immunol 15(3):172–184. doi:10.1038/nri3814
6. Casola S, Cattoretti G, Uyttersprot N, Koralov SB, Seagal J, Hao Z, Waisman A, Egert A, Ghitza D, Rajewsky K (2006) Tracking germinal center B cells expressing germ-line immunoglobulin gamma1 transcripts by conditional gene targeting. Proc Natl Acad Sci U S A 103 (19):7396–7401. doi:10.1073/pnas.0602353103
7. Toellner KM, Gulbranson-Judge A, Taylor DR, Sze DM, MacLennan IC (1996) Immunoglobulin switch transcript production in vivo related to the site and time of antigen-specific B cell activation. J Exp Med 183(5):2303–2312
8. Caganova M, Carrisi C, Varano G, Mainoldi F, Zanardi F, Germain PL, George L, Alberghini F, Ferrarini L, Talukder AK, Ponzoni M, Testa G, Nojima T, Doglioni C, Kitamura D, Toellner KM, Su IH, Casola S (2013) Germinal center dysregulation by histone methyltransferase EZH2 promotes lymphomagenesis. J Clin Invest 123(12):5009–5022. doi:10.1172/JCI70626
9. Calado DP, Zhang B, Srinivasan L, Sasaki Y, Seagal J, Unitt C, Rodig S, Kutok J, Tarakhovsky A, Schmidt-Supprian M, Rajewsky K (2010) Constitutive canonical NF-kappaB activation cooperates with disruption of BLIMP1 in the pathogenesis of activated B cell-like diffuse large cell lymphoma. Cancer Cell 18 (6):580–589. doi:10.1016/j.ccr.2010.11.024

10. Hao Z, Duncan GS, Seagal J, Su YW, Hong C, Haight J, Chen NJ, Elia A, Wakeham A, Li WY, Liepa J, Wood GA, Casola S, Rajewsky K, Mak TW (2008) Fas receptor expression in germinal-center B cells is essential for T and B lymphocyte homeostasis. Immunity 29 (4):615–627. doi:10.1016/j.immuni.2008.07.016
11. Hikida M, Casola S, Takahashi N, Kaji T, Takemori T, Rajewsky K, Kurosaki T (2009) PLC-gamma2 is essential for formation and maintenance of memory B cells. J Exp Med 206 (3):681–689. doi:10.1084/jem.20082100
12. Klein U, Casola S, Cattoretti G, Shen Q, Lia M, Mo T, Ludwig T, Rajewsky K, Dalla-Favera R (2006) Transcription factor IRF4 controls plasma cell differentiation and class-switch recombination. Nat Immunol 7(7):773–782. doi:10.1038/ni1357
13. Sander S, Chu VT, Yasuda T, Franklin A, Graf R, Calado DP, Li S, Imami K, Selbach M, Di Virgilio M, Bullinger L, Rajewsky K (2015) PI3 kinase and FOXO1 transcription factor activity differentially control B cells in the germinal center light and dark zones. Immunity 43(6):1075–1086. doi:10.1016/j.immuni.2015.10.021
14. Zhang J, Dominguez-Sola D, Hussein S, Lee JE, Holmes AB, Bansal M, Vlasevska S, Mo T, Tang H, Basso K, Ge K, Dalla-Favera R, Pasqualucci L (2015) Disruption of KMT2D perturbs germinal center B cell development and promotes lymphomagenesis. Nat Med 21 (10):1190–1198. doi:10.1038/nm.3940
15. Calado DP, Sasaki Y, Godinho SA, Pellerin A, Kochert K, Sleckman BP, de Alboran IM, Janz M, Rodig S, Rajewsky K (2012) The cell-cycle regulator c-Myc is essential for the formation and maintenance of germinal centers. Nat Immunol 13(11):1092–1100. doi:10.1038/ni.2418
16. Sander S, Calado DP, Srinivasan L, Kochert K, Zhang B, Rosolowski M, Rodig SJ, Holzmann K, Stilgenbauer S, Siebert R, Bullinger L, Rajewsky K (2012) Synergy between PI3K signaling and MYC in Burkitt lymphomagenesis. Cancer Cell 22(2):167–179. doi:10.1016/j.ccr.2012.06.012
17. Crouch EE, Li Z, Takizawa M, Fichtner-Feigl S, Gourzi P, Montano C, Feigenbaum L, Wilson P, Janz S, Papavasiliou FN, Casellas R (2007) Regulation of AID expression in the immune response. J Exp Med 204 (5):1145–1156. doi:10.1084/jem.20061952
18. Dogan I, Bertocci B, Vilmont V, Delbos F, Megret J, Storck S, Reynaud CA, Weill JC (2009) Multiple layers of B cell memory with different effector functions. Nat Immunol 10 (12):1292–1299. doi:10.1038/ni.1814
19. Kwon K, Hutter C, Sun Q, Bilic I, Cobaleda C, Malin S, Busslinger M (2008) Instructive role of the transcription factor E2A in early B lymphopoiesis and germinal center B cell development. Immunity 28(6):751–762. doi:10.1016/j.immuni.2008.04.014
20. Robbiani DF, Bothmer A, Callen E, Reina-San-Martin B, Dorsett Y, Difilippantonio S, Bolland DJ, Chen HT, Corcoran AE, Nussenzweig A, Nussenzweig MC (2008) AID is required for the chromosomal breaks in c-myc that lead to c-myc/IgH translocations. Cell 135(6):1028–1038. doi:10.1016/j.cell.2008.09.062
21. Muramatsu M, Sankaranand VS, Anant S, Sugai M, Kinoshita K, Davidson NO, Honjo T (1999) Specific expression of activation-induced cytidine deaminase (AID), a novel member of the RNA-editing deaminase family in germinal center B cells. J Biol Chem 274 (26):18470–18476
22. Mao C, Jiang L, Melo-Jorge M, Puthenveetil M, Zhang X, Carroll MC, Imanishi-Kari T (2004) T cell-independent somatic hypermutation in murine B cells with an immature phenotype. Immunity 20(2):133–144
23. Béguelin W, Popovic R, Teater M, Jiang Y, Bunting KL, Rosen M, Shen H, Yang SN, Wang L, Ezponda T, Martinez-Garcia E, Zhang H, Zheng Y, Verma SK, McCabe MT, Ott HM, Van Aller GS, Kruger RG, Liu Y, McHugh CF, Scott DW, Chung YR, Kelleher N, Shaknovich R, Creasy CL, Gascoyne RD, Wong KK, Cerchietti L, Levine RL, Abdel-Wahab O, Licht JD, Elemento O, Melnick AM (2013) EZH2 is required for germinal center formation and somatic EZH2 mutations promote lymphoid transformation. Cancer Cell 23(5):677–692. doi:10.1016/j.ccr.2013.04.011
24. Madisen L, Zwingman TA, Sunkin SM, Oh SW, Zariwala HA, Gu H, Ng LL, Palmiter RD, Hawrylycz MJ, Jones AR, Lein ES, Zeng H (2010) A robust and high-throughput Cre reporting and characterization system for the whole mouse brain. Nat Neurosci 13 (1):133–140. doi:10.1038/nn.2467
25. Muzumdar MD, Tasic B, Miyamichi K, Li L, Luo L (2007) A global double-fluorescent Cre reporter mouse. Genesis 45(9):593–605. doi:10.1002/dvg.20335
26. Plummer NW, Evsyukova IY, Robertson SD, de Marchena J, Tucker CJ, Jensen P (2015) Expanding the power of recombinase-based labeling to uncover cellular diversity. Development 142(24):4385–4393. doi:10.1242/dev.129981

27. Novak A, Guo C, Yang W, Nagy A, Lobe CG (2000) Z/EG, a double reporter mouse line that expresses enhanced green fluorescent protein upon Cre-mediated excision. Genesis 28 (3–4):147–155
28. Srinivas S, Watanabe T, Lin CS, William CM, Tanabe Y, Jessell TM, Costantini F (2001) Cre reporter strains produced by targeted insertion of EYFP and ECFP into the ROSA26 locus. BMC Dev Biol 1:4
29. Snippert HJ, van der Flier LG, Sato T, van Es JH, van den Born M, Kroon-Veenboer C, Barker N, Klein AM, van Rheenen J, Simons BD, Clevers H (2010) Intestinal crypt homeostasis results from neutral competition between symmetrically dividing Lgr5 stem cells. Cell 143(1):134–144. doi:10.1016/j.cell.2010.09.016
30. Casola S (2010) Mouse models for miRNA expression: the ROSA26 locus. Methods Mol Biol 667:145–163. doi:10.1007/978-1-60761-811-9_10
31. Mullins CD, Su MY, Hucthagowder V, Chu L, Lu L, Kulkarni S, Novack D, Vij R, Tomasson MH (2013) Germinal center B-cells resist transformation by Kras independently of tumor suppressor Arf. PLoS One 8(6):e67941. doi:10.1371/journal.pone.0067941
32. Schenten D, Gerlach VL, Guo C, Velasco-Miguel S, Hladik CL, White CL, Friedberg EC, Rajewsky K, Esposito G (2002) DNA polymerase kappa deficiency does not affect somatic hypermutation in mice. Eur J Immunol 32(11):3152–3160. doi:10.1002/1521-4141(200211)32:11<3152::AID-IMMU3152>3.0.CO;2-2
33. Esposito G, Texido G, Betz UA, Gu H, Muller W, Klein U, Rajewsky K (2000) Mice reconstituted with DNA polymerase beta-deficient fetal liver cells are able to mount a T cell-dependent immune response and mutate their Ig genes normally. Proc Natl Acad Sci U S A 97(3):1166–1171
34. Cain DW, Sanders SE, Cunningham MM, Kelsoe G (2013) Disparate adjuvant properties among three formulations of "alum". Vaccine 31(4):653–660. doi:10.1016/j.vaccine.2012.11.044
35. Victora GD, Schwickert TA, Fooksman DR, Kamphorst AO, Meyer-Hermann M, Dustin ML, Nussenzweig MC (2010) Germinal center dynamics revealed by multiphoton microscopy with a photoactivatable fluorescent reporter. Cell 143(4):592–605. doi:10.1016/j.cell.2010.10.032
36. Weiss U, Rajewsky K (1990) The repertoire of somatic antibody mutants accumulating in the memory compartment after primary immunization is restricted through affinity maturation and mirrors that expressed in the secondary response. J Exp Med 172(6):1681–1689

Chapter 18

Development of Mouse Model Systems of Germinal Center Lymphomas

Eleni Kabrani and Sandrine Sander

Abstract

Germinal centers (GC) are the predominant origin of human B cell lymphomagenesis. Transgenic mice in which gene expression is altered specifically in GC B cells have broadened our knowledge about the mechanisms of malignant transformation. However, extensive resources are needed due to the genetic complexity of these mouse models. Thus, bone marrow (BM)-derived chimerism is an attractive approach to study GC B cell derived lymphomagenesis, as it allows for an efficient allocation of resources and reduces the number of animals used.

Key words Lymphomagenesis, Mouse model, BM chimerism, SHM assay

1 Introduction

The B cell receptor (BCR) is essential for B cell survival and function. Diversity in the BCR repertoire is achieved in early and late stages of B cell development: in the bone marrow (BM) the immunoglobulin (Ig) variable (V), joining (J), and diversity (D) gene segments of B cell progenitors are joined by site-specific recombination processes [1]. Upon antigen activation mature B cells expand in germinal centers (GC) and GC B cells introduce DNA mutations in their Ig genes by class switch recombination (CSR) and somatic hypermutation (SHM) [2, 3]. Both mechanisms require the recruitment of the enzyme AID (Activation Induced cytidine Deaminase) to rearranged V(D)J genes and switch regions of the Ig loci. At hotspot sites AID targets cytosine residues and produces uracils. During SHM DNA mismatches (U:G) can be repaired by the DNA replication machinery resulting in transition (C → T and G → A depending on the deaminated DNA strand) or transversion (C → A or G and G → C or T) substitutions. In contrast, CSR is based on DNA double strand breaks which are introduced by uracil excisions from neighboring U:G mismatches in selected gene regions.

Dinis Pedro Calado (ed.), *Germinal Centers: Methods and Protocols*, Methods in Molecular Biology, vol. 1623,
DOI 10.1007/978-1-4939-7095-7_18, © Springer Science+Business Media LLC 2017

Infidelity in both processes increases the risk for chromosomal translocations and somatic mutations in non-Ig genes [4]. These aberrations are frequently detected in GC B cell derived human lymphomas resulting in abnormal cell proliferation and death. By engineering mice in which oncogenic pathways are modulated, tumor promoting events can be recapitulated in vivo. Conditional approaches are highly favorable compared to global ones in the analyses of GC B cell transformation: restricted oncogene activation and/or tumor suppressor gene inactivation in the presumed cells of tumor origin circumvents the risk for malignant transformation of pre-GC B cells and non-lymphocytes.

Thus, the generation of preclinical lymphoma models benefits from loxP-mediated gene control and GC B cell specific expression of Cre recombinase (such as in the AID-cre and Cγ1-cre mouse strains [5, 6]). Due to the multistep process of cancerogenesis, complex intercrossings are needed to create sophisticated multi-allelic mouse strains, an undertaking typically associated with high costs and time consuming experimental settings.

BM based transfer systems are suitable tools to overcome these limitations. Large cohorts of tumor bearing mice are generated by BM reconstitution of irradiated recipients: at least ten animals can be transplanted with BM cells of a single adult mouse, thereby limiting the number of donor mice carrying complex genotypes. BM transfer in mice lacking lymphocytes due to Rag2 gene ablation is especially attractive to study lymphoma development as transplant derived B and T cells will not compete with the recipient's lymphocytes. In the absence of the interleukin 2 receptor gamma chain (Il2rg) gene and natural killer cells, donor derived lymphocytes repopulate secondary lymphoid organs after BM reconstitution of Rag2/Il2rg double-knockout animals and lymphocytes participate in antigen-dependent immune responses. In GC B cells Cre-mediated recombination allows the expression of donor specific genetic modifications thereby promoting B cell malignancies [7].

The analysis of the individual Ig gene rearrangement and its mutation load in tumors arising in chimeric animals is highly recommended because it allows to (1) discriminate between clonal processes and reactive polyclonal expansion of transgene expressing cells i.e., tumor clonality; (2) prove the GC B cell origin of the lymphoma cells, including SHM based aberrations; and (3) exclude the presence of preexisting tumor cells within the BM transplant; i.e., the occurrence of unique tumor VDJ rearrangements in various recipients of a single BM.

2 Materials

2.1 Isolation of Bone Marrow Cells

1. Forceps, scissors, mortar and pestle, or syringes (1 ml) and needles (29G × ½).
2. Cell strainers (50 or 70 μm mesh size).
3. Plastic disposables for tissue culture.
4. B cell medium: Dulbecco's Modified Eagle's Medium (DMEM) with 4.5 g/l D-glucose, without L-glutamine, without sodium pyruvate, 10% fetal calf serum (FCS), 1% penicillin–streptomycin (10.000 U/ml), 1% MEM, NEAA (100×), 1% HEPES (1 M), 1% sodium pyruvate 100 mM (100×), 1% L-glutamine 200 mM (100×), 0.05 M 2-ß-mercaptoethanol (prepare stock solution; *see* **Note 1**).
5. 70% ETOH.
6. Hemolytic Gey's solution: Prepare stock solutions A, B, and C according to *Selected Methods in Cellular Immunology* by B.B. Mishell and S.M. Shiigi [8]. Store the stocks at 4 °C and prepare the working solution immediately before usage: mix 7 ml water, 2 ml stock A, 0.5 ml stock B, and 0.5 ml stock C per 10 ml Gey's working solution.
7. Hemocytometer and Trypan Blue solution (or automated cell counter).
8. CD90.2 magnetic beads, magnet, and separation columns.
9. Optional: cryogenic vials and freezing medium: 10% DMSO in FCS, prepared freshly.

2.2 Irradiation of Recipient Animals

1. Neomycin: 20% stock solution in water. Strain the solution through a 0.2-μm filter to remove undissolved particles. Prepare freshly.
2. γ-Irradiation source (either ^{137}Cesium or ^{60}Cobalt or high-energy X-rays).
3. Acidified water (pH 2.5–2.8).

2.3 Bone Marrow Transplantation

1. Optional: water bath (37 °C).
2. B cell medium (*see* Subheading 2.1).
3. 1×PBS without Ca^{2+} and Mg^{2+}.
4. Cell strainers (50 or 70 μm mesh size).
5. Hemocytometer and Trypan Blue solution (or automated cell counter).
6. Syringes (1 ml) and needles (27–30G).

2.4 Molecular Biology

1. Reagents for genomic DNA isolation (phenol–chloroform extraction or mini-column based purification).
2. High-fidelity DNA polymerase (e.g., KOD Hot start DNA polymerase) and appropriate buffer (*see* **Note 2**).
3. Primers: listed primer sequences are suitable for SHM analysis of the heavy chain (H) in C57BL/6 animals [9]. Universal reverse primer: JH4 intronic: CTCCACCAGACCTCTCTAGACAGC; individual forward primers: VHA (J558/VH1): ARGCCT GGGRCTTCAGTGAAG, VHB (Q52/VH2): CTCACAGA GCCTGTCCATCAC, VHC (36–60/VH3): TCTCAGWCT CTGTCYCTCACC, VHD (X24/VH4): CTCGAGTCTGG AGGTGGCCTG, VHE (7183/VH5): GTGGAGTCTGGGG GAGGCTTA, VHF (J606/VH6, S107/VH7, X24/VH5): WCTGGAGGAGGCTTGGTGCAA, VHG (GAM3/VH9): GGAGAGACAGTCAAGATCTCC.
4. Agarose, ethidium bromide, running buffer 1× Tris–acetate–EDTA (TAE): 2.0 M Tris acetate, 0.05 M EDTA, pH 8.2–8.4 (at 25 °C); loading dye and DNA size marker.
5. Razor or scalpel blade.
6. DNA gel extraction kit.
7. Plasmid vector kit for blunt DNA cloning.
8. Chemically competent bacteria for plasmid transformation.
9. LB agar plates (petri dishes and 96-well plates) containing the proper antibiotic for selection.
10. Optional: isopropyl-β-D-thiogalactopyranoside (IPTG), 5-bromo-4-chloro-3-indolyl-beta-D-galactoside (X-Gal).
11. Incubator (37 °C) for bacteria.
12. Plasmid DNA isolation kit.
13. Primers for insert sequencing in recombinant plasmid (e.g., T3 and T7 primers).
14. Sequence alignment software.

3 Methods

Carry out all procedures in a laminar air flow, except for molecular biology experiments such as SHM analysis. Keep the cells always on ice at 4 °C. Use sterile material, including medium, disposables, and scissors, throughout the experiment.

3.1 Isolation of Bone Marrow Cells

1. Euthanize a 8–12-week-old donor mouse according to the institutional regulations. Isolate the femur and tibia bones (both legs) and collect the bones in a petri dish filled with B cell medium. Store the samples on ice.

2. Briefly soak the intact bones in 70% ETOH and transfer them to a new petri dish filled with B cell medium.
3. Isolate BM cells in B cell medium by flushing the marrow content or grinding the bones using a mortar and pestle (*see* **Note 3**). Prepare a single cell suspension by vigorously pipetting and straining the solution through a 50- or 70-μm filter.
4. Spin the cell suspension at 200 × *g* for 10 min (4 °C). Discard the supernatant and resuspend the cell pellet in 1 ml B cell medium (transfer the cell suspension in a 15 ml tube).
5. Lyse red blood cells (RBC) by adding 5 ml of Gey's working solution and incubate on ice for 3 min. Stop RBC lysis by filling up the 15-ml tube with B cell medium. Spin the cell suspension at 200 × *g* for 10 min (4 °C) and discard supernatant.
6. Wash cells once with 10 ml B cell medium. After spinning resuspend the cells in an appropriate amount of B cell medium and take out an aliquot for cell count analysis. Use a hemocytometer and Trypan Blue (or automated cell counter) to determine the number of living cells.
7. Enrich the stem cell population in the BM sample by T cell depletion using magnetic beads coupled to an antibody against the pan-T cell marker CD90.2 (*see* **Note 4**). Perform the depletion according to the manufacturer's protocol. Collect the cells on ice.
8. Spin the CD90.2 negative (= non-expressing) cells at 200 × *g* for 10 min (4 °C). Discard the supernatant, resuspend the cell pellet in B cell medium and take out an aliquot for cell count analysis.
9. BM cells can be used for injections into recipients or for long-term storage at N_2 after transferring them in freezing medium.

3.2 Irradiation of Recipient Animals

1. Dilute the neomycin stock solution 1:100 in the animals' acidified drinking water (0.2% final concentration). Start the neomycin treatment of animals 3 days before irradiation. The antibiotic should be added to the water for 14 days after irradiation. Refresh the water at least once per week (*see* **Note 5**).
2. Irradiate recipient animals (6–12 weeks of age) sublethally (5–6 Gy γ-irradiation).
3. Transfer BM cells 24 h after irradiation and monitor mice for signs of sickness in the subsequent weeks and months (*see* **Note 6**).

3.3 Bone Marrow Transplantation

1. Optional: thaw the frozen BM cells in a 37 °C water bath and wash the cells twice using 10 ml B cell medium for each washing step.
2. Spin the cell suspension at 200 × *g* for 10 min (4 °C) and discard the supernatant.

3. Add 10 ml 1×PBS, resuspend the cell pellet and filter the solution through a 50 or 70 μm mesh.
4. Spin the cells at 200 × *g* for 10 min (4 °C) and discard the supernatant.
5. Resuspend the cells in an appropriate amount of 1×PBS and take out an aliquot for cell count analysis (*see* **Note 7**). Use a hemocytometer and Trypan Blue (or automated cell counter) to determine the number of living cells.
6. Resuspend the cells in 1×PBS to obtain a final concentration of 5×10^6 cells/ml.
7. Inject 200 μl cell suspension per mouse (via tail vein or retro-orbital venous sinus; *see* **Note 8**).
8. Validate BM reconstitution around 6 weeks after transplantation by analyzing lymphocytes in the peripheral blood.
9. Monitor animal health for signs of tumor development.

3.4 DNA Isolation and PCR Amplification of VDJ Rearrangements

1. Prepare genomic DNA from (FACS sorted) B cell lymphomas using standard methods (phenol–chloroform extraction or mini-column based purification).
2. Identify the tumor specific VDJ rearrangement by PCR amplification using high-fidelity DNA polymerase (*see* **Note 9**).
3. Use 10 ng of genomic DNA per 25 μl PCR reaction for an explorative gel (1% agarose; *see* Fig. 1).
4. Include a single forward primer and the JH4 reverse primer per PCR reaction (*see* **Note 10**).
5. After identification of the tumor specific V gene, repeat the PCR reaction with the previously verified primer pair using 50 ng of genomic DNA in a 50 μl total reaction volume.
6. Load the PCR mix on a 1% agarose gel and include a DNA size marker.
7. Cut the band of interest out of the gel using a razor or scalpel blade (*see* **Note 11**).
8. Purify the DNA using a DNA gel extraction kit.

3.5 Clonality Assessment via SHM Analysis

1. Clone the blunt-end PCR products of the VDJ amplification into a plasmid vector for subsequent Sanger sequencing (*see* **Note 12**).
2. Transform competent cells with the recombinant plasmid and inoculate a LB agar filled petri dish containing the proper antibiotic.
3. Incubate at 37 °C overnight.
4. Pick single colonies; typically 30–40 colonies per tumor.

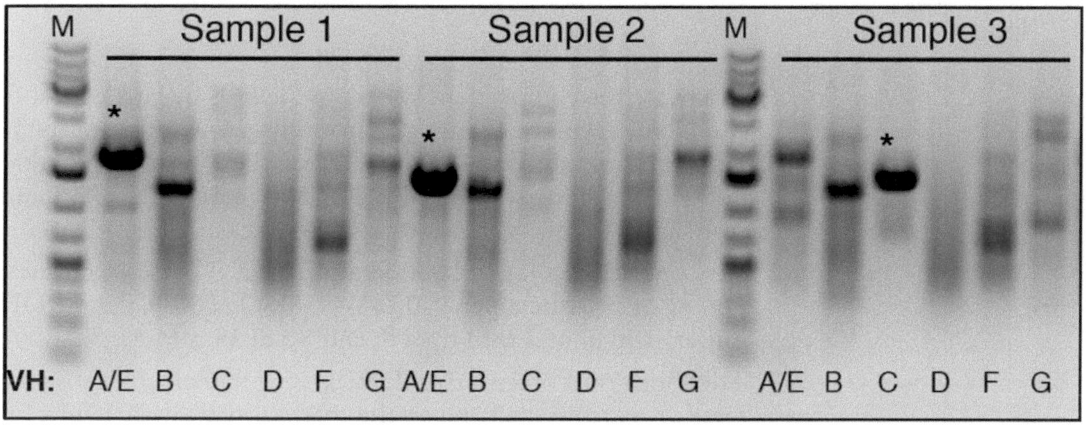

Fig. 1 Gel for V gene analysis. Example of explorative gel for V gene analysis in three individual mouse lymphomas (sample 1–3). Tumor specific B cell clones (*asterisk*) were detected in all samples. The VDJ assembly was subsequently confirmed by Sanger sequencing which also verified the non-Ig origin of other PCR products (= unspecific bands). Individual primer combinations are indicated by the VH primer usage. M = size marker (1 kb plus DNA ladder)

5. Inoculate 96-well plates filled with LB agar and antibiotic.
6. Incubate at 37 °C overnight.
7. Purify DNA using a plasmid DNA isolation kit.
8. Submit samples for Sanger sequencing (*see* **Note 13**).
9. Using a DNA sequence analysis software tool exclude sequences of poor quality in terms of for example signal strength, persistence, and overlapping peaks.
10. Analyze each sequence in IgBlast (http://www.ncbi.nlm.nih.gov/igblast/) in order to: (1) identify the tumor specific germ line VDJ segment; (2) discriminate between productive and nonproductive rearrangements (optional); and (3) calculate the mutation load, including replacement/silent mutation ratio, in the CDRs (complementarity determining regions) and FWRs (framework regions) (optional).
11. Align all sequences of an individual tumor with the control JH germ line sequence (C57BL/6 background) using a software which allows multiple alignments (*see* **Note 14**).
12. Cellular origin and tumor clonality is reflected in the mutation pattern, e.g., nucleotide replacements in (nearly) all sequences is indicative of malignant transformation of a single GC B cell (*see* **Notes 15–17**).

4 Notes

1. Inactivate the complement system in the FCS by heat exposure (56 °C for 30 min), aliquot and store at −20 °C. Aliquot the L-glutamine and store at −20 °C. Prepare the 2-ß-mercaptoethanol (ß-ME) stock solution: 70 μl ß-ME in 10 ml 1×PBS and store at 4 °C; renew the stock every 4 weeks.
2. KOD buffer (2×): per 1000 μl add: 20 μl DNAse-free water; 400 μl of 5 M betaine; 100 μl of DMSO; 200 μl 10× KOD buffer; 200 μl of 2 mM dNTP; and 80 μl 25 mM $MgSO_4$.
3. Trim both ends of femurs and tibias using scissors to expose the interior marrow shaft. Flush the marrow shaft with 1 ml of B cell medium using a 1-ml insulin syringe with a 29G × ½ needle. Repeat the procedure with additional B cell medium until the bones appear white. Collect the bone contents into a sterile 15 or 50-ml centrifuge tube.
4. In CD90.2 expressing mouse strains, hematopoietic stem cell (HSC) capacity is found in the CD90.2 low and negative HSC fraction [10]. Thus, lymphocyte reconstitution using 1 × 10e6 CD90.2 depleted BM cells is routinely achieved.
5. To reduce the bacterial contamination of water bottles and prevent Pseudomonas aeruginosa infections animals are routinely housed with autoclaved water acidified with hydrochloric acid (pH 2.5–2.8) [11]. Neomycin can be added to the water without reduction in its effectiveness.
6. Between day 5 and day 14 mice are prone to infections due to irradiation-mediated BM injury and myeloid suppression. During this time animal illness is indicative for improper sanitation procedures and not related to lymphoma development. Several weeks after irradiation the animal's fur color might change; typically black hair will turn gray.
7. Cells should not be stored for long periods of time in 1×PBS. Alternatively, cells may be kept in B cell medium during counting. Add 1×PBS to the maximum of the tube and spin the cell suspension at 200 × *g* for 10 min (4 °C). Estimate 10–15% cell loss due to the additional centrifugation step.
8. Carefully check the cell suspension before injection for clogs of dead cells. Include an additional filter step immediately before injection, if necessary.
9. Define the PCR conditions according to your polymerase; for example hot start polymerases will require an initialization step for activation. Typically, 30–35 cycles are performed and include template denaturation: 15 s at 95 °C; primer annealing: 30 s at 59 °C; and primer extension: 1–2.5 min at 68 °C.

10. To use small amounts of tumor DNA, VHE and VHA primers can be analyzed in a single PCR reaction: mix both forward primers at equal amounts and add the reverse primer JH4 intronic.
11. In order to minimize DNA exposure to mutagenic radiation work quickly. J gene usage can be predicted by the size of the amplification product: JH1 = 2040 bp; JH2 = 1740 bp; JH3 = 1300 bp; JH4 = 800 bp.
12. To increase the frequency of successfully recombined clones use either a plasmid that allows blue–white screening and inoculate the transformed bacteria on X-Gal soaked petri dishes. Or work with a vector system that encodes a suicide gene (such as ccdB) whose expression is disrupted after insert integration.
13. Select primers that bind in the vector backbone. For longer PCR products (>900 bp) sequencing from both directions might be necessary.
14. The quantification of the intronic SHM load reassures that BCR selection processes during the GC reaction will not influence the analysis. In addition, both productive and nonproductive

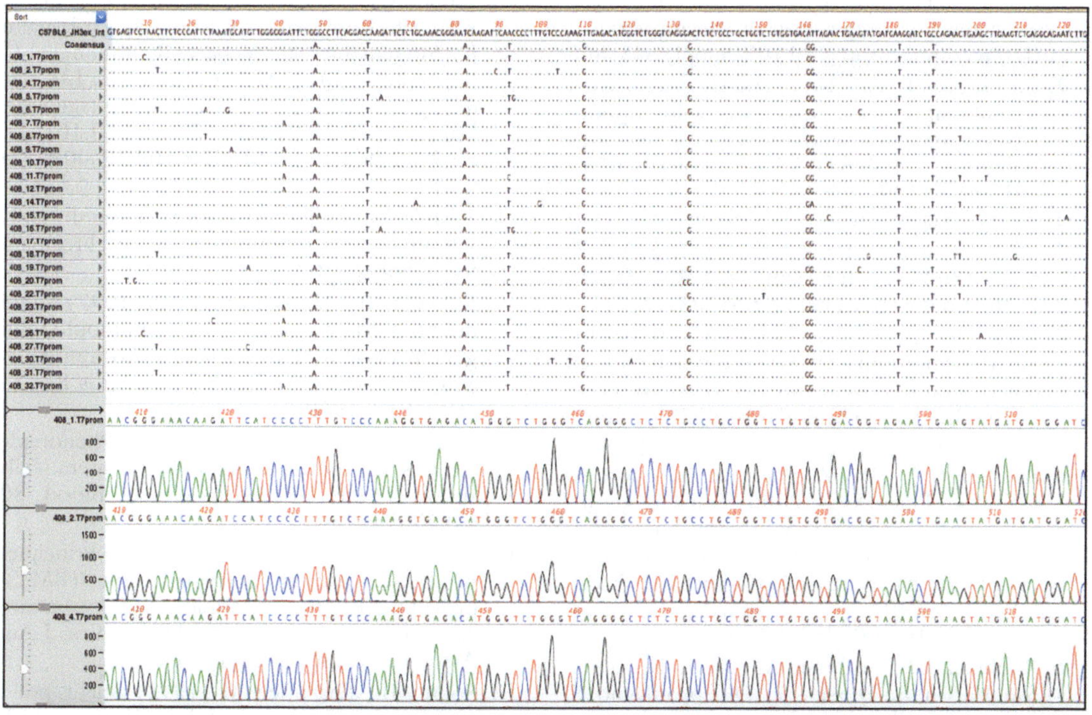

Fig. 2 Sequence alignment. Example of tumor specific sequences aligned with the control (C57BL/6 derived intronic region of JH3). The analysis revealed highly mutated and clonally related tumor sequences indicative of a monoclonal GC B cell derived lymphoma

VDJ rearrangements can be taken into account. Germ line sequences for J genes have been previously described [12].

15. To exclude mouse strain related SNPs multiple tumors are analyzed in parallel and identical mutations occurring in different individual tumors are excluded from the analysis. In addition, the analysis of nonmutated B cell populations, e.g., follicular B cells in tumor bearing animals will identify background noise in the SHM analysis. To calculate mutational frequency, SNPs corresponding to different background strains should be excluded.
16. The output of AID-mediated deamination of cytosine residues in nonmalignant GC B cells is biased toward transition (C → T and G → A) mutations (*see* Fig. 2).
17. Ongoing SHM after tumor development will result in tumor heterogeneity and the acquisition of mutations in subpopulations of the tumor bulk. Clone sizes and their pedigrees can be estimated by the percentage of shared mutations.

References

1. Bassing CH, Swat W, Alt FW (2002) The mechanism and regulation of chromosomal V(D)J recombination. Cell 109(Suppl):S45–S55
2. Chaudhuri J, Alt FW (2004) Class-switch recombination: interplay of transcription, DNA deamination and DNA repair. Nat Rev Immunol 4(7):541–552. doi:10.1038/nri1395. nri1395 [pii]
3. Di Noia JM, Neuberger MS (2007) Molecular mechanisms of antibody somatic hypermutation. Annu Rev Biochem 76:1–22. doi:10.1146/annurev.biochem.76.061705.090740
4. Nussenzweig A, Nussenzweig MC (2010) Origin of chromosomal translocations in lymphoid cancer. Cell 141(1):27–38. doi:10.1016/j.cell.2010.03.016. S0092-8674(10)00289-8 [pii]
5. Casola S, Cattoretti G, Uyttersprot N, Koralov SB, Seagal J, Hao Z, Waisman A, Egert A, Ghitza D, Rajewsky K (2006) Tracking germinal center B cells expressing germ-line immunoglobulin gamma1 transcripts by conditional gene targeting. Proc Natl Acad Sci U S A 103(19):7396–7401. doi:10.1073/pnas.0602353103. 0602353103 [pii]
6. Robbiani DF, Bothmer A, Callen E, Reina-San-Martin B, Dorsett Y, Difilippantonio S, Bolland DJ, Chen HT, Corcoran AE, Nussenzweig A, Nussenzweig MC (2008) AID is required for the chromosomal breaks in c-myc that lead to c-myc/IgH translocations. Cell 135(6):1028–1038. doi:10.1016/j.cell.2008.09.062. S0092-8674(08)01310-X [pii]
7. Sander S, Calado DP, Srinivasan L, Kochert K, Zhang B, Rosolowski M, Rodig SJ, Holzmann K, Stilgenbauer S, Siebert R, Bullinger L, Rajewsky K (2012) Synergy between PI3K signaling and MYC in Burkitt lymphomagenesis. Cancer Cell 22(2):167–179. doi:10.1016/j.ccr.2012.06.012. S1535-6108(12)00261-9 [pii]
8. Mishell BB, Shiigi SM (1980) Selected methods in cellular immunology. W.H. Freeman, San Francisco, CA
9. Jolly CJ, Klix N, Neuberger MS (1997) Rapid methods for the analysis of immunoglobulin gene hypermutation: application to transgenic and gene targeted mice. Nucleic Acids Res 25(10):1913–1919
10. Spangrude GJ, Brooks DM (1992) Phenotypic analysis of mouse hematopoietic stem cells shows a Thy-1-negative subset. Blood 80(8):1957–1964
11. Duran-Struuck R, Dysko RC (2009) Principles of bone marrow transplantation (BMT): providing optimal veterinary and husbandry care to irradiated mice in BMT studies. J Am Assoc Lab Anim Sci 48(1):11–22
12. Solin ML, Kaartinen M (1992) Allelic polymorphism of mouse Igh-J locus, which encodes immunoglobulin heavy chain joining (JH) segments. Immunogenetics 36(5):306–313

Chapter 19

The AID-Cre-ERT2 Model: A Tool for Monitoring B Cell Immune Responses and Generating Selective Hybridomas

Simon Le Gallou, Takuya Nojima, Daisuke Kitamura, Jean-Claude Weill, and Claude-Agnès Reynaud

Abstract

Expression of activation-induced cytidine deaminase (AID) is the hallmark of B cells engaged in an immune response in germinal centers. We designed an inducible fate-mapping reporter mouse in which AID-expressing B cells could be timely and irreversibly marked, by knockin at the *Aicda* locus of a tamoxifen-inducible Cre recombinase. This mouse model allows notably for the long-term follow-up of memory B cells and plasma cells engaged in an immune response. We describe here a protocol to generate hybridomas from small memory subsets that can be easily traced and identified in this mouse line through Cre-activated fluorescent reporters.

Key words Memory B cells, Fluorescent reporter, Germinal center, Fate mapping, *Aicda*

1 Introduction

T-dependent immune responses trigger the activation of cognate B cells that drives them into reactive structures called germinal centers in which two major events contributing to diversification of the B-cell receptor occur: somatic hypermutation, which allows the emergence and selection of B cell clones with higher affinity for the immunizing antigen, and class switch recombination, which allows B cells to acquire new effector functions by expressing an immunoglobulin class distinct from the IgM isotype that the responding B cell initially harbored. Both events are mediated by an enzyme, activation-induced cytidine deaminase (AID, encoded by the *Aicda* locus), which is specifically activated during the germinal center reaction and triggers these diversification events through cytidine deamination targeted at the Ig loci, followed by mobilization of error-prone repair pathways [1]. AID expression is thus the hallmark of the B-cell immune response.

Dinis Pedro Calado (ed.), *Germinal Centers: Methods and Protocols*, Methods in Molecular Biology, vol. 1623,
DOI 10.1007/978-1-4939-7095-7_19, © Springer Science+Business Media LLC 2017

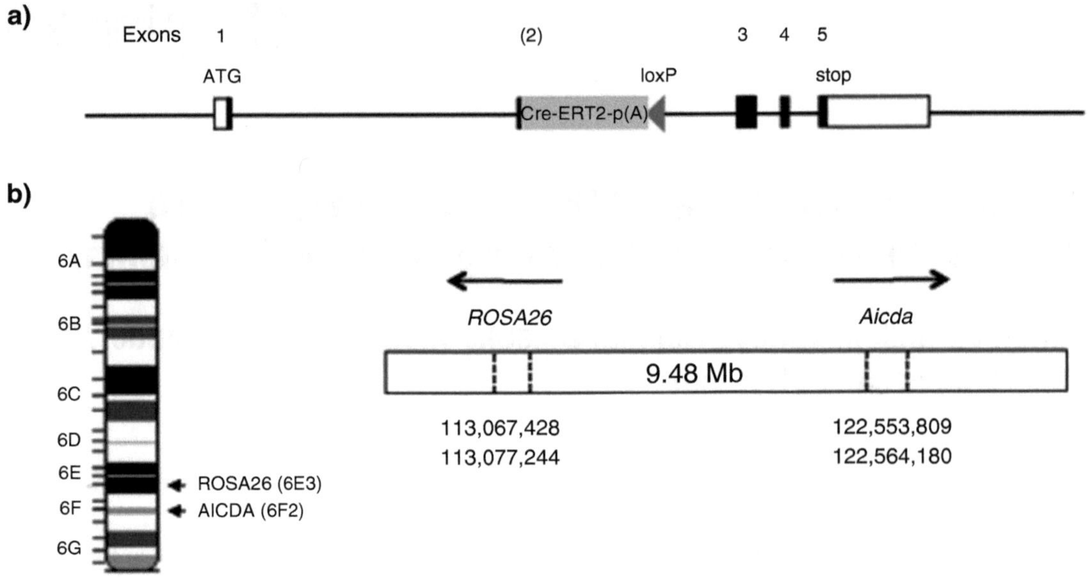

Fig. 1 The *Aicda-Cre-ERT2* knocked-in locus on mouse chromosome 6. (**a**) The tamoxifen-inducible Cre-ERT2 sequence (a gift from D. Metzger, Strasbourg) was inserted in place of exon 2 of the *Aicda* gene, in fusion with the first four amino acids of AID (encoded by 12 nucleotides in exon 1 and 4 nucleotides in exon 2) and followed by an SV40 poly(A) addition site (not drawn on scale) [2]. A neo^R gene flanked by loxP sites was used to select the transfected construct in ES cells, and further excised by transient Cre transfection that left one loxP site downstream of the poly(A) addition site. The splicing acceptor site of exon 3 was mutated to avoid skipping of the Cre-ERT2 exon. *Black boxes* represent coding exons and *white boxes* untranslated exons. (**b**) The *Aicda* and Gt(ROSA)26Sor (hereafter named ROSA26) loci are 9.48 Mb apart on chromosome 6 (represented on the *left* with its karyotype bands). A mouse line in which the ROSA26-loxP-stop-loxP-EYFP reporter [6] and *Aicda* loci recombined and segregated on the same chromosome has been generated (numbering refers to the 5′ and 3′ limits of the two genes as referenced in the annotation release 105 of the mouse genome, with their orientation represented by the *horizontal arrow*)

We thus generated a "fate-mapping" reporter mouse line, based on inducible Cre expression under the control of the AID promoter, in which all cells engaged in an immune reaction would acquire an irreversible fluorescent labeling [2]. The tamoxifen-inducible Cre-ERT2 enzyme was inserted within the second exon of the *Aicda* gene, fused with the first four amino acids of AID (Fig. 1a). *Aicda* and *ROSA26* loci are linked on chromosome 6 (9.48 Mb apart), and the line used was isolated from a recombination event (occurring at the frequency of approximately 1/20 animals) that segregated both reporters on the same chromosome (Fig. 1b). Such a configuration facilitates further breeding. Alternatively, the Aid-Cre-ERT2 line is also bred isolated to allow for other reporter choices and is available from the EMMA consortium.

The *Aicda* locus was disrupted by Cre-ERT2 insertion. Heterozygous AID-Cre reporter mice are thus usually generated by crossing AID-Cre-ERT2/ROSA26-loxP-EYFP males with wild-type females (hereafter referred to as AID-Cre-EYFP mice), to avoid AID deficiency in the pregnant females and the lack of maternal IgG in the offspring. Interestingly, homozygous animals, which are AID-deficient, can be used to follow cells having tentatively expressed AID in the absence of affinity maturation and isotype switch [3].

We used this mouse line to describe the diversity of mouse memory subsets [2] and, more recently, to study persistent B cell subsets generated in the absence of deliberate immunization (Le Gallou et al., in preparation). While the sensitivity of this model is particularly appropriate for studying long-term immune memory in which small numbers of antigen-specific B cells persist, it has been recently used with success for the analysis of clonal diversity in short-term germinal center reactions as well [3].

This method will more specifically focus on the use of the AID-Cre-ERT2 mouse line for generating hybridomas from limited amounts of specific memory B-cell populations, an approach that should facilitate the identification of antigenic specificities in defined memory subsets. An asset of hybridoma technology, compared to single cell V_H-V_L amplification, cloning and transfection strategies, is notably its capacity to assess the specificity of IgM in their pentameric form (Le Gallou et al., in preparation).

2 Materials

2.1 Cells and Reagents

1. FACS purified $EYFP^+$ B cell subsets from AID-Cre-EYFP mice.
2. Sp2/0 myeloma cell line.
3. Balb/c 3T3 cell line transfected with mouse CD40L and BAFF (40LB) [4].
4. Mitomycin C.
5. Recombinant IL-4.
6. Tamoxifen Citrate, e.g., Nolvadex (AstraZeneca).
7. ClinOleic (Baxter).
8. Ethylenediaminetetraacetic acid, EDTA.
9. Syringes.
10. Cell culture plasticware.
11. 0.22-μm vacuum filter units.
12. Gavage needles.
13. 16-gauge blunt-end needles.
14. Polymerase chain reaction (PCR) reagents.
15. Agarose.

2.2 Media and Solutions

1. Phosphate buffered saline, PBS.
2. ClonaCell™-HY Hybridoma Cloning Kit (STEMCELL Technologies, *see* **Notes 1** and **2**), containing Media A, B, C, D, and E.
3. Sp2/0 medium: RPMI-1640 supplemented with 10% FCS and 1% penicillin–streptomycin.
4. 40LB medium: DMEM, 10% FCS, and 1% penicillin–streptomycin.
5. Medium to culture EYFP$^+$ B cell subsets (MB medium): RPMI-1640 supplemented with 10% FCS, 5.5×10^{-5} M 2-mercaptoethanol, 10 mM HEPES, 1 mM sodium pyruvate, 1% penicillin–streptomycin.

3 Methods

AID-Cre-EYFP is an efficient model for hybridoma isolation from as little as a few tens of thousands EYFP$^+$ B cells, after in vitro expansion in the culture system of Nojima et al. [4].

3.1 Genotyping of the AID-Cre-ERT2 Mouse Line

1. For Cre detection, AID-fwd, GTAGGTCCAGCCATCAG CAG; AID-rev, CGAAGGTGGCCGAAGTCCAG; Cre-rev, AGGTTCTGCGGGAAACCATTTCCG (94 °C, 2 min, 35 cycles of 30 s at 92 °C, 1 min at 60 °C, 1.5 min at 72 °C, followed by 2 min at 72 °C. The wild-type *Aicda* locus is diagnosed by a 250-bp band and the Cre-targeted locus by a 300-bp band.
2. For EYFP detection, ROSA26-fwd, AAGGGAGCTGCAGT GGAGTA; ROSA26-rev, CCGAAAATCTGTGGGAAGTC; EYFP-fwd, ACATGGTCCTGCTGGAGTTC (94 °C 2 min, 38 cycles of 20 s at 94 °C, 20 s at 60 °C and 20 s at 72 °C, followed by 2 min at 72 °C. The wild-type *ROSA26* locus is diagnosed by the presence of a 300-bp band and the ROSA26-EYFP allele by a 500-bp band.

3.2 Tamoxifen Injection for Cre Activation and EYFP Expression In Vivo

1. To prepare 12 doses of tamoxifen, take four 30 mg Nolvadex tablets into 6 mL of 20% Clinoleic.
2. Rotate tube at 4 °C for 2 h or until the tablets are completely dissolved.
3. Using a syringe and a gavage needle, inject 500 μL of the solution into the mouse by gavage, which corresponds to 10 mg of tamoxifen (*see* **Note 3**).

3.3 Culture of Sp2/0 Cells for Hybridoma Generation

1. Seed Sp2/0 cells at a density of approximately 5×10^4 cells/mL.
2. Passage cells every 2 days.
3. Calculate the growth rate at each passage. Cell density should not exceed 4×10^5 cells/mL (*see* **Note 4**).

3.4 In Vitro Culture and Expansion of EYFP+ Memory B Cells

1. Culture the 40LB feeder cells in 40LB Medium.
2. Passage the cells every 3 days.
3. Twelve hours prior to coculture with EYFP+ B cells, treat 40LB cells with 10 μg/mL final concentration of mitomycin C in culture medium.
4. Incubate for 2.5–3 h at 37 °C in humidified incubator with 5% CO_2.
5. Wash the cells three times in PBS and then add trypsin for 5 min at 37 °C.
6. Centrifuge cells at 400 × *g* at RT for 10 min.
7. Count cells and reseed 3×10^5 40LB cells per well in 3 mL of 40LB Medium in a 6-well plate.
8. Resuspend 5×10^4 of FACS purified EYFP+ memory B cells from spleen or other lymphoid tissue in 8 mL of MB medium.
9. Seed cells into the well with mitomycin-treated 40LB.
10. Add 1 ng/mL rIL-4 to the culture.
11. On day 2 of EYFP+ memory B cells/40LB coculture, remove 4 mL of the medium and replace with fresh medium and IL4 (*see* **Note 5**). Day of fusion to generate hybridomas takes place at day 4 or 5 of culture (*see* **Note 6**).

3.5 Generation of Hybridoma from EYFP+ Memory B Cells

1. The day before the fusion, count the Sp2/0 cells and dilute them in order to obtain 2×10^5 cells/mL for the next day.
2. On the day of fusion, place a 9 mL aliquot (*see* **Note 2**) of ClonaCell™-HY Medium D at 2–8 °C and thaw overnight.
3. On the day of fusion; day 4 or 5 of EYFP+ memory B cells/40LB coculture; harvest the EYFP+ memory B cells from the 40LB culture supernatant.
4. Incubate the remaining cells in warm PBS with 2 mM EDTA for 5 min, (*see* **Note 5**), and then detach the whole cells by pipetting.
5. Centrifuge at 400 × *g* at RT for 10 min and discard supernatant.
6. Count cells in medium or PBS at RT or 37 °C (*see* **Note 7**).
7. Wash EYFP cells with 20 mL of ClonaCell™-HY Medium B.
8. Centrifuge at 400 × *g* at RT for 10 min and discard supernatant.
9. Count Sp2/0 cells and centrifuge a number of Sp2/0 cells in order to obtain a 1:1 EYFP+ memory B cells:Sp2/0 cell ratio.
10. Wash Sp2/0 cells with 20 mL of ClonaCell™-HY Medium B.
11. Centrifuge at 400 × *g* at RT for 10 min and discard supernatant.

12. Resuspend the Sp2/0 and EYFP+ memory B cells together in 20 mL of ClonaCell™-HY Medium B in a 50-mL conical tube.
13. Centrifuge at 400 × *g* at RT for 10 min and discard supernatant.
14. Repeat wash with 20 mL of ClonaCell™-HY Medium B.
15. Centrifuge at 400 × *g* at RT for 10 min and discard supernatant.
16. Resuspend the cell pellet by gently tapping the bottom of the tube to disperse cells for optimal fusion.
17. Add 37 °C 100 μL of ClonaCell™-HY PEG Solution (PEG) to the pellet dropwise using a 100 μL pipette, over a period of 1 min without stirring.
18. Stir cells continuously but gently, with the pipette tip, over the next minute.
19. Add 0.4 mL, 2 × 200 μL with a 200-μL pipette, of ClonaCell™-HY Medium B to the fusion mix, stirring continuously but gently as before, over a period of 4 min.
20. Slowly add 1 mL of ClonaCell™-HY Medium B to the cells with a 1000 μL pipette.
21. Incubate for 15 min in a water bath at 37 °C (*see* **Note 8**).
22. Slowly add 3 mL of ClonaCell™-HY Medium A.
23. Centrifuge cells at 400 × *g* at RT for 7 min and discard supernatant.
24. Wash cells with 4 mL of ClonaCell™-HY Medium A to ensure that all PEG is removed.
25. Slowly resuspend the cell pellet in 1 mL of ClonaCell™-HY Medium C.
26. Transfer the cell suspension to a well of a 12-well plate containing 2 mL of ClonaCell™-HY Medium C, for a total culture volume = 3 mL.
27. Incubate for 24 h at 37 °C in 5% CO_2 atmosphere.
28. The day after the fusion, shake vigorously the thawed 9 mL aliquot of ClonaCell™-HY Medium D to mix contents well and let warm to room temperature, do not warm the solution to 37 °C.
29. Transfer fused cell suspension into the 15 mL.
30. Centrifuge for 10 min at 400 × *g* at RT or 37 °C, and discard supernatant.
31. Resuspend the cells in ClonaCell™-HY Medium C to a total volume of 1 mL.
32. Transfer the 1 mL cell suspension into the tube containing 9 mL of ClonaCell™-HY Medium D.

33. Mix thoroughly by gently inverting the tube several times.
34. Let the tube sit for 15 min at 37 °C to allow the bubbles to come to the top.
35. Using a 12 mL syringe and a 16-gauge blunt-end needle, aseptically plate out 9.5 mL of cell suspension medium into a 100 mm Petri plate.
36. Tilt the plate to evenly distribute the medium to cover the bottom of the plate. Avoid the introduction of bubbles during plating.
37. Incubate the plate at 37 °C in 5% CO_2 atmosphere
38. Do not disturb plates for 14 days to prevent colony dispersion.
39. Fourteen days after, examine the plates for the presence of colonies visible by eye (*see* **Note 9**).
40. Pick isolated colonies from the plate using a pipette set to 10 μL using sterile pipette tips.
41. Pipette each clone into individual wells of a 96-well tissue culture plate containing 200 μL of ClonaCell™-HY Medium E, changing tip between each clone.
42. With a multichannel pipette set at 150 μL, pipette the entire contents of the well several times to resuspend the colony. Ensure a new sterile tip is used for each clone.
43. Incubate the plates at 37 °C in 5% CO_2 for 4 days without feeding.
44. Gently resuspend all hybridomas with a multichannel pipette.
45. Transfer 100 μL of cells to two wells of a 24-well plate, containing 1 mL of ClonaCell™-HY Medium E.
46. When cells have grown to approximately 4×10^5 cells/mL, freeze the cells from one well and expand the remaining positive clones in a 25 cm^2 tissue culture flask containing 5 mL of ClonaCell™-HY Medium A and 5 mL of Medium E.
47. When cells have grown to a suitable density, freeze the cells.
48. Aliquot the supernatants of each hybridoma.
49. Analyze the supernatants by ELISA for immunoglobulin secretion and by appropriate assays for antigen specificity, including ELISA, flow cytometry, peptide microarrays, or western blot, (*see* **Note 10**).

4 Notes

1. Filter A and B media of the ClonaCell™-HY Hybridoma Cloning Kit on a 0.22 μm vacuum filter unit before using them for fusion.

2. Each medium of the ClonaCell™-HY kit can be thawed once, aliquoted and refrozen before use. Aliquots of 9 mL in 15-mL conical tubes for media C and D and 50 mL in 50 mL conical tubes for media A, B, and E.
3. We estimated that 20% of splenic germinal center B cells are labeled through such tamoxifen gavage, and Victora and colleagues reported 50% efficiency for lymph nodes [3]. The duration of tamoxifen action has been estimated between 3 and 5 days (*see* Jarjour et al., 2014, Fig. S3) [5]. Multiple feedings are usually performed during the immune response to increase the number of labeled B cells. Alternatively, a tamoxifen diet of 400 mg tamoxifen citrate/kg can be used. We observed that a 2-week diet was approximately equivalent for EYFP B cell labeling efficiency as a three-course gavage 5 days apart. Dry tamoxifen food is the only protocol we found adapted to germ-free mouse studies, as such food can easily be sterilized by irradiation. We failed to produce tamoxifen solution for gavage that presented the required sterility for long-term maintenance of an axenic status.
4. Sp2/0 cells must be kept in early/mid log phase growth prior to fusion (maximum density of 8×10^5 cells/mL). If cells grow beyond 8×10^5 cells/mL, passage them at least twice to return them to early/mid log phase growth prior to fusion.
5. All solutions and media should be prewarmed to 37 °C prior to use for fusion and EYFP$^+$ memory B cells/40LB coculture.
6. The number of EYFP$^+$ memory B cells increases around tenfold between day 0 and day 4 of coculture.
7. For counting, EYFP$^+$ memory B cells and 40LB can be distinguished by size under microscope, with the 40LB cells displaying a larger size.
8. Fusion events can be observed under microscope after 15 min.
9. Colonies start to be visible by naked eye around 10 days of culture. A typical fusion will produce around 100 colonies per plate.
10. Immunoglobulin secretion dosage by ELISA is performed from 25-cm^2 tissue culture flask due to the large amount of supernatant available. It is also possible to do it at the 24-well culture step. Concerning 96-well supernatants, all of them will be immunoglobulin positive due to large amount of immunoglobulin secretion carried over from the 100 mm Petri plate, generating false-positive colonies.

Acknowledgments

This work was supported by an ERC Advanced Grant (Memo-B), by the Ligue Nationale contre le Cancer and by the Fondation Princesse Grace de Monaco. We thank Sandra Weller for valuable comments and advices.

References

1. Di Noia JM, Neuberger MS (2007) Molecular mechanisms of antibody somatic hypermutation. Annu Rev Biochem 76:1–22. doi:10.1146/annurev.biochem.76.061705.090740
2. Dogan I, Bertocci B, Vilmont V, Delbos F, Megret J, Storck S, Reynaud CA, Weill JC (2009) Multiple layers of B cell memory with different effector functions. Nat Immunol 10 (12):1292–1299. doi:10.1038/ni.1814. ni.1814 [pii]
3. Tas JM, Mesin L, Pasqual G, Targ S, Jacobsen JT, Mano YM, Chen CS, Weill JC, Reynaud CA, Browne EP, Meyer-Hermann M, Victora GD (2016) Visualizing antibody affinity maturation in germinal centers. Science 351 (6277):1048–1054. doi:10.1126/science.aad3439. science.aad3439 [pii]
4. Nojima T, Haniuda K, Moutai T, Matsudaira M, Mizokawa S, Shiratori I, Azuma T, Kitamura D (2011) In-vitro derived germinal centre B cells differentially generate memory B or plasma cells in vivo. Nat Commun 2:465. doi:10.1038/ncomms1475. ncomms1475 [pii]
5. Jarjour M, Jorquera A, Mondor I, Wienert S, Narang P, Coles MC, Klauschen F, Bajenoff M (2014) Fate mapping reveals origin and dynamics of lymph node follicular dendritic cells. J Exp Med 211(6):1109–1122. doi:10.1084/jem.20132409. jem.20132409 [pii]
6. Srinivas S, Watanabe T, Lin CS, William CM, Tanabe Y, Jessell TM, Costantini F (2001) Cre reporter strains produced by targeted insertion of EYFP and ECFP into the ROSA26 locus. BMC Dev Biol 1:4

Chapter 20

Determining the Origin of Human Germinal Center B Cell-Derived Malignancies

Marc Seifert and Ralf Küppers

Abstract

Most human B cell lymphomas originate from germinal center (GC) B cells. This is partly caused by the high proliferative activity of GC B cells and the remodeling processes acting at the immunoglobulin (Ig) loci of these cells, i.e., somatic hypermutation and class-switching. Mistargeting of these processes can cause chromosomal translocations, and the hypermutation machinery may also target non-Ig genes. As somatic hypermutation is exclusively active in GC B cells, the presence of somatic mutations in rearranged IgV genes is a standard criterium for a GC or post-GC B cell origin of lymphomas. Beyond this, ongoing somatic hypermutation during lymphoma clone expansion indicates that the lymphoma has an active GC B cell differentiation program. The proto-oncogene *BCL6* is specifically expressed in GC B cells and also acquires somatic mutations as a physiological by-product of the somatic hypermutation process, albeit at a lower level than IgV genes. Thus, detection of *BCL6* mutations is a further genetic trait of a GC experience of a B cell lymphoma. Typically, B cell lymphomas retain key features of their specific cells of origin, including a differentiation stage-specific gene expression pattern. This is at least partly due to genetic lesions, which "freeze" the lymphoma cells at the differentiation stage at which the transformation occurred. Therefore, identification of the normal B cell subset with the most similar gene expression pattern to a particular type of B cell lymphoma has been instrumental to deduce the precise cell of origin of lymphomas.

We present here protocols to analyze human B cell lymphomas for a potential origin from GC B cells by determining the presence of mutations in rearranged IgV genes and the *BCL6* gene, and by comparing the gene expression pattern of lymphoma cells with those of normal B cell subsets by genechip or RNA-sequencing analysis.

Key words B cells, BCL6, Gene expression profiling, Germinal center, Immunoglobulin genes, Somatic hypermutation

1 Introduction

1.1 B Cell Development

B cells are generated in the bone marrow in a multistep developmental process from hematopoietic stem cells [1]. A hallmark of this process is the rearrangement of the immunoglobulin (Ig) genes to equip each B cell with a functional B cell antigen receptor (BCR). Rearrangement at the Ig heavy chain locus occurs first [1]. The V region gene of the heavy chain is composed of three gene segments,

Dinis Pedro Calado (ed.), *Germinal Centers: Methods and Protocols*, Methods in Molecular Biology, vol. 1623,
DOI 10.1007/978-1-4939-7095-7_20, © Springer Science+Business Media LLC 2017

V_H, D_H, and J_H. In the human, at the pro B cell stage one of 27 D_H gene segments is rearranged to one of six J_H gene segments [2]. Then, one of 38–46 (depending on the haplotype) functional V_H gene segments is recombined to the D_HJ_H joint [2]. The resulting $V_HD_HJ_H$ rearrangement is tested for functionality, and if it can be expressed as a pre-BCR together with a surrogate light chain at the pre-B cell stage, the cell initiates light chain rearrangement. This is done first at the Igκ locus, and in case no functional $V_\kappa J_\kappa$ rearrangement can be generated, rearrangement of the Igλ locus takes place [3]. Rearrangement diversity is not only achieved by the random assortment of the V, (D) and J segments, but also by junctional diversity, i.e., hairpin structures at the ends of rearranging gene segments are randomly incised and the overhanging nucleotides may be removed or randomly extended by insertion of the so-called N nucleotides before the joining sites are ligated. As a consequence, each B cell carries a unique BCR, which is therefore an ideal clonal marker. A further consequence of the random V(D)J recombination process is that many rearrangements are nonfunctional because of reading frameshifts or the generation of nonsense codons at the joining sites. Consequently, often rearrangements of the second allele occur, when the first attempt was unsuccessful. Hence, a fraction of B cells carries two IgH rearrangements, one productive and one nonfunctional. The same principles apply to the light chain loci. After successful generation and expression of a functional, non-self-reactive BCR, transitional or mature naive B cells leave the bone marrow and circulate through the body.

1.2 Germinal Center Reaction

When mature B cells encounter cognate antigen and T cell help is available, the antigen-activated B and T cells interact in primary foci in secondary lymphoid organs, such as lymph nodes [4, 5]. These activated B cells proliferate and mostly differentiate into short-lived plasma cells. A fraction of the activated B cells migrates into B cell follicles, differentiates and thereby establishes histological structures called germinal centers (GC) [5]. In the GC, B cells vigorously proliferate in the dark zone, so that large B cell clones can be generated (Fig. 1). The proliferating GC B cells, named centroblasts, activate the process of somatic hypermutation (SHM), which introduces point mutations and occasionally also deletions or duplications specifically into the heavy and light chain V region genes at a very high rate (10^{-3}–10^{-4}/bp/cell division) [6–8]. The enzyme activation-induced cytidine deaminase (AID) is the key factor for SHM and catalyzes the deamination of C to U [6]. As U pairs with A instead of G and is recognized as a base not belonging into DNA, this induces error-prone repair processes which often result in mutations at the altered C and neighboring positions [6]. Both productive as well as nonproductive rearrangements accumulate mutations. After several cell divisions and the potential acquisition of one or even several mutations, centroblasts become resting

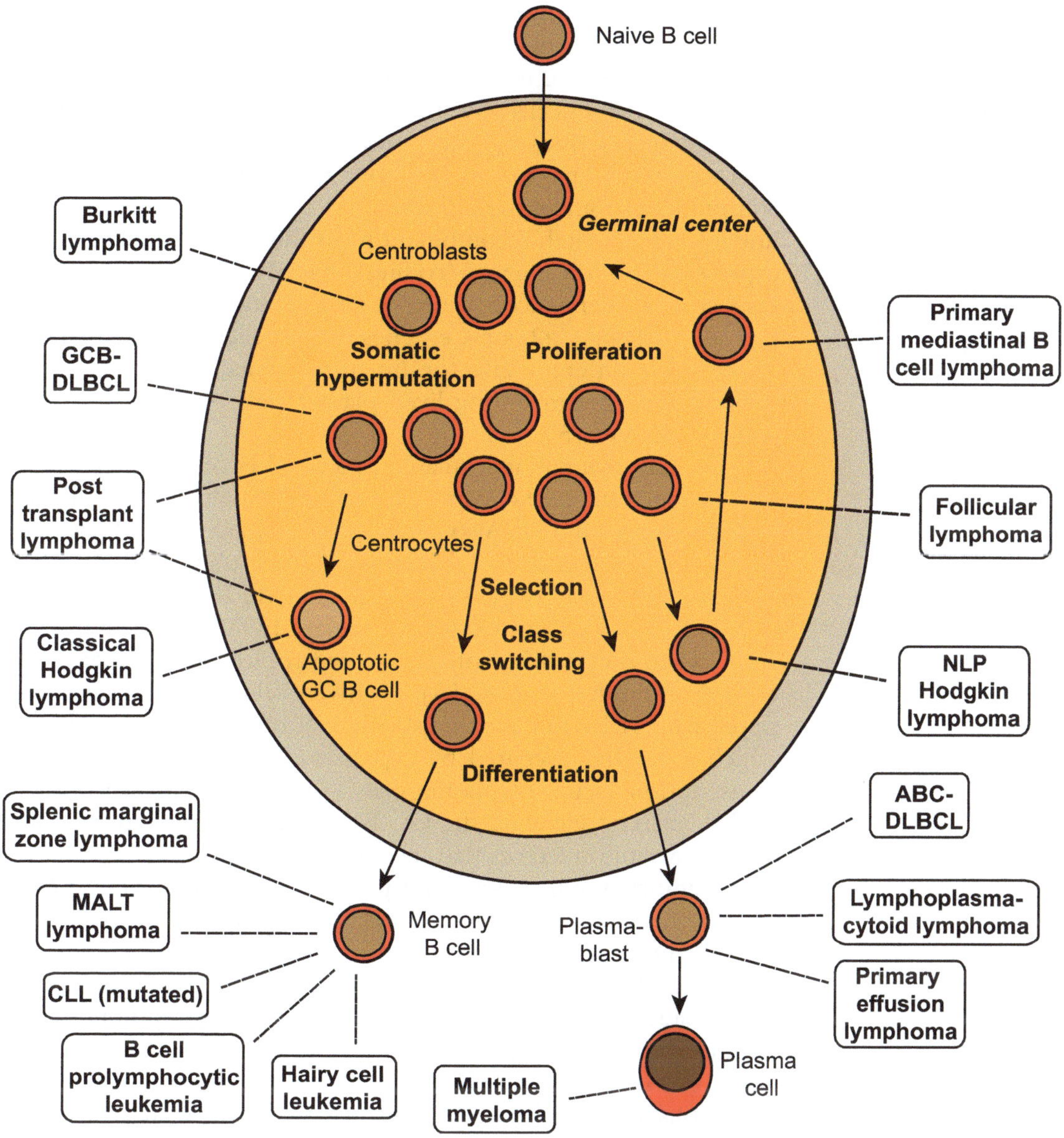

Fig. 1 B cell differentiation in the GC and origin of GC B cell-associated lymphomas. Antigen-activated mature, naive B cells are driven into a GC reaction when T cell help is available. GC B cells undergo massive clonal expansion in the dark zone of the GC. These proliferating GC B cells are called centroblasts. They activate the process of SHM that introduces mutations at a very high rate into IgV region genes, and at a lower rate also in some non-Ig genes, including *BCL6*. After one or several cell divisions, centroblasts differentiate into nonproliferating centrocytes that are mainly found in the light zone of the GC. The centrocytes are then selected for affinity-increasing IgV gene mutations by interaction with follicular dendritic cells and T follicular helper cells (not shown). Positively selected centrocytes will typically return to the dark zone for further rounds of proliferation, mutation and selection before they differentiate into memory B cells or plasmablasts and exit the GC. The majority of GC B cells will acquire disadvantagous mutations and undergo apoptosis. Many GC B cells perform class switch recombination in the light zone. Most B cell lymphomas are derived from GC B cells or from post-GC B cells. Also in the latter types of lymphomas, decisive transforming events have presumably

centrocytes and move into the light zone of the GC [5]. The centrocytes are then selected for affinity-increasing mutations by competition for the interaction with follicular dendritic cells and follicular T helper cells. GC B cells that are not positively selected undergo apoptosis as a default program [9]. Positively selected B cells typically return to the dark zone for further rounds of proliferation, mutation, and selection (Fig. 1) [5]. Some GC B cells also undergo class switch recombination. In this process, the originally expressed Cμ and Cδ genes (coding for IgM and IgD heavy chains, respectively) are replaced by one of seven 3′ located CH genes, i.e., Cγ1–4, Cα1 or Cα2, or Cε. These CH genes encode Ig heavy chains with different effector functions. Finally, selected GC B cells differentiate into long-lived plasma cells or memory B cells and exit the GC [5].

The GC program is regulated by a number of master transcription factors. One of the main factors is BCL6, which is essential for GC development [5]. Also IRF8 [10] and NF-κB [11, 12] are important for GC B cell formation. Other transcription factors expressed by (subsets of) GC B cells include IRF4, MEF2B, and BACH2 [5, 13]. Centrocytes upregulate FOXO1 and transiently MYC when they redifferentiate into centroblasts for further proliferation and mutation [14, 15]. Positively selected centrocytes that upregulate IRF4 initiate the plasma cell differentiation program, which is further executed by expression of PRDM1, the master transcription factor of plasma cells [5]. PRDM1 downregulates BCL6, thereby silencing the GC B cell gene expression program. The transcription factors that mediate differentiation of GC B cells into memory B cells are still poorly understood. NF-κB activation in centrocytes seems to play a role in this differentiation process [11].

The process of SHM is not entirely specific for rearranged IgV genes, as a few other genes may also be targeted physiologically. A well-characterized off-target is *BCL6*, which acquires mutations at about 1/50 of the frequency of IgV genes [16, 17]. Consequently, about one third of GC B cells and memory B cells carry one or a few mutations in the major mutation cluster (MMC), which is a region

Fig. 1 (Continued) occurred in premalignant GC B cell precursors of these lymphomas. About half of the cases of chronic lymphocytic leukemia (CLL) carry mutated IgV genes. These cases are likely derived from a small subset of $CD5^+$ memory B cells [29]. Most mantle cell lymphomas carry unmutated IgV genes and are presumably derived from ($CD5^+$) mantle zone B cells. However, 20–30% of these lymphomas carry mutated V genes, suggesting a GC experience (not shown). Primary mediastinal large B cell lymphoma is likely derived from GC-experienced thymic B cells. A considerable fraction of splenic marginal zone B cell lymphomas carries mutated IgV genes and hence likely derives from GC-experienced marginal zone B cells. The tumor cells of classical Hodgkin lymphoma and some post-transplant lymphomas carry destructive IgV gene mutations, indicating a derivation from pre-apoptotic GC B cells. ABC-DLBCL, activated B cell-like diffuse large B cell lymphoma; GCB-DLBCL, GC B cell-like diffuse large B cell lymphoma. MALT, mucosa-associated lymphatic tissue; NLP, nodular lymphocyte predominant about 1 kb downstream of the *BCL6* promoter [16, 17]. Importantly, SHM is dependent on active transcription of the target genes [6, 18, 19], so that the detection of *BCL6* mutations is a genetic trait of the GC origin of these mutations, as AID and BCL6 are together expressed only in GC B cells. B cells potentially undergoing SHM outside of the GC, where no *BCL6* is transcribed, cannot acquire *BCL6* mutations.

There is a controversial discussion whether SHM might also take place outside of GCs in humans, and that in particular $IgM^+IgD^+CD27^+$ B cells carrying mutated IgV genes, may be generated independent of a GC reaction [20], so that the detection of mutated IgV genes in IgM^+ B cell lymphomas would not be an unequivocal sign for their GC experience. However, there is now strong evidence that at least in adults, most if not all $IgM^+IgD^+CD27^+$ B cells are generated in GC reactions [16, 21].

1.3 Lymphoma Pathogenesis

An important concept of lymphoma pathogenesis is that transformed lymphocytes typically retain key features of the differentiation stage of their cell of origin [22]. This is different to most solid cancers, where it is assumed that these either derive from tissue stem or precursor cells, or from terminally differentiated cells that underwent a partial redifferentiation to regain proliferative potential. Despite their mature and differentiated character, lymphocytes intrinsically have a high proliferative potential, a property that is often abused by lymphoma cells. For example, several B cell lymphomas derive from GC B cells that were "frozen"in their highly proliferative state by particular transforming events [23, 24]. Examples of this are translocations of *BCL6* causing its constitutive expression, or inactivating mutations of *BLIMP1*, preventing a differentiation of GC B cells into resting plasma cells [23]. As B cell leukemias and lymphomas typically retain much of the differentiation program of their cell of origin [23, 24], gene expression profiling of lymphomas in comparison to the main normal B cell subsets is a valuable and frequently used approach to identify the cellular origin of a B cell malignancy [25–30]. Please note that transforming events often accumulate in lymphoma precursor cells over several subsequent stages of B cell development, before the fully malignant clone evolves (e.g., in the GC B cell-derived follicular lymphoma, the typical t(14;18) IgH-*BCL2* translocation already occurs at the pro- or pre-B cell stage [31]), and that we refer here to the stage at which the final transformation occurred. Overall, most types of human B cell lymphomas are derived from GC B cells (Fig. 1) [22–24].

1.4 Approaches to Study Lymphomas for a GC Derivation

We present here protocols for three molecular biological approaches to study the potential origin of a human lymphoma from GC B cells: (1) IgV gene mutation analysis, (2) *BCL6* gene mutation analysis, and (3) gene expression profiling of lymphoma and normal B cells to identify the normal B cell subset with the most similar gene expression pattern. Basic insights into the cellular origin of lymhomas are also obtained by classical pathological evaluation, including morphology, histology, and immunohistochemical analysis. For example, the morphology of the lymphoma cells in Burkitt lymphoma resembles normal centroblasts, and the histological picture in follicular lymphoma resembles GC, as the lymphoma cells are found in networks of follicular dendritic cells and are intermingled with GC-type T_H cells [22, 24]. Further indication for a GC B cell identity of lymphoma cells is the expression of typical GC B cell differentiation markers, such as BCL6 [32]. Although class switching mostly takes place in the GC, the expression of IgG or IgA by a lymphoma is not a strong argument for a GC experience of the lymphoma precursor cell, as class switching also occurs during T cell independent immune responses not involving GC reactions (*see* for example Ref. [33]).

More recently, it has been revealed that also epigenetic studies, and in particular analyses of DNA methylation patterns, can be used to infer the cellular origin of lymphomas, because normal B cells undergo specific and quite extensive epigenetic changes during their development and differentiation [34, 35].

2 Materials

2.1 IgV and BCL6 Gene Sequencing and Mutation Analysis

1. DNA isolation Kit for blood, cells or tissue (e.g., Gentra Puregene).
2. Optional: 100 mg/ml RNase A.
3. 70% ethanol.
4. 100% isopropanol.
5. High Fidelity PCR System, (e.g., Expand High Fidelity PCR System, Roche).
6. 2 mM dNTPs.
7. V_H primers: *see* Table 1; 2.5 μM working dilution.
8. J_H primers: *see* Table 1; mix same volumes of 2.5 μM working dilutions.
9. V_κ primers: see Table 1; 2.5 μM working dilution.
10. J_κ primers: see Table 1; mix each J_κ primer to a 2.5 μM working dilution.
11. *BCL6* primers: *BCL6*, F1 forward primer: 5′-CGCTCTTGCC AAATGCTTTGGC-3′ (first round); *BCL6*, Rl reverse primer: 5′-CTCTCGTTAGGAAGATCACGGC-3′ (first round); *BCL6*, F2 forward primer: 5′-CGCTCTTGCCAAATGCT TTG-3′ (second round); *BCL6*, R2 reverse primer: 5′-GAC ACGATACTTCATCTCATC -3′ (second round).

Table 1
Primer sequences to amplify rearranged human IgH and Igκ V genes

Name	Sequence (5′–3′)
VH1	CAG-TCT-GGG-GCT-GAG-GTG-AAG-A
VH2	GTC-CTR-CGC-TGG-TGA-AAC-CCA-CAC-A
VH3	GGG-GTC-CCT-GAG-ACT-CTC-CTG-TGC-AG
VH4	GAC-CCT-GTC-CCT-CAC-CTG-CRC-TGT-C
VH5	AAA-AAG-CCC-GGG-GAG-TCT-CTG-ARG-A
VH6	ACC-TGT-GCC-ATC-TCC-GGG-GAC-AGT-G
JH1.2.4.5	ACC-TGA-GGA-GAC-GGT-GAC-CAG-GGT
JH3	ACC-TGA-AGA-GAC-GGT-GAC-CAT-TGT
JH6	ACC-TGA-GGA-GAC-GGT-GAC-CGT-GGT
Vk1	GAC-ATC-CRG-WTG-ACC-CAG-TCT-CCW-TC
Vk2	CAG-WCT-CCA-CTC-TCC-CTG-YCC-GTC-A
Vk3	TTG-TGW-TGA-CRC-AGT-CTC-CAG-SCA-CC
Vk4	AGA-CTC-CCT-GGC-TGT-GTC-TCT-GGG-C
Vk5	CAG-TCT-CCA-GCA-TTC-ATG-TCA-GCG-A
Vk6	TTT-CAG-TCT-GTG-ACT-CCA-AAG-GAG-AA
Jk1.2.4	ACT-CAC-GTT-TGA-TYT-CCA-SCT-TGG-TCC
Jk3	GTA-CTT-ACG-TTT-GAT-ATC-CAC-TTT-GGT-CC
Jk5	GCT-TAC-GTT-TAA-TCT-CCA-GTC-GTG-TCC

The numbers in the V primer designations indicate the respective V gene families. The numbers in the J primers indicate that J segments that are preferentially amplified by these primers
R: A&G; W: A&T; Y: C&T; S: G&C

12. Standard agarose.
13. Nucleic acid stain, 10,000× in water (e.g., GelRed, Biotium).
14. TAE: 40 mM Tris–acetate, 1 mM EDTA, pH 8.
15. 6× gel loading dye.
16. 100-bp DNA ladder.
17. Kit for extraction of DNA fragments from Agarose (e.g., Innu-Prep Doublepure Extraction Kit, Analytik Jena).
18. DNA mass ladder.
19. Nuclease-free water.
20. Mineral oil.

21. Big Dye Terminator v3.1 Cycle Sequencing Kit from Applied Biosystems.
22. Na-acetate–dextran blue: 3 M Na-acetate (pH 5–5.5), 12.5 mg/ml dextran blue.
23. Ethanol 100%, 70%.
24. Nuclease-free water.
25. Formamide.
26. Sequence analysis software (e.g., SeqScape software, Applied Biosystems).

2.2 DNA Fragment Cloning

1. For addition of 3′ adenosine overhangs: 2 mM dATP; 25 mM $MgCl_2$; *Taq* DNA polymerase (5 U/μl); 10× PCR buffer.
2. TA cloning Kit (e.g., pGemTeasy, Promega).
3. Competent cells (e.g., XL-1 blue, Promega).
4. SOC medium.
5. Agar-agar.
6. LB medium.
7. IPTG/XGAL.
8. Ampicilin.
9. Petri dishes with 10 cm diameter.
10. 2 and 1.5 ml reaction tubes.
11. 13 ml reaction tubes with ventilated cap.
12. Plasmid Miniprep Kit (e.g., Metabion).
13. *EcoRI* restriction enzyme and 10× digestion buffer.
14. Nuclease-free water.
15. Standard agarose.
16. 1-kb DNA ladder.
17. 6× orange loading dye.

2.3 Cell and RNA Isolation

1. Steel mesh (ca. 1 mm pore size).
2. Phosphate buffered saline–5% w/v bovine serum albumin (PBS–5% BSA).
3. Filter (45 μm pore size) for lymphocytes (e.g., Partec).
4. Density gradient solution (1.077 g/ml), (e.g., Amersham Biosciences).
5. Appropriate magnetic bead coupled antibodies for surface antigens, e.g., CD19 for positive B cell enrichment from Miltenyi Biotech, or antibody cocktails marking multiple cell populations (T cells, monocytes, NK cells, etc.) for depletion of

unwanted cell types and negative enrichment of desired population (e.g., Stem Cell).

6. Magnets and columns (e.g., Miltenyi Biotech, Stem Cell).
7. RNA isolation kit (e.g., RNeasy Mini kit, Qiagen).
8. 70% EtOH.
9. RNAse-free water.
10. RNA 6000 Nano kit from Agilent.
11. 2100 Bioanalyzer from Agilent.

2.4 Gene Expression Profiling by Oligonucleotide Arrays

1. Magnetic Stand-96 Agencourt SPRI(R)Plate Super Magnet Plate (Beckman Coulter).
2. Magnetic Stand-96 (e.g., Ambion, ThermoFisher).
3. 96-Well Magnetic-Ring Stand (e.g., ThermoFisher).
4. Microcentrifuge.
5. UV spectrophotometer (e.g., NanoDrop, ThermoFisher).
6. 2100 Bioanalyzer from Agilent.
7. RNA 6000 Nano kit from Agilent.
8. GeneChip® Hybridization, Wash, and Stain Kit from Affymetrix.
9. GeneChip® Whole Transcript Expression Arrays from Affymetrix.
10. GeneChip® Hybridization Oven 640 from Affymetrix.
11. GeneChip® Fluidics Station 450 from Affymetrix.
12. GeneChip® Scanner 3000 (or higher) from Affymetrix.
13. Affymetrix Gene Expression Console: http://www.affymetrix.com/support/technical/software_downloads.affx.
14. GeneSpringGX software from Agilent, or similar.
15. ComBat software: http://www.bu.edu/jlab/wp-assets/ComBat/, or similar.

2.5 Gene Expression Profiling by RNA-Sequencing (RNA-Seq)

1. TruSeq® RNA Sample Preparation Kit v2 from Illumina.
2. 96-Well thermal cycler.
3. Microplate shaker.
4. Agencourt AMPure XP kit from Beckman Coulter.
5. Reverse transcriptase (e.g., SuperScript II, Invitrogen)
6. Agilent DNA 1000 kit from Agilent Technologies.
7. Illumina Cluster station.
8. Illumina HiSeq2000 or GAIIx Systems.
9. Illumina Sequence Analysis Viewer: https://support.illumina.com/downloads/sequencing-analysis-viewer-software-v1-10-2.html.

10. FastQC: http://www.bioinformatics.bbsrc.ac.uk/projects/fastqc.
    BWA: https://sourceforge.net/projects/bio-bwa/files (or similar).

3 Methods

3.1 IgV Gene Mutation Analysis

3.1.1 DNA Isolation

DNA is isolated from 10^5–10^7 peripheral blood mononuclear cells or cell suspensions from lymphoma biopsies, or from 5–10 sections of frozen biopsies (*see* **Note 1**). An RNA digestion step with RNase A is recommended to remove nucleic acids that may interfere with the PCR reaction. The isolated DNA is diluted in 10 mM Tris or nuclease-free water and stored at 4 °C.

3.1.2 V_H PCR

1. Prepare six master mixes (one for each of the six V_H families; the V_H1 primer also binds to members of the V_H7 family) on ice, each consisting of 5 μl dNTPs, 5 μl 10× PCR buffer, 2.5 μl of the respective V_H primer, 2.5 μl J_H primer mix, 3 μl 25 mM $MgCl_2$ (5 μl for V_H1), and 30.5 μl H_2O (28.5 μl for V_H1) (*see* **Note 2**) (Table 2).
2. Prepare aliquots of 48.5 μl master mix for each sample in PCR tubes.
3. Add 50–100 ng of template DNA (*see* Subheading 3.1.1).
4. Add 0.5 μl of High fidelity DNA polymerase mixture after the first denaturation step (hot start) in a PCR cycler with heated lid. PCR program: 95 °C 5 min, 68 °C break (polymerase addition), 72 °C 1 min, 40 × (95 °C 50 s, 61 °C 30 s, 72 °C 60 s), 72 °C 5 min, 10 °C pause (*see* **Note 3**).

Table 2
Family-specific VH leader exon primers

Name	Sequence (5′–3′)
VH1L	CTC-ACC-ATG-GAC-TGG-ACC-TGG-AG
VH2L	TGC-TCC-ACR-CTC-CTG-CTR-CTG-A
VH3L	ACC-ATG-GAG-TTT-GGG-CTG-AGC-TG
VH3.2L	ACC-ATG-GAA-CTG-GGG-CTC-CGC-TG
VH4L	CTC-CTG-GTG-GCA-GCT-CCC-AGA-T
VH5L	ATC-ATG-GGG-TCA-ACC-GCC-ATC-CT

Two primers are used to cover the members of the V_H3 family. No V_H leader primer was designed for the single gene family V_H6
R: A&G

3.1.3 V_κ PCR

1. Prepare six master mixes (one for each of the six V_κ families) each consisting of 5 μl dNTPs, 5 μl 10× PCR buffer, 2.5 μl of the respective V_κ primer, 2.5 μl J_κ primer mix, 5 μl 25 mM $MgCl_2$, and 28.7 μl H_2O on ice (*see* **Note 4**).
2. Prepare aliquots of 48.5 μl master mix for each sample in PCR tubes.
3. Add 50–100 ng of template DNA (*see* Subheading 3.1.1) and start PCR program in a cycler with heated lid.
4. Add 0.5 μl of High fidelity DNA polymerase mixture after the first denaturation step. PCR program: 95 °C 5 min, 68 °C pause (polymerase addition), 40 × (95 °C 50 s, 61 °C 30 s, 72 °C 60 s), 72 °C 5 min, 10 °C pause (*see* **Note 5**).

3.1.4 PCR Product Purification

Since primer dimers and further by-products are often generated during the PCR, it is highly recommended to purify the PCR products by agarose gel electrophoresis before sequencing. The amount of purified product is estimated from an agarose gel using a DNA mass ladder to estimate the DNA concentration (*see* **Note 6**).

1. PCR products are separated and visualized by gel electrophoresis (2% agarose gel with nucleic acid stain) by UV excitation.
2. To retrieve sufficient PCR product for Sanger sequencing, the second round PCR is performed in 2–3 independent reactions of 50 μl volume each, which are finally pooled and separated by preparative gel electrophoresis (2% agarose gel).
3. Desired PCR products (about 350 bp in size) are cut out of the gel (UV excitation is needed), purified with the DNA gel Extraction Kit and eluted in 30 μl nuclease-free water.
4. An aliquot of the eluted DNA is used for estimation of the DNA concentration by analytical gel electrophoresis using a DNA mass ladder.

3.1.5 Sequencing

1. Mix the following components on ice: 5–20 ng PCR product (depending on the size of the fragment), 3.75 μl 5× sequencing buffer, 0.75 μl primer (2.5 μM; e.g., the V primer), 0.5 μl Big Dye reaction mix plus x μl H_2O to obtain a final volume of 20 μl.
2. Run sequence reaction in a PCR cycler with heated lid: 96 °C 5 min, 24 times (95 °C 15 s, 50 °C 10 s, 60 °C 4 min), cool to 10 °C.
3. Precipitation (room temperature): Add 40 μl H_2O, 6 μl 3 mM Na-acetate–dextran blue, and 150 μl 100% ethanol (1.5× volume). Mix thoroughly (vortexer) and centrifuge at 12,000 × g for 30 min. Remove supernatant and wash the sediment with 400 μl 70% ethanol. Centrifuge at 12,000 × g for 10 min,

remove supernatant, and let the pellet air-dry. Samples may be stored at 4 °C.

4. For further processing, resolve the sediment in 10 µl formamide at room temperature.
5. For analysis on an DNA sequencer (e.g., ABI 3130), the samples are transferrred to a 96 well sequencing plate followed by a denaturation step for 2 min at 95 °C.

3.1.6 Cloning of Amplificates

1. For cloning of the PCR product (*see* **Note 7**) into a plasmid with a 1-bp thymidine nucleotide overhang at the 3′ ends, it is recommended to add an adenine nucleotide to the 3′ ends of the amplificate; in contrast to conventional Taq DNA polymerases, the high fidelity enzyme mixture produces adenine nucleotide overhangs rather inefficiently. The protocol does not involve precipitation steps to avoid product loss. The amount of DNA should be quantified (*see* Subheading 3.1.4). Mix the following components on ice: 1 µl of 10× PCR buffer, 1 µl dATPs (2 mM), 0.6 µl $MgCl_2$, 1 µl *Taq* DNA polymerase and add purified DNA to a final volume of 10 µl. Incubate for 30 min at 72 °C. The products can be stored at 4 °C.
2. To perform fragment ligation and transformation, 20–50 ng (depending on the size of the product) of the purified PCR product is ligated into the TA cloning Kit vector according to manufacturer's instructions and stored at 4 °C overnight.
3. The bacteria are treated with 1.7 µl of ß-mercaptoethanol per 100 µl cells (provided with competent cells) for 10 min on ice.
4. Three microliters of the ligation reaction are added to 20 µl of competent cells and incubated for 30 min on ice.
5. Heat-pulse the samples in a 42 °C water bath for exactly 50 s (the duration of this heat shock is a critical step).
6. Briefly put the tubes back on ice (2 min).
7. Add 180 µl of preheated (37 °C) SOC-medium and incubate the tubes at 37 °C for 50 min with shaking at 240 rounds per minute (rpm).
8. Distribute 100 µl of the transformation mixture on LB agar plates containing 100 µg/ml ampicilline and IPTG/X-gal for blue-white color screening.
9. Incubate the plates at 37 °C overnight (color development of the blue-white screening may be enhanced by subsequent incubation for 2–24 h at 4 °C). Colonies containing plasmids with inserts will show white colour, colonies containing plasmids without inserts will be blue.

10. Pick plasmid-containing colonies (white) and inoculate them in 3–4 ml LB medium containing 100 μg ampicillin per ml in 13-ml reaction tubes with vent cap.
11. Culture bacteria for 16 h at 37 °C with shaking at 240 rpm.
12. Perform the plasmid isolation with the mini plasmid preparation kit according to manufacturer's instructions.
13. Elute plasmid DNA in 30 μl nuclease-free water (*see* **Note 8**).
14. Perform the sequencing as described under Subheading 3.1.5, using 2 μl of plasmid DNA solution.

3.1.7 Sequence Evaluation

V gene rearrangement sequences can be evaluated for gene segment homology and somatic mutations by using the IMGT database and the DNA plot software (http://www.imgt.org/IMGT_vquest/share/textes/ or http://vbase.mrc-cpe.cam.ac.uk/index.php?&MMN_position=1:1) or the Igblast tool (http://www.ncbi.nlm.nih.gov/igblast/) (*see* **Notes 9** and **10**).

3.2 BCL6 Mutation Analysis

3.2.1 DNA Isolation

DNA is isolated from cell suspensions or frozen tissue sections as described (*see* Subheading 3.1.1).

3.2.2 BCL6 PCR

As in our experience the MMC of *BCL6* is often difficult to amplify with sufficient yield in one round of PCR, we recommend a semi-nested PCR strategy to amplify the MMC of *BCL6*. The resulting amplificate has a length of 744 bp.

1. The first round of amplification is carried out in 2.0 mM $MgCl_2$, 200 μM dNTPs, 0.1 μM each *BCL6* first round primer and 1.75 U High Fidelity DNA polymerase.
2. Prepare a master mix including 5 μl dNTPs (2 mM), 5 μl 10× PCR buffer, 2.5 μl each of *BCL6* first round primers, 4 μl 25 mM $MgCl_2$ and 29.5 μl H_2O on ice.
3. Distribute 48.5 μl master mix to an appropriate number of PCR reaction vials.
4. Add 1–10 ng of the desired DNA template (*see* Subheading 3.2.1) and overlay the reaction mix with mineral oil to avoid evaporation and contaminations.
5. Add 0.5 μl High fidelity enzyme solution after the first denaturation step (hot start). PCR program: 95 °C 4 min, 68 °C pause (add enzyme), 72 °C 1 min, 30 × (95 °C 50 s, 59 °C 30 s, 72 °C 60 s), 72 °C 5 min. As a two-rounded PCR is particularly prone to contamination issues, please consider **Note 3**.
6. For the second round of amplification, 1.0 μl of the first round reaction is used as template. PCR conditions are 2.0 mM $MgCl_2$, 0.2 mM dNTPs, 5 μl 5× PCR buffer, 2.5 μM *BCL6*

second round primers (see above) and 1.6 U *Taq* DNA polymerase.

7. Run PCR in a cycler with heated lid. PCR program: 95 °C 4 min, 30 × (95 °C 50 s, 56 °C 30 s, 72 °C 60 s), 72 °C 5 min, 10 °C pause.

3.2.3 PCR Product Purification

Amplificates are purified with the DNA Gel extraction kit. If by-products are detectable by analytical 2% agarose gel electrophoresis, the desired product (size 744 bp) has to be isolated as described (*see* Subheading 3.1.4).

3.2.4 BCL6 MMC Sequencing (Sanger)

Sequencing is best performed for both DNA strands with second round amplification primers. Follow the sequence protocol as described under Subheading 3.1.5 but use 2 μl of Big Dye reaction mix for the sequence reaction if using the ABI Cycle Sequencing Kit, version 3.1 on an ABI3130 sequencer. Sequence analysis can be performed with the SeqScape software version 2.5 and the GenBank data library.

3.2.5 Cloning of PCR Products

The *BCL6* amplificates can also be cloned into the TA cloning kit vector and transformed into competent cells as described (Subheading 3.1.6). Use 2 μl of the isolated plasmid DNA for sequencing (*see* Subheading 3.1.5) (*see* **Note 11**).

3.3 Differential Gene Expression

Gene expression profiling (GEP) is a method to quantify a multitude of RNA transcripts in parallel within a biological sample, aiming at the determination of transcriptome patterns that may help to identify active biological processes and/or understand similarities or differences between individual samples. GEP technologies are highly diverse, ranging from comprehensive transcriptome profiling of single cells to the specialized comparison of selected transcripts among different cell types [36]. The use of GEP for the determination of the origin of GC-derived malignancies requires transcriptome-wide analyses and comparisons of normal and malignant populations. We focus on two different techniques, i.e., oligonucleotide array analysis, and next generation sequencing of RNA (RNA-Seq), using samples of several thousand sort-purified lymphocytes. We consider these approaches as the current state-of-the-art to comprehensively reveal the degree of relationship as well as individual expression patterns of normal and malignant lymphocyte populations. We further focus on standard reagents and protocols for both, RNA-Seq and oligonucleotide array technologies, anticipating that putative normal precursors as well as malignant cells are available in sufficient amount and high quality, i.e., several thousand sort-purified, viable cells. Finally, we discuss advantages and limitations of both technical approaches (*see* **Note 12**) and we describe general pitfalls in GEP analysis (*see* **Note 13**), which may be largely avoided by careful experimental design (*see* **Note 14**).

3.3.1 Cell Isolation

Both normal and malignant lymphocyte populations should be prepared for sort purification by depletion of unwanted cells and cell debris. This is carried out best by pre-enrichment of leukocytes by density gradient centrifugation of cell suspensions prepared from peripheral blood (PB) or dispersed lymphoid tissue.

1. Prepare cell suspensions from tissues, e.g., by slicing, mincing, and pressing through steel meshes (ca. 1 mm pore size).
2. Dilute and wash with PBS–5% BSA.
3. Filter cell suspension through filters with appropriate pore sizes, e.g., 45 μm for lymphocytes.
4. Density gradient centrifugation is carried out by overlaying appropriate polysaccharide solutions (density 1.077 g/ml) with the cell suspension and subsequent centrifugation with a swing rotor at 400 × *g* at 4 °C for 30 to 45 min.
5. Optional: Further enrichment of desired cell types may be achieved by magnetic activated cell sorting (MACS) with bead-coupled antibodies against appropriate surface antigens, e.g., CD19 for positive B cell enrichment, or antibody cocktails marking non-B cells for depletion of unwanted cell types (negative enrichment).
6. Optional: Flow cytometric sort purification is carried out via direct or indirect labeling of desired normal and malignant cell-types with fluorescent antibodies against an appropriate combination of surface molecules, e.g., CD77, CD10, $CD38^{int}$ or $CD20^{high}$ for human GC B cells. Cell sorting systems with high throughput, high optical resolution performance and sample cooling systems are recommended for purification and purity control. Sort purity should be determined by aliquot reanalysis. Samples with purity below 99% should be purified repeatedly or excluded from the analysis. For small samples (<10,000 cells), direct sort into RNA lysis buffer is possible, however, large cell samples should be sedimented and the supernatant removed after sorting for storage (−80 °C) or further processing.

3.3.2 RNA Isolation

RNA isolation is carried out by RNeasy Mini kit, according to manufacturer's instructions. After a mandatory DNAseI digestion and sample washing, RNA elution is carried out in two subsequent steps with each 25 μl of RNAse-free water. RNA quality should be determined by measurement on an Agilent Bioanalyzer. The RNA integrity numbers should exceed 8.5.

3.3.3 GEP by Oligonucleotide Arrays

For the preparation of oligonucleotide arrays, we focus here on the standard GeneChip® WT PLUS Reagent kit manual target preparation procedure for analysis with GeneChip® Whole Transcript Expression arrays. In short, 50–500 ng of high quality total RNA are reverse transcribed with random primers containing a T7 promotor sequence at the 5′ end, generating DNA/RNA hybrid molecules. For the second strand synthesis, RNaseH and DNA polymerase treatment act together to digest the RNA and generate double stranded cDNAs containing a T7 promotor sequence 5′ of the initial transcript sequence. In a subsequent in vitro transcription of the cDNA, antisense RNA (aRNA or cRNA for complementary RNA) is synthesized according to the Eberwine protocol, thus linearly amplifying the initial transcript information. The aRNA is purified by magnetic beads, quantified and quality controlled by Agilent 2100 Bioanalyzer analysis, using a RNA 6000 Nano Kit, loading at least 200 ng of aRNA per sample to obtain reliable results. High quality aRNA samples show a distribution profile of sizes from 50 to 4500 nucleotides, with a maximum at 200 to 2000 nucleotides. The sense-strand cDNA, necessary for array hybridization, is synthesized by the reverse transcription of 15 μg aRNA using second cycle random primers and a dNTP mixture including a fixed ratio of dUTP to dTTP. The single stranded cDNA from the second cycle is purified with magnetic beads and analysed for yield and sample quality with an Agilent 2100 Bioanalyzer. High quality samples show a homogenous distribution profile with median sizes of about 400 nucleotides. Subsequently, 5.5 μg of second cycle cDNA are fragmented by uracil-DNA glycosylase treatment and covalently labeled with biotin by terminal deoxynucleotidyl transferase treatment using the Affymetrix proprietary DNA labeling reagent. Finally, the sample is hybridized on the oligonucleotide arrays for 16 h at 45 °C, followed by washing and staining in the GeneChip® Fluidics Station 450 and scanned in the GeneChip Scanner as recommended by Affymetrix.

3.3.4 Library Preparation and RNA-Seq

A broad range of RNA-Seq library construction kits as well as different strategies for next generation sequencing are commercially available, even more variety in technological approaches is given when RNA amplification prior to RNA-Seq—e.g., due to small samples—is required [37, 38]. As mentioned above, we focus here on the polyA-based TruSeq V2 library construction kit without prior RNA amplification and subsequent Illumina sequencing. Library preparation is carried out according to manufacturer's instructions. In short, the polyA-containing mRNA molecules from 100 ng to 1 μg of total RNA are enriched using oligo-dT-attached magnetic beads using two rounds of purification. During the second elution, the polyA$^+$ RNA is fragmented and primed for first strand cDNA synthesis. In the first strand cDNA synthesis, the cleaved polyA RNA fragments are primed with random hexamers and reverse transcribed. In the second strand synthesis, the RNA template is removed and a replacement strand is generated, resulting in double strand cDNA, subsequently purified with AMPure XP beads. The next steps involve a cDNA end repair, removing 3′- and refilling 5′-overhangs, introduced by the fragmentation process, as well as an adenylation of 3′ blunt ends, thereby avoiding concatemer formation. Next, indexed sequencing adapters are added through ligation and after clean up, the library is amplified in 15 PCR cycles. After magnetic purification of PCR products, these are quantified by Qubit fluorometric quantitation and quality controlled by a DNA chip analysis on an Agilent 2100 Bioanalyzer. Finally, libraries with appropriately selected indexes are pooled and samples are sequenced in a HiSeq 2000 sequencing run with Illumina PhiX controls, at an initial depth of four to five libraries per lane, which may be sufficient for initial analyses and deepened in subsequent sequencing runs of the same libraries. In many cases, single-read sequencing with 50-cycles kits will suffice for quantification of mRNAs. However, more sophisticated analyses, such as variant identification and splicing analysis may require sequencing of 100 bp reads in the paired-end mode. After the run is finished, a demultiplexing step is performed, which assigns the individual sequenced reads based on index sequence determined in the index read back to each sample present in the lanes of the flowcell. The resulting FASTQ raw sequence reads are suitable for many different alignment tools and contain base quality information attached to each position in the reads. For detailed information, refer to the respective manual (http://support.illumina.com/content/dam/illumina-support/documents/documentation/chemistry_documentation/samplepreps_truseq/truseqrna/truseq-rna-sample-prep-v2-guide-15026495-f.pdf).

3.3.5 Data Analysis

GEP data mining is a time intensive process that requires appropriate use and adaptation of computational algorithms, provided by numerous sources. It is possible to combine multiple software modules that are available as freeware or rather rely on commercially available analysis software packages that usually include convenient guidelines for data processing. For the unexperienced user, the latter approach is highly recommended and usually sufficient for a solid analysis of cellular relationships. However, it should be noted that the analysis of the cellular origin of GC B cell-derived malignancies includes the evaluation of immunologically relevant patterns among normal cell types, as well as the evaluation of deregulated gene expression among malignant cell types and finally, the integration of both findings to deduce the putative normal precursor. The definition of the normal precursor and the evaluation of deregulated gene patterns in the corresponding tumor may require revisions, depending on additional cellular subsets or further normal or malignant expression patterns described. To be on a safe side, the identification and interpretation of patterns, as comprehensive and as consistent as possible is necessary. In the following, we provide an overview of a typical possible evaluation pathway, however, we admit that there may be more suitable solutions for individual data sets.

First of all, before further processing of the RNA-Seq as well as array data sets, quality control should be performed. In the case of array data this begins with an inspection of the array images followed by reviewing various parameters, such as percentage of present calls, average fluorescence intensity of present calls, background, noise, scaling factors, housekeeping gene 3′/5′ fluorescence intensity ratio, hybridization control and polyA spike-in control fluorescence intensities and signal distribution patterns. Important parameters to be reviewed in RNA-Seq sets involve the global error rate as revealed by the PhiX spike-in library and number of sequence reads obtained per library. After alignment to the reference genome, e.g., with Tophat2, mapping efficiency, rRNA content, duplication rates, GC bias, transcript coverage, 3′/5′ bias, and count of detectable transcripts. Samples representing severe outliers in any of these parameters should be treated with caution or excluded.

Second, the experimental design for tumor precursor analyses may easily exceed sample numbers that cannot be covered by commercially available kit sizes or simply cannot be processed in a suitable time period in parallel. This issue generates an often underestimated problem called batch-effect, meaning a systematic, but inhomogenous technical bias between independently processed sample batches that may exceed and/or blur biological variation. To avoid batch effects and to minimize their influence, the initial experimental design is important as well as the use of appropriate algorithms, adjusting the raw data sets (*see* **Note 13**).

Third, to enable comparison of independent samples, data sets must be normalized on sample level. These normalization procedures account for variations between data sets, introduced by physical or chemical bias in individually processed samples (e.g., sample concentration). A broad range of normalization methods are available, ranging from simple scaling factors to sophisticated error models including transcript nucleotide composition and abundance. Normalization methods may significantly influence results. To find out which normalization suits best to your individual data set, start with one method and determine the consistency of your results with other normalization approaches.

Before transcript specific analyses can be performed, the RNA-Seq approach requires transcript annotation, especially to determine alternative splice variants or noncoding transcripts (provided by a constantly growing number of different software tools), whereas oligonucleotide arrays are much more convenient in that they are predesigned for specific transcripts and annotation is provided by the respective company.

To enable better visualization of the usually very large and often unseizable data points within and between samples, a gene-to-gene normalization may be performed, allowing convenient presentation of multiple data points in colour schemes like heat maps or hierarchical clusters.

Next, for the determination of normal precursers of GC B cell-derived malignancies, unsupervised comparisons are required. This means that clustering algorithms (hierarchical clustering, principal component analysis, classification algorithms) are used to depict similarity and distinctness among unclassified samples, thereby revealing relative relationships. As each sample includes false positive/negative data points and considerable amounts of background noise, a prefiltering of transcripts according to unsupervised metrics is necessary. Such a filtering step may be relative transcript abundance or minimum standard deviation among the entire data set. Suitable quality control metrics for clustering algorithms are minimal distances among replicate samples in combination with maximal transcript numbers underlying these calculations.

The identification of tumor-specific expression patterns is desired to comprehensively and exactly understand the transforming events leading to its pathobiology. These analyses are performed supervised, i.e., resemble a quantification of statistically significant differences, e.g., based on parameters like fold-change, probability of error and multiple testing correction, of previously defined sample subsets (e.g., normal versus tumor). The quantification of such differences is versatile, ranging from simply counting the number of differentially expressed genes to pattern recognition software like gene set enrichment analysis [39] or ingenuity pathway analysis. Also here, consistency among different approaches is usually a good indicator for a reliable determination of biological relevance.

Finally, it should be mentioned that a GEP analysis, no matter how deep sequenced, how pure the sample or how stringent the cutoffs chosen, can never represent a scientific proof, but should be rather considered as basis for the formulation of hypotheses, to be verified (or falsified) in independent tests, e.g., by analysing the observed patterns on protein level or, if possible, performing functional experiments.

4 Notes

1. The protocol given here is not meant for formalin-fixed and paraffin-embedded biopsy specimens. If only such material is available, one may try to perform the PCR after appropriate DNA isolation from the paraffin-embedded material with more cycles (40–45), to compensate for the partial degradation of

DNA in such material, or use a protocol with IgV_H primers binding to framework region (FR) II or III, so that shorter products are amplified [40].

2. The protocol described here uses V_H and V_κ family-specific primers binding to the FR1 of the respective genes together with mixtures of J segment primers. We also established PCR strategies for the amplification of V_λ gene rearrangements [41]. Also in this approach, V gene family-specific primers binding to sequences in FRI are used together with primers for the J gene segments. For the amplification of V_H gene rearrangements, V_H family-specific primers binding to sequences in the leader exon may be used instead of the FRI primers [42]. With the V_H leader primers, amplificates of about 500–550 bp size are obtained. There are usually less somatic mutations in the leader exon than in FRI. Therefore, usage of the leader primers is advantageous when highly mutated B cells are studied, because there is less chance for leader primers than for FRI primers that they cannot bind efficiently to the V genes due to somatic mutations at their binding sites. A further advantage of usage of V_H leader primers is that the complete V_H gene sequence is obtained. However, when the DNA quality is not optimal, the longer amplificates with V_H leader primers may be amplified with lower efficiency.
3. PCR contamination is a major concern. Therefore, precautions should be followed to obtain reliable results: Pre and post PCR work should be performed in separate rooms. Gloves should be worn and frequently changed. Separate pipettes, reaction tubes, aerosol resistant pipette tips and other equipment should be dedicated only for the pre PCR work. It is mandatory to run in parallel negative controls for each V gene PCR.
4. Since V_κ genes of the $V_\kappa5$ and $V_\kappa6$ families are only rarely used, the V_κ analysis may be restricted to families 1–4.
5. Since 60% of human B cells use Igκ light chains, the V_H and V_κ primer combination is the preferred one to analyse human B cells for clonality and the presence of somatic mutations. However, one has to keep in mind that most Igλ light chain-expressing B cells carry $V_\kappa J_\kappa$ joints, and that these joints are usually inactivated by rearrangements of a kappa deleting element (KDE) that is located downstream of the Cκ gene [3]. Rearrangement of the KDE to a sequence in the Jκ-Cκ intron deletes not only the Cκ gene but also the two Igκ enhancers [3]. However, $V_\kappa J_\kappa$ joints remain on the chromosome and can be amplified. As the Ig enhancers are needed for somatic hypermutation activity, such $V_\kappa J_\kappa$ joints remain unmutated when the other Ig loci in the cell acquire mutations.

6. The PCR products amplified with the V gene primers binding to the FR1 of human V_H and V_κ genes and the corresponding J segment primers are usually 300–350 bp long. However, occasionally also longer or shorter PCR products are obtained. These should not be disregarded, as they may represent V gene rearrangements, too. On the one hand, one has to keep in mind that the J gene segments are located close to each other in the IgH and Igκ loci (200–600 bp distances; Ref. [43, 44]) and that J primer mixes are used that bind to all J_H and J_κ segments. Hence, longer PCR products may represent amplificates ranging from the V primer binding site to a J gene segment downstream of the one that is rearranged. If a somatic mutation in the rearranged J segments hampers amplification from this J segment, the product from the next downstream J gene segment may be the predominant one. On the other hand, somatic hypermutation not only introduces point mutations, but also deletions or duplications [7]. Overall, about 40% of mutated out-of-frame rearrangements from human B cells have been found to carry deletions and/or duplications, which may range in length from single nucleotides to several hundred basepairs [7].
7. If it is only of interest to determine whether the clonal V region genes of a lymphoma are somatically mutated, the PCR products can be directly sequenced. However, if it is also of interest to determine whether there is intraclonal V gene mutation diversity as a trait of ongoing somatic hypermutation, the PCR products need to be cloned into plasmids and numerous plasmids should be sequenced. In that case, it is important to use a high fidelity DNA polymerase (mixture), so that PCR-introduced errors do not interfere with a reliable identification of low level intraclonal diversity. If subclonal mutations are seen in several plasmid sequences, and/or if a genealogical tree with several levels of intraclonal diversification can be established from the sequences, these are strong arguments for ongoing hypermutation during lymphoma clone expansion.
8. Before sequencing, one may test the plasmids for the presence of an insert of the expected size by performing a restriction enzyme digestion and agaraose gel electrophoresis. With the plasmid used here, the insert can be cut out of the plasmid with 0.5 U *EcoRI* restriction enzyme in a final volume of 10 μl. Put the complete digest on a 2% agarose gel and determine the positive clones. Positive colonies give raise to two bands, open vector and insert (3 kb and about 350 bp, respectively), whereas insert-less plasmids show just the size of the linearized vector.
9. For the determination of somatic mutations in the rearranged IgV genes, the J_H segments are normally disregarded, as there

are only few bases of J_H 5′ of the J_H primer. D_H segments are also disregarded, because D_H segments are typically trimmed during the V(D)J recombination process at both ends, so that it is difficult to distinguish true mutations in a D_H segment from variability generated through the concerted action of nucleases and terminal deoxynucleotidyl transferase, which adds one or several non-germline encoded bases to the ends of rearranging gene segments. Similary, mutations near the 3′ end of the V_H (or V_L) segments should only be counted if mutations are followed by at least three homologus bases.

10. Normally, a V gene rearrangement is considered as mutated if it carries at least 1% mutations. For rearrangements with only one or two bases mismatch to the germline sequence, rare germline polymorphisms as source of the mismatch have to be considered, or a rare mutation happening independent of conventional somatic hypermutation. An exception is CLL, where typically a border of 2% mutation frequency is used to distinguish mutated from unmutated CLL [45, 46]. The reason for this is that in CLL, cases with less than 2% *IGVH* gene mutation frequency clinically behave similar as CLL with 100% germline homology [45, 46].
11. When sequencing the BCL6 PCR product directly, a mutation on one allele will appear as a “double peak” in the electropherogram. However, if the DNA used for the analysis does not originate from purified lymphoma cells, but, for example, from tissue sections with less than 90% lymphoma cell content, the peak representing the mutation might be rather small. Therefore, to validate the presence of *BCL6* mutations, it is recommended in such an instance to clone the amplificate and sequence 10–20 plasmids. If several plasmids carry identical mutations, these mutations are reliable.
12. Advantages and limitations of RNA-Seq and oligonucleotide array analysis: Generally speaking, it is likely that in the future RNA-Seq analysis will replace currently available microarray technologies, due to several reasons: standard microarrays only allow the relative measurement and quantification of known transcripts, irrespective of their posttranscriptional variations (e.g., alternative splicing). Although modern microarray technologies also allow the analysis of alternatively spliced transcripts, the RNA-Seq analysis has the advantage of sequence information retention, thereby allowing a more precise quantification of single transcripts (especially, when unique molecular identifiers are included), the detection of unknown transcripts and fusion transcripts resulting from chromosomal rearrangements. Moreover, additional information on allelic variants, polymorphisms or mutations may be extracted from the data (Please note that the latter sequence informations are

not necessarily important for a sufficient uncovering of the cellular origin, but of course represent attractive sources of further information). However, RNA-Seq as the more detailed and comprehensive GEP technique is much more computationally intensive and usually requires sustainment from experienced users or bioinformaticians. The latter circumstance, as well as the fact that microarry technologies are faster and more easy to evaluate via a broad range of available tools (commercially available and open source), makes the choice of microarrays over RNA-Seq still a suitable approach for analyses that do not require in-depth evaluations.

13. Pitfalls of GEP approaches: The standard GEP approach requires the input of a minimum amount of RNA, which is only covered by the analysis of larger cell populations (unless single cell-based approaches are used which are technical highly demanding and produce more technical noise). The analysis of cell populations, however, is limited by the fact that individual data sets will always represent mixtures of cells, varying at least in terms of activation state, signaling activity or intensity of homeostatic functions. Even more variation in the respective composition of different cell types may be derived from the age of the donor, the available sample size, the phenotypical definition of the cell population, the cell purification approach used, or other parameters. These (mostly) biological parameters together will seriously affect the quality of the respective GEP analysis as they contribute considerable variation (noise), especially among low-level transcripts. A similar source for biological variation are putative contaminations with unwanted cell types: even the analysis of 99% pure B lymphocyte populations include the existence of one non-B cell among every 100 cells, so that strongly transcribed genes of these contaminating cell types (often represented by thousands of individual transcripts per single cell) may lead to substantial expression levels of false positive signals among the population of interest. A further pitfall in GEP analysis is the fact that usually more different cell types and replicates are required for a study than are possible to be processed in one experiment, due to limitations in assay sizes, laboratory capacities, or timely or manual feasibility. This results in the necessity to split the total study into more or less independent batches. Each of these batches, however, must be considered as an independent experiment, as usually none of the above mentioned technical or manual parameters or laboratory conditions can be kept entirely identical and reproducible for the total number of batches. Thus, each batch will include a batch-specific signature that will most likely affect the GEP of each batch-associated data set similarly, but differ substantially from the

batch signature of independent batches. These "batch-effect "issues critically enhance technical noise, or may even hamper the evaluability of the entire study, if not considered appropriately in the experimental design. Strong batch effects are usually identified by unsupervised clustering analyses that separate experimental conditions (generation date) rather than biological or technical replicates.

Finally, the analysis of the cellular origin of GC-derived B cell lymphomas by comparative GEP may help to identify the most similar normal counterpart of the respective tumor entity only from the cell types that have been included in the study. As similarity is a relative measure, the existence of even more closely related cell types should be ruled out from the conclusions by further, independent validation studies, once the most similar normal entities have been identified.

14. Experimental design: The appropriate experimental design of a GEP study may help to exclude typical pitfalls. First of all, the cell isolation strategy for both, tumor and normal cell samples should be considered carefully. Keep in mind that too few markers will lead to the inclusion of unwanted cell types, whereas too many markers will select a more and more specialized subpopulation. Thus, both extremes will considerably skew the data set. Moreover, typical experimental controls of the cell isolation procedure, e.g., by FACS-based reanalysis of MACS-enriched and sort-purified cell aliquots, cell numbers, the RNA and cDNA amount and quality by appropriate techniques are mandatory and allow exclusion of inappropriate samples before integration into the whole data set. Consider the use of biological replicates to account for sample-specific variations. The more even the distribution of biological replicates among different batches, the less impact will be observed from each batch on the sample specific expression patterns; however, more batches will always mean more technically induced variation. The latter effect may be encountered by using technical replicates, processed evenly distributed among the independent batches. Finally, the use of spike-in controls may be considered as an independent measure for estimating RNA-Seq library or microarray performance quality.

Acknowledgments

We thank Kerstin Heise and Ludger Klein-Hitpass for help in gathering the protocols. Our own work on lymphoma pathogenesis and GEP is supported by the Deutsche Forschungsgemeinschaft through the grants Ku1315/10-1, TRR60/A2, and SE1885/2-1, the Wilhelm Sander Foundation (2014.136.1), the Deutsche Krebshilfe (70112112), and by the Hariy Cell Leukemia Foundation.

References

1. Rajewsky K (1996) Clonal selection and learning in the antibody system. Nature 381 (6585):751–758
2. Matsuda F, Honjo T (1996) Organization of the human immunoglobulin heavy-chain locus. Adv Immunol 62:1–29
3. Bräuninger A, Goossens T, Rajewsky K, Küppers R (2001) Regulation of immunoglobulin light chain gene rearrangements during early B cell development in the human. Eur J Immunol 31(12):3631–3637
4. Jacob J, Kelsoe G (1992) In situ studies of the primary immune response to (4-hydroxy-3-nitrophenyl)acetyl. II. A common clonal origin for periarteriolar lymphoid sheath-associated foci and germinal centers. J Exp Med 176 (3):679–687
5. De Silva NS, Klein U (2015) Dynamics of B cells in germinal centres. Nat Rev Immunol 15 (3):137–148. doi:10.1038/nri3804
6. Di Noia JM, Neuberger MS (2007) Molecular mechanisms of antibody somatic hypermutation. Annu Rev Biochem 76:1–22
7. Goossens T, Klein U, Küppers R (1998) Frequent occurrence of deletions and duplications during somatic hypermutation: implications for oncogene translocations and heavy chain disease. Proc Natl Acad Sci U S A 95:2463–2468
8. Küppers R, Zhao M, Hansmann ML, Rajewsky K (1993) Tracing B cell development in human germinal centres by molecular analysis of single cells picked from histological sections. EMBO J 12(13):4955–4967
9. Liu YJ, Joshua DE, Williams GT, Smith CA, Gordon J, MacLennan IC (1989) Mechanism of antigen-driven selection in germinal centres. Nature 342(6252):929–931
10. Xu H, Chaudhri VK, Wu Z, Biliouris K, Dienger-Stambaugh K, Rochman Y, Singh H (2015) Regulation of bifurcating B cell trajectories by mutual antagonism between transcription factors IRF4 and IRF8. Nat Immunol 16(12):1274–1281. doi:10.1038/ni.3287
11. Basso K, Klein U, Niu H, Stolovitzky GA, Tu Y, Califano A, Cattoretti G, Dalla-Favera R (2004) Tracking CD40 signaling during germinal center development. Blood 104 (13):4088–4096. doi:10.1182/blood-2003-12-4291
12. Heise N, De Silva NS, Silva K, Carette A, Simonetti G, Pasparakis M, Klein U (2014) Germinal center B cell maintenance and differentiation are controlled by distinct NF-kappaB transcription factor subunits. J Exp Med 211 (10):2103–2118. doi:10.1084/jem.20132613
13. Dominguez-Sola D, Kung J, Holmes AB, Wells VA, Mo T, Basso K, Dalla-Favera R (2015) The FOXO1 transcription factor instructs the germinal center dark zone program. Immunity 43 (6):1064–1074. doi:10.1016/j.immuni.2015.10.015
14. Calado DP, Sasaki Y, Godinho SA, Pellerin A, Kochert K, Sleckman BP, de Alboran IM, Janz M, Rodig S, Rajewsky K (2012) The cell-cycle regulator c-Myc is essential for the formation and maintenance of germinal centers. Nat Immunol 13(11):1092–1100. doi:10.1038/ni.2418
15. Dominguez-Sola D, Victora GD, Ying CY, Phan RT, Saito M, Nussenzweig MC, Dalla-Favera R (2012) The proto-oncogene MYC is required for selection in the germinal center and cyclic reentry. Nat Immunol 13 (11):1083–1091. doi:10.1038/ni.2428
16. Seifert M, Küppers R (2009) Molecular footprints of a germinal center derivation of human IgM+(IgD+)CD27+ B cells and the dynamics of memory B cell generation. J Exp Med 206 (12):2659–2669
17. Pasqualucci L, Migliazza A, Fracchiolla N, William C, Neri A, Baldini L, Chaganti RSK, Klein U, Küppers R, Rajewsky K, Dalla-Favera R (1998) BCL-6 mutations in normal germinal center B cells: evidence of somatic hypermutation acting outside Ig loci. Proc Natl Acad Sci U S A 95:11816–11821
18. Fukita Y, Jacobs H, Rajewsky K (1998) Somatic hypermutation in the heavy chain locus correlates with transcription. Immunity 9:105–114
19. Yang SY, Fugmann SD, Schatz DG (2006) Control of gene conversion and somatic hypermutation by immunoglobulin promoter and enhancer sequences. J Exp Med 203 (13):2919–2928. doi:10.1084/jem.20061835
20. Weill JC, Weller S, Reynaud CA (2009) Human marginal zone B cells. Annu Rev Immunol 27:267–285

21. Budeus B, Schweigle de Reynoso S, Przekopowitz M, Hoffmann D, Seifert M, Küppers R (2015) Complexity of the human memory B-cell compartment is determined by the versatility of clonal diversification in germinal centers. Proc Natl Acad Sci U S A 112(38): E5281–E5289. doi:10.1073/pnas.1511270112
22. Küppers R, Klein U, Hansmann M-L, Rajewsky K (1999) Cellular origin of human B-cell lymphomas. N Engl J Med 341:1520–1529
23. Klein U, Dalla-Favera R (2008) Germinal centres: role in B-cell physiology and malignancy. Nat Rev Immunol 8(1):22–33
24. Küppers R (2005) Mechanisms of B-cell lymphoma pathogenesis. Nat Rev Cancer 5 (4):251–262
25. Brune V, Tiacci E, Pfeil I, Döring C, Eckerle S, van Noesel CJM, Klapper W, Falini B, von Heydebreck A, Metzler D, Bräuninger A, Hansmann M-L, Küppers R (2008) Origin and pathogenesis of nodular lymphocyte-predominant Hodgkin lymphoma as revealed by global gene expression analysis. J Exp Med 205(10):2251–2268
26. Tiacci E, Döring C, Brune V, van Noesel CJ, Klapper W, Mechtersheimer G, Falini B, Küppers R, Hansmann ML (2012) Analyzing primary Hodgkin and Reed-Sternberg cells to capture the molecular and cellular pathogenesis of classical Hodgkin lymphoma. Blood 120 (23):4609–4620. doi:10.1182/blood-2012-05-428896
27. Basso K, Liso A, Tiacci E, Benedetti R, Pulsoni A, Foa R, Di Raimondo F, Ambrosetti A, Califano A, Klein U, Dalla Favera R, Falini B (2004) Gene expression profiling of hairy cell leukemia reveals a phenotype related to memory B cells with altered expression of chemokine and adhesion receptors. J Exp Med 199 (1):59–68. doi:10.1084/jem.20031175
28. Klein U, Tu Y, Stolovitzky GA, Mattioli M, Cattoretti G, Husson H, Freedman A, Inghirami G, Cro L, Baldini L, Neri A, Califano A, Dalla-Favera R (2001) Gene expression profiling of B cell chronic lymphocytic leukemia reveals a homogeneous phenotype related to memory B cells. J Exp Med 194 (11):1625–1638
29. Seifert M, Sellmann L, Bloehdorn J, Wein F, Stilgenbauer S, Dürig J, Küppers R (2012) Cellular origin and pathophysiology of chronic lymphocytic leukemia. J Exp Med 209 (12):2183–2198. doi:10.1084/jem.20120833
30. Alizadeh AA, Eisen MB, Davis RE, Ma C, Lossos IS, Rosenwald A, Boldrick JC, Sabet H, Tran T, Yu X, Powell JI, Yang L, Marti GE, Moore T, Hudson J Jr, Lu L, Lewis DB, Tibshirani R, Sherlock G, Chan WC, Greiner TC, Weisenburger DD, Armitage JO, Warnke R, Levy R, Wilson W, Grever MR, Byrd JC, Botstein D, Brown PO, Staudt LM (2000) Distinct types of diffuse large B-cell lymphoma identified by gene expression profiling. Nature 403(6769):503–511. doi:10.1038/35000501
31. Küppers R, Dalla-Favera R (2001) Mechanisms of chromosomal translocations in B cell lymphomas. Oncogene 20(40):5580–5594
32. Fukuda T, Yoshida T, Okada S, Hatano M, Miki T, Ishibashi K, Okabe S, Koseki H, Hirosawa S, Taniguchi M, Miyasaka N, Tokuhisa T (1997) Disruption of the Bcl6 gene results in an impaired germinal center formation. J Exp Med 186(3):439–448
33. He B, Qiao X, Cerutti A (2004) CpG DNA induces IgG class switch DNA recombination by activating human B cells through an innate pathway that requires TLR9 and cooperates with IL-10. J Immunol 173(7):4479–4491
34. Kulis M, Merkel A, Heath S, Queiros AC, Schuyler RP, Castellano G, Beekman R, Raineri E, Esteve A, Clot G, Verdaguer-Dot N, Duran-Ferrer M, Russinol N, Vilarrasa-Blasi R, Ecker S, Pancaldi V, Rico D, Agueda L, Blanc J, Richardson D, Clarke L, Datta A, Pascual M, Agirre X, Prosper F, Alignani D, Paiva B, Caron G, Fest T, Muench MO, Fomin ME, Lee ST, Wiemels JL, Valencia A, Gut M, Flicek P, Stunnenberg HG, Siebert R, Küppers R, Gut IG, Campo E, Martin-Subero JI (2015) Whole-genome fingerprint of the DNA methylome during human B cell differentiation. Nat Genet 47(7):746–756. doi:10.1038/ng.3291
35. Oakes CC, Seifert M, Assenov Y, Gu L, Przekopowitz M, Ruppert AS, Wang Q, Imbusch CD, Serva A, Koser SD, Brocks D, Lipka DB, Bogatyrova O, Weichenhan D, Brors B, Rassenti L, Kipps TJ, Mertens D, Zapatka M, Lichter P, Döhner H, Küppers R, Zenz T, Stilgenbauer S, Byrd JC, Plass C (2016) DNA methylation dynamics during B cell maturation underlie a continuum of disease phenotypes in chronic lymphocytic leukemia. Nat Genet 48 (3):253–264. doi:10.1038/ng.3488
36. Kawasaki ES (2006) The end of the microarray Tower of Babel: will universal standards lead the way? J Biomol Tech 17(3):200–206
37. Wang Z, Gerstein M, Snyder M (2009) RNA-Seq: a revolutionary tool for transcriptomics. Nat Rev Genet 10(1):57–63. doi:10.1038/nrg2484
38. Shanker S, Paulson A, Edenberg HJ, Peak A, Perera A, Alekseyev YO, Beckloff N, Bivens NJ, Donnelly R, Gillaspy AF, Grove D, Gu W,

Jafari N, Kerley-Hamilton JS, Lyons RH, Tepper C, Nicolet CM (2015) Evaluation of commercially available RNA amplification kits for RNA sequencing using very low input amounts of total RNA. J Biomol Tech 26(1):4–18. doi:10.7171/jbt.15-2601-001

39. Subramanian A, Tamayo P, Mootha VK, Mukherjee S, Ebert BL, Gillette MA, Paulovich A, Pomeroy SL, Golub TR, Lander ES, Mesirov JP (2005) Gene set enrichment analysis: a knowledge-based approach for interpreting genome-wide expression profiles. Proc Natl Acad Sci U S A 102(43):15545–15550
40. van Dongen JJ, Langerak AW, Bruggemann M, Evans PA, Hummel M, Lavender FL, Delabesse E, Davi F, Schuuring E, Garcia-Sanz R, van Krieken JH, Droese J, Gonzalez D, Bastard C, White HE, Spaargaren M, Gonzalez M, Parreira A, Smith JL, Morgan GJ, Kneba M, Macintyre EA (2003) Design and standardization of PCR primers and protocols for detection of clonal immunoglobulin and T-cell receptor gene recombinations in suspect lymphoproliferations: report of the BIOMED-2- Concerted Action BMH4-CT98-3936. Leukemia 17(12):2257–2317
41. Bräuninger A, Küppers R, Spieker T, Siebert R, Strickler JG, Schlegelberger B, Rajewsky K, Hansmann ML (1999) Molecular analysis of single B cells from T cell-rich B-cell lymphoma shows the derivation of the tumor cells from mutating germinal center B cells and exemplifies means by which immunoglobulin genes are modified in germinal center B cells. Blood 93:2679–2687
42. Braeuninger A, Küppers R, Strickler JG, Wacker HH, Rajewsky K, Hansmann ML (1997) Hodgkin and Reed-Sternberg cells in lymphocyte predominant Hodgkin disease represent clonal populations of germinal center-derived tumor B cells. Proc Natl Acad Sci U S A 94(17):9337–9342
43. Hieter PA, Maizel JV Jr, Leder P (1982) Evolution of human immunoglobulin kappa J region genes. J Biol Chem 257(3):1516–1522
44. Ravetch JV, Siebenlist U, Korsmeyer S, Waldmann T, Leder P (1981) Structure of the human immunoglobulin mu locus: characterization of embryonic and rearranged J and D genes. Cell 27(3 Pt 2):583–591
45. Damle RN, Wasil T, Fais F, Ghiotto F, Valetto A, Allen SL, Buchbinder A, Budman D, Dittmar K, Kolitz J, Lichtman SM, Schulman P, Vinciguerra VP, Rai KR, Ferrarini M, Chiorazzi N (1999) Ig V gene mutation status and CD38 expression as novel prognostic indicators in chronic lymphocytic leukemia. Blood 94 (6):1840–1847
46. Hamblin TJ, Davis Z, Gardiner A, Oscier DG, Stevenson FK (1999) Unmutated Ig V(H) genes are associated with a more aggressive form of chronic lymphocytic leukemia. Blood 94(6):1848–1854

Chapter 21

Tracking B-Cell Repertoires and Clonal Histories in Normal and Malignant Lymphocytes

Nicola J. Weston-Bell, Graeme Cowan, and Surinder S. Sahota

Abstract

Methods for tracking B-cell repertoires and clonal history in normal and malignant B-cells based on immunoglobulin variable region (*IGV*) gene analysis have developed rapidly with the advent of massive parallel next-generation sequencing (mpNGS) protocols. mpNGS permits a depth of analysis of *IGV* genes not hitherto feasible, and presents challenges of bioinformatics analysis, which can be readily met by current pipelines. This strategy offers a potential resolution of B-cell usage at a depth that may capture fully the natural state, in a given biological setting. Conventional methods based on RT-PCR amplification and Sanger sequencing are also available where mpNGS is not accessible. Each method offers distinct advantages. Conventional methods for *IGV* gene sequencing are readily adaptable to most laboratories and provide an ease of analysis to capture salient features of B-cell use. This chapter describes two methods in detail for analysis of *IGV* genes, mpNGS and conventional RT-PCR with Sanger sequencing.

Key words *IGHV* gene, Massive parallel next-generation sequencing, Germinal center, B-cell repertoire, Tumor origins, Multiple myeloma, B-cell malignancy

1 Introduction

Maturation of normal B-cells underpins humoral immunity and available naïve and antigen-experienced B-cell repertoires can be tracked by analyzing immunoglobulin variable (*IGV*) region genes, encoding the antigen-binding variable domain of the B-cell receptor (BCR). Signals via the BCR regulate differentiation in response to antigen. In malignant B-cells, *IGV* gene analysis provides an important tool to identify the cell of origin and to track its clonal history in disease pathogenesis [1].

We present two different and distinct methodologies for the analysis of the *IGV* gene repertoires in normal and malignant

Nicola J. Weston-Bell and Graeme Cowan contributed equally to this work.

Dinis Pedro Calado (ed.), *Germinal Centers: Methods and Protocols*, Methods in Molecular Biology, vol. 1623,
DOI 10.1007/978-1-4939-7095-7_21, © Springer Science+Business Media LLC 2017

B-cells: first, using massive parallel next generation sequencing (mpNGS), and second using conventional RT-PCR followed by Sanger sequencing. The two differ vastly in scope of depth of analysis of *IGV* genes, and their use will be dictated by needs of the investigation to be undertaken. With the advent of mpNGS, it has become possible to study the composition and dynamics of the BCR repertoires on a scale and level of detail that surpasses conventional analysis massively [2]. However, where mpNGS is not available as an analytical tool, conventional Sanger based sequencing of *IGV* genes remains a highly informative method of analysis.

1.1 Germline Immunoglobulin Variable Region Gene Repertoire and Diversity

The immunoglobulin molecule consists of two identical heavy (IGH) glycoprotein chains paired with either two identical kappa (IGK) or lambda (IGL) light chains. In these the variable domains that recognize antigen are encoded very early in B-cell development by rearranged variable (*V*), diversity (*D*) (*H* chain only), and joining (*J*) gene segments selected from a *germline* genomic repertoire, and combined uniquely in individual B-cells to generate the first tier of the naïve B-cell repertoire [3, 4]. The human *IGV* germline gene database, ImMunoGeneTics (IMGT; www.imgt.org), lists 55 *IGHV*, 23 *IGHD*, and 6 *IGHJ* genes with functional alleles that can be combined and paired with rearranged 41 *IGKV* and 5 *IGKJ* or 33 *IGLV* and 5 *IGLJ* genes with functional alleles. Further to the germline combinatorial diversity, naïve B-cell encounter with antigen in germinal centers (GC) of secondary lymph nodes triggers a further diversity by somatic hypermutation (SHM) of *IGV* genes to improve BCR fit for antigen, and generate the second tier of the B-cell repertoire [5].

1.2 Relevance

High-throughput NGS methods have been used, for example in the study of the normal B-cell repertoire by assessing bias in *IGHV* gene usage [6–8] and in D segment reading frame [9]. *IGV* mutational status differentiates antigen-naïve B-cells from their antigen-experienced memory B-cell counterparts and these characteristics can be exploited to evaluate the B-cell of origin in B-cell malignancy and its evolution in relation to the GC [1, 10]. To illustrate, mpNGS of tumor-derived *IGHV* genes reveal that reentry to the dark zone of the GC most likely occurs in the cell of origin giving rise to monoclonal gammopathy of undetermined significance (MGUS) and multiple myeloma, and that this cell of origin appears to be a pre-plasma cell committed to terminal differentiation but still undergoing SHM [11].

1.3 mpNGS IGV Analysis

A number of different experimental approaches exist, with particular methodologies being advantageous for different experimental questions. Broadly, the genes encoding the variable region of the BCR are amplified by PCR either from genomic DNA or cDNA.

The use of genomic DNA may yield more accurate transcript quantitation since mRNA levels vary greatly between different B-cell populations; however, greater numbers of nonproductive arrangements are also obtained with DNA as template [12]. Methods relying on the amplification of the *IGV* genes from cDNA mostly utilize a reverse primer situated in a conserved region of the gene encoding the immunoglobulin constant domain (*IGHC; IGHA, IGHD, IGHE, IGHG, IGHM*) and owing to the intrinsic variability of the variable region, pools of primers, often degenerate, are used to cover known *IGV* gene alleles. An alternative approach for BCR repertoire sequencing is the use of the SMART cDNA synthesis system [13], which employs a phenomenon known as template switching to attach an oligonucleotide of known sequence to all 5′ cDNA ends. This approach is most suited to studies of the 3′ end of the *IGV*-region, including CDR3, since with current sequencing limitations (2 × 300 bp paired ends) the sequence of the 5′ region of the V segment may be missing or error-prone.

This protocol selectively describes the generation of immunoglobulin G (IgG) and immunoglobulin M (IgM) encoding B-cell *IGV* repertoire sequencing data using total RNA extracted from human B-cell populations as source material. It can be readily adapted for analysis of other isotype repertoires by change of reverse primer in the constant region *IGHV* gene. Owing to the potential for PCR or sequencing errors, in the analysis of sequence variants it is advisable to include technical replicates (analysis of aliquots of the same biological sample). Antibody sequence variants found in multiple libraries are only likely to occur at similar frequencies where they are true variants present in each of the starting aliquots [11].

1.4 Conventional *IGV* Analysis

Conventional *IGHV* sequence analysis utilizes degenerate primers targeted to the *V* regions of *IGH* or *IGK/L* genes in the RT-PCR step, with cloning prior to Sanger sequencing providing a means to measure variation in the B-cell repertoire.

The cloning method can also be used to identify tumor-related clonal sequence in analysis of intraclonal variation (ICV), or explore the normal B-cell repertoire, albeit with less depth than high-throughput methods. Where a tumor clonal sequence is the aim, knowledge of the surface Ig isotype is of great benefit as the RT-PCR reactions to be performed can be limited to the appropriate *IGHC* antisense primer combined with the chosen V region-specific sense primer mix.

Care should be taken to obtain a sufficiently pure B-cell population depending on the question to be addressed prior to nucleic acid extraction, as the data produced will only be as clean as the source sample.

2 Materials

2.1 Samples

Sample purity should be appropriate for type of repertoire information required. When total B-cells from a normal sample are being investigated, purification to >90% (ideally >95%) becomes critical when subsets of normal B-cells are to be compared, as all will contain *IGHV-D-J* rearrangements. Flow cytometry based sorting (FACSAria) can achieve B-cell populations of 95–99% purity using a two-step approach with an initial "yield" precision sort for CD19+ cells, followed by a second "purity" precision sort using stringent gates for specific subsets [11, 14]. Obtaining >99% pure bulk T-cells for use as a paired germline control can be achieved with a single "purity" sort to assess background error rates [11, 14]. Tumor clonal sequences can be identified using the conventional method with total RNA from whole blood or bone marrow mononuclear cells but cloning of amplification product prior to Sanger sequencing is essential (clonal tumor-derived sequences are identified by identical or shared CDR3 motifs when present at a frequency of ~35% or greater). Purifying specific tumor populations enables a more detailed analysis of clonal cells.

The most common sample types are peripheral blood or bone marrow, for which procedures are outlined below:

1. Peripheral blood (PB); obtain 10–20 mL (volume dependent on expected cellular yield for disease type, *see* **Note 1**) peripheral blood by venepuncture and put into lithium-heparin (LH) or ethylenediamine tetraacetic acid (EDTA) tubes, for processing on the same day.
2. Bone marrow (BM) aspirates; volume dependent on expected cellular yield for disease type (*see* **Note 1**).
3. PB or BM mononuclear cells (PBMC/BMMC) are obtained by density centrifugation using Lymphoprep according to manufacturer's protocol.
4. PBMC/BMMC can be stored as pellets (approx. 5×10^6) at −80 °C or in DMSO-containing FCS (10% DMSO, 90% FCS) for RNA.
5. For samples stored in DMSO, a wash step using sterile RPMI 1640 medium will be required before the extraction of RNA.
6. For DNA, cells are best stored in DMSO-containing FCS.

2.2 mpNGS *IGV* Analysis

2.2.1 cDNA Synthesis

1. Total RNA extracted from B-cell population (*see* **Note 2**).
2. Heat block or thermal cycler for 0.2 μL PCR tubes.
3. PCR workstation (*see* **Note 3**).
4. oligo(dT)18 (e.g., Thermo Fisher Scientific), 20 μM.

5. 0.2 μL thin-wall PCR tubes (alternatively 96-well PCR plates may be used).
6. DNase/RNase-free water.
7. Recombinant RNase inhibitor (e.g., Clontech).
8. SMARTScribe reverse transcriptase and 5× first strand buffer (Clontech).
9. 20 mM DTT (supplied with SMARTscribe transcriptase).
10. dNTP mix (10 mM).

2.2.2 V-Region PCR

1. Phusion Flash Master Mix (Thermo Fisher Scientific).
2. PCR primers (Tables 1 and 2).
3. PCR tubes or PCR plate with sealing foil.
4. PCR workstation (*see* **Note 3**).
5. Thermal cycler.

2.2.3 Sample Pooling and Preparation for Sequencing

1. Propan-2-ol, molecular biology grade.
2. Ethanol, molecular biology grade.
3. 3 M sodium acetate.
4. DNase/RNase-free water.
5. NanoDrop spectrophotometer or similar device for DNA quantitation (e.g., Thermo Fisher Scientific).
6. eGel SizeSelect gel apparatus and 2% SizeSelect gels (Thermo Fisher Scientific).
7. 50 bp DNA ladder (e.g., Thermo Fisher Scientific).
8. 0.5 mL screw-cap microcentrifuge tubes.

2.3 Conventional IGV Analysis

2.3.1 Extraction and Complimentary Deoxyribonucleic Acid Preparation Reagents

1. Lymphoprep (Stemcell Technologies).
2. (Optional) RNeasy Mini Kit and RNase-Free DNase Set (Qiagen).
3. Isopropanol, molecular biology grade.
4. RPMI 1640 medium.
5. Fetal calf serum.
6. DMSO (e.g., Sigma-Aldrich).
7. DNeasy Blood & Tissue Kit and 100 mg/mL RNase A (e.g., Qiagen)
8. Ethanol, molecular biology grade.
9. GoScript Reverse Transcription System (Promega).

2.3.2 Primers

1. Desalt purity custom oligos (e.g., Sigma-Aldrich), (*see* Table 4).

Table 1
Oligonucleotide sequences of PCR primers

Forward amplification primers		
V0	hIGHV0.0	AATGATACGGCGACCACCGAGATCTACACGACATAGACCGGT**TCAGTGAAGGTYTCCTGC**
V1	hIGHV1.0	AATGATACGGCGACCACCGAGATCTACACCGCCGGAAGCAGT**GCTGAGGTGAAGAAGCCT**
	hIGHV1.1	AATGATACGGCGACCACCGAGATCTACACCGCCGGAAGCAGT**YCAGGACTGGTGAAGCCT**
	hIGHV1.2	AATGATACGGCGACCACCGAGATCTACACCGCCGGAAGCAGT**SCAGGACTGTTGAAGCCT**
V2	hIGHV2.0	AATGATACGGCGACCACCGAGATCTACACGACATGACTTCGT**CCCTCTCACTCACCTGTG**
	hIGHV2.1	AATGATACGGCGACCACCGAGATCTACACGACATGACTTCGT**CCCTKAGACTCTCCTGTG**
	hIGHV2.2	AATGATACGGCGACCACCGAGATCTACACGACATGACTTCGT**CCCTGARACTCTCCTGTG**
V3	hIGHV3.0	AATGATACGGCGACCACCGAGATCTACACGACACTACAGCGT**AGACCCTCACRCTGACCT**
V4	hIGHV4.0	AATGATACGGCGACCACCGAGATCTACACCATGTGCCTGTGT**GTCCCTGAGACTCTCCTG**
	hIGHV4.1	AATGATACGGCGACCACCGAGATCTACACCATGTGCCTGTGT**GTCTCTGARGATCTCCTG**
V5	hIGHV5.0	AATGATACGGCGACCACCGAGATCTACACCTTAGAGTCACGT**CTCCTGCAAGGYTTCTGG**
Reverse amplification primers		
hIGGHrevINDEX1		CAAGCAGAAGACGGCATACGAGAT*AGCTCT*AGTCAGTCAGCC**ATGGGCCCTTGGTGGA**
hIGMHrevINDEX1		CAAGCAGAAGACGGCATACGAGAT*AGCTCT*AGTCAGTCAGCC**GGCGGATGCACTCCC**

Where multiple primers are present within an individual reaction, a pool (marked as V0–V5) is created containing an equimolar mix of the individual primers to achieve the same overall final primer concentration (10 μM). Gene specific portions are shown in *bold* and the index component of the reverse primers is written in *italics*. This portion is changed to the reverse complement of each index sequence (*see* Table 2)

Table 2
Oligonucleotide index sequences

Index	Sequence/Reverse
1	AGAGCT/AGCTCT
2	AGGATC/GATCCT
3	AGCTAG/CTAGCT
4	AGTCGA/TCGACT
5	ACACTG/CAGTGT
6	ACGTCA/TGACGT
7	ACCAGT/ACTGGT
8	ACTGAC/GTCAGT
9	ATCGTA/TACGAT
10	GAAGTC/GACTTC
11	GAGACT/AGTCTC
12	GACTGA/TCAGTC
13	GATCAG/CTGATC
14	GGAAGG/CCTTCC
15	GGCCTT/AAGGCC
16	GGTTCC/GGAACC
17	GCATAT/ATATGC
18	GTACCA/TGGTAC
19	GTGTTG/CAACAC
20	GTCAAC/GTTGAC

Index sequences are used to allow multiplexing of multiple samples on a single sequencing run. This table shows 20 individual index sequences systematically generated to each differ by at least two nucleotides and to prevent serial repeats of greater than two bases. Sequences should be substituted into the index position of the reverse primers shown in Table 1 to facilitate sample multiplexing. Many hundred additional indices may be generated according to these rules but are not shown here for space purposes

2.3.3 PCR Reagents

1. GoTaq Hot Start Polymerase (Promega).
2. dNTPs—prepare a mix with each dNTP at 10 mM (e.g., Promega; Set of dATP, dCTP, dGTP, dTTP).
3. Agarose (e.g., Fisher Scientific).
4. Webgreen Advance DNA Stain (Web Scientific).
5. QIAquick Gel Extraction Kit (Qiagen).
6. Isopropanol, molecular biology grade.

2.3.4 Cloning Reagents

1. pGEM-T Vector System I (Promega).
2. *E. coli* strain XL-1 Blue competent cells (Agilent).
3. Ampicillin (e.g., Merck Millipore).
4. LB and LB Agar (e.g., Sigma).
5. X-gal; 5-bromo-4-chloro-3-indoyl-ß-D-galactopyranoside, 50 mg/mL (e.g., Promega).
6. IPTG; isopropyl-ß-D-thio-galactopyranoside, 0.1 M (e.g., Cambio).
7. QIAprep Spin Miniprep kit (Qiagen).
8. Ethanol, molecular biology grade.

2.3.5 DNA Sequence Analysis

1. BigDye Terminator v1.1 Cycle Sequencing Kit (Thermo Fisher Scientific).
2. ABI 3130xl DNA Sequencer (Thermo Fisher Scientific).
3. Sodium acetate (3 M, pH 5.2 [glacial acetic acid]) (e.g., Fisher Scientific).
4. Ethanol, molecular biology grade.
5. HiDi Formamide (Thermo Fisher Scientific).

2.3.6 Software

1. MacVector (MacVector Inc).
2. IMGT V-QUEST (http://www.imgt.org/IMGT_vquest/vquest?livret=0&Option=humanIg).

3 Methods

3.1 mpNGS IGV Analysis

3.1.1 cDNA Synthesis

1. Dilute 100 ng total RNA with RNase-free water to a volume of 4.5 μL in a 0.2 μL PCR tube (*see* **Note 2** regarding suitable methods for total RNA purification).
2. Add 1 μL oligo(dT)18 primer to the PCR tube, mix briefly by pipetting.
3. Incubate sample at 72 °C for 3 min on a heat block or thermal cycler.
4. During incubation make up the cDNA synthesis mix below. Volumes should be increased proportionately according to the number of biological samples or technical replicates to be performed.

2 μL	5× First-strand buffer
0.25 μL	DTT (20 mM)
1 μL	dNTP mix (10 mM)
0.25 μL	Recombinant RNase inhibitor
1 μL	SMARTscribe reverse transcriptase

5. Transfer sample tubes to a block set to 42 °C or set thermal cycler to 42 °C.
6. Add 4.5 μL of cDNA synthesis mix to each reaction tube and briefly mix by pipetting.
7. Incubate samples at 42 °C for 1 h, then terminate the reaction by heating to 70 °C for 15 min.
8. Store cDNA at −70 °C or below until ready to proceed to V-region amplification reactions.

3.1.2 V-Region PCR

1. For each B-cell sample, set up six individual PCR reactions as detailed below, each containing a single forward primer pool (V0–V5). Each individual B-cell sample or technical replicate should use a reverse primer with a unique barcode for sample identification following sequencing (*see* Table 2).

Phusion Flash Master Mix	10 μL
cDNA (from 100 ng RNA reaction)	0.75 μL
Fwd Primer (10 μM) V0–V5	1 μL
IgG or IgM Reverse Primer (10 μM)	1 μL
DNase/RNase-free water	7.25 μL

Optional: SYBR-green I may be added to water at a dilution of 0.05% v/v (1 in 2000 dilution) to enable real-time tracking of PCR reactions using a quantitative thermocycler. This can be used to optimize the number of PCR cycles to the lowest number of cycles and to confirm that product has been produced for all reactions when many samples are being simultaneously processed.

2. Collect tube contents by centrifugation then place tubes on the thermocycler block and perform PCR using the cycling program below:

98 °C	5 m	
98 °C 57 °C 72 °C	10 s 60 s 30 s	6 cycles
98 °C 65 °C 72 °C	10 s 15 s 10 s	28 cycles
72 °C	2 m	

3. Remove tubes from thermal cycler.
4. Analyze 4 μL of each individual reaction by DNA agarose gel electrophoresis. The expected dominant band size is 450 bp (*see* **Note 4** for trouble shooting).

3.1.3 Sample Pooling and Preparation for Sequencing

1. Combine all PCRs within each sample and determine DNA concentration of each pool using a NanoDrop spectrometer or similar quantification device.
2. Combine equal quantities of each individually barcoded sample pool generated in **step 1** into a single tube.
3. Add 0.1 volumes 3 M sodium acetate and 1 volume propan-2-ol (where 1 volume represents the starting DNA volume).
4. Centrifuge the sample immediately at 10,000–15,000 × *g* for 20 min at 4 °C.
5. Decant the supernatant without disturbing the pellet.
6. Wash the DNA pellet by adding 1 volume of room-temperature 70% ethanol.
7. Centrifuge at 10,000–15,000 × *g* for 10 min at 4 °C.
8. Carefully decant the supernatant without disturbing the pellet, remove any remaining supernatant carefully with a fine pipette.
9. Air-dry the pellet for 5 min at room temperature.
10. Redissolve the DNA in 20 μL DNase/RNase-free water.
11. Determine the quantity of pooled DNA using a NanoDrop spectrophotometer or similar.
12. Load 100 ng of pooled DNA sample to each sample well on a 2% SizeSelect eGel and load 20 μL 50-bp ladder into the central ladder well.
13. Add 20 μL DNase/RNase-free water to each collection well on the eGel.
14. Run the eGel system on the 2% Size-Select program and monitor the progression of DNA bands towards the collection wells.

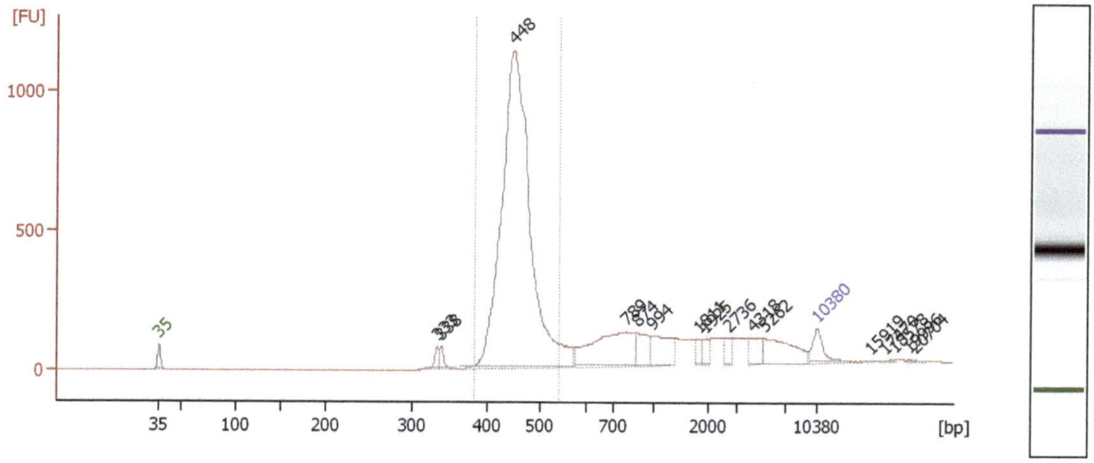

Fig. 1 Bioanalyzer trace of gel-purified amplicon pool ready for Illumina MiSeq sequencing. A dominant peak with mean fragment length of approximately 450 bp is expected; smaller bands representing primers and primer dimers should be absent

15. When the 450-bp band reaches the collection wells, collect the gel band and place in a 0.5 mL screw-cap microcentrifuge tube.
16. Analyze the gel purified DNA sample using a Bioanalyzer to determine DNA quantity and size distribution prior to sequencing. A sample trace is shown in Fig. 1.
17. Store DNA sample at −70 °C or below until sequencing.

3.1.4 Sequencing Requirements

Samples should be sequenced using Illumina MiSeq V2 (or above), 2 × 250 bp reads (or longer). Custom sequencing primers should be used as detailed in Table 3. A schematic diagram of the sequencing process is shown in Fig. 2.

3.2 Conventional IGV Analysis

3.2.1 Extraction of RNA and DNA

1. Extract RNA from approx. 2–5 × 10^6 PBMC/BMMCs (*see* **Notes 1** and **5**) using the RNeasy mini kit in accordance with manufacturer's protocol for animal cells. RNA can be eluted in as little as 30 μL nuclease-free water. For DNA-free RNA, include the optional DNase step (*see* **Note 6**). Store RNA at −80 °C
2. Extract DNA from approx. 2–5 × 10^6 PBMC/BMMCs (*see* **Notes 1** and **5**) using the DNeasy Blood & Tissue kit in accordance with manufacturer's protocol for animal blood or cells. Include the optional RNase step. Elute in 100–200 μL buffer AE. DNA eluted in buffer AE can be stored at 4 °C for a few weeks or longer term at −20 °C. Avoid repeated freeze–thaw cycles.

Table 3
Sequencing primers

Read 1	hIGHV0.0SEQ	GACATAGACCGGTTCAGTGAAGGTYTCCTGC
	hIGHV1.0SEQ	CGCCGGAAGCAGTGCTGAGGTGAAGAAGCCT
	hIGHV1.1SEQ	CGCCGGAAGCAGTYCAGGACTGGTGAAGCCT
	hIGHV1.2SEQ	CGCCGGAAGCAGTSCAGGACTGTTGAAGCCT
	hIGHV2.0SEQ	GACATGACTTCGTCCCTCTCACTCACCTGTG
	hIGHV2.1SEQ	GACATGACTTCGTCCCTKAGACTCTCCTGTG
	hIGHV2.2SEQ	GACATGACTTCGTCCCTGARACTCTCCTGTG
	hIGHV3.0SEQ	GACACTACAGCGTAGACCCTCACRCTGACCT
	hIGHV4.0SEQ	CATGTGCCTGTGTGTCCCTGAGACTCTCCTG
	hIGHV4.1SEQ	CATGTGCCTGTGTGTCTCTGARGATCTCCTG
	hIGHV5.0SEQ	CTTAGAGTCACGACTCCTGCAAGGYTTCTGG
Index	IgGH_index	TCCACCAAGGGCCCATGGCTGACTGACT
	IgMH_index	GGGAGTGCATCCGCCGGCTGACTGACT
Read2	IgGH_read2	AGTCAGTCAGCCATGGGCCCTTGGTG
	IgMH_read2	AGTCAGTCAGCCGGCGGATGCACTC

All sequencing primer mixes contain equimolar quantities of each primer in the pool. Where only individual isotypes are being sequenced, the remaining isotype may be omitted from the mix. The first two nucleotides of reverse sequencing reads will contain amplification primer-encoded bases to allow identification of the IGHC isotype

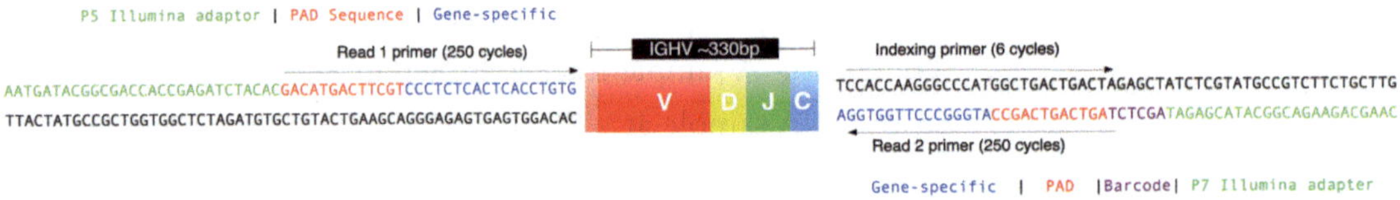

Fig. 2 PCR amplicon sequence and sequencing strategy, illustrated for an IgG heavy chain amplicon using primers of index 1. Nontemplate encoded sequences are incorporated into amplicons during PCR to facilitate direct Illumina sequencing of amplicons. PAD sequences extend the available length for sequencing primers, thereby increasing annealing temperatures. Read 1 primer, Indexing primer and Read 2 primer locations are indicated by the *black arrows*

3.2.2 Preparation of cDNA

1. Prepare cDNA using the GoScript Reverse Transcription System in accordance with manufacturer's protocol using oligo $(dT)_{15}$ primers. We do not generally include the optional RNase inhibitor.
2. Store at −20 °C.

3.2.3 Amplification of IGHV Genes

For PCR amplification of *IGV* genes, a number of protocols can be used depending on the template source. For cDNA, the following primers can be used: (a) Mix of leader-specific primers (sense) and constant region primer (antisense): for *IGHV* genes, leader primers are for *IGHV* families 1–6 (*IGHV7* family genes are amplified by the *IGHV1* primer) and an *IGHC* primer matching the tumor Ig isotype (Table 4). This mix of leader primers can also be used with

Table 4
PCR primers for conventional *IGHV* sequencing

Primer	Location	Orientation	Sequence (5′–3′)
IGHV-specific primers			
IGHV 1/7 Leader	Leader	Sense	CTC ACC ATG GAC TGG ACC TGG AG
IGHV 2 Leader	Leader	Sense	ATG GAC ATA CTT TGT TCC ACG CTC
IGHV 3 Leader	Leader	Sense	CCA TGG AGT TTG GGC TGA GCT GG
IGHV 4 Leader	Leader	Sense	ACA TGA AAC AYC TGT GGT TCT TCC
IGHV 5 Leader	Leader	Sense	ATG GGG TCA ACC GCC ATC CTY G
IGHV 6 Leader	Leader	Sense	ATG TCT GTC TCC TTC CTC ATC TTC
IGHV 1/7 FR1	FR1	Sense	CAG GTG CAG CTG GTG CAR YCT G
IGHV 2 FR1	FR1	Sense	CAG RTC ACC TTG AGG GAG TCT G
IGHV 3 FR1	FR1	Sense	GAG GTG CAG CTG GTG SAG TCY G
IGHV 4a FR1	FR1	Sense	CAG STG CAG CTG CAG GAG TCS G
IGHV 4b FR1	FR1	Sense	CAG GTG CAG CTA CAR CAG TGG G
IGHV 5 FR1	FR1	Sense	GAG GTG CAG CTG KTG CAG TCT G
IGHV 6 FR1	FR1	Sense	CAG GTA CAG CTG CAG CAG TCA G
IGHJ-specific primers			
IGHJ1	FR4	Antisense	TGA GGA GAC GGT GAC CAG GGT GCC CTG
IGHJ2	FR4	Antisense	TGA GGA GAC AGT GAC CAG GGT GCC ACG
IGHJ3	FR4	Antisense	TGA AGA GAC GGT GAC CAT TGT CCC TTG
IGHJ4	FR4	Antisense	TGA GGA GAC GGT GAC CAG GGT TCC CTG
IGHJ5	FR4	Antisense	TGA GGA GAC GGT GAC CAG GGT TCC CTG
IGHJ6	FR4	Antisense	TGA GGA GAC GGT GAC CGT GGT CCC TTG
IGHJ consensus	FR4	Antisense	TGA GGA GAC GGT GAC CAG GAT CCC TTG GCC CCA G
IGHC-specific primers			
IGHM (104–123)	IGHC1	Antisense	GGA GAA AGT GAT GGA GTC GG
IGHG (110–130)	IGHC1	Antisense	CTG AGT TCC ACG ACA CCG TCA
IGHA (108–129)	IGHC1	Antisense	TTC GCT CCA GGT CAC ACT GAG T
IGHD (52–72)	IGHC1	Antisense	CAC AGG GCT GTT ATC CTT TGG
IGHM (16–35)	IGHC1	Antisense	ACG AGG GGG AAA AGG GTT GG
IGHG (16–33)	IGHC1	Antisense	CAG GGG GAA GAC CGA TGG
IGHA (13–33)	IGHC1	Antisense	CAG CGG GAA GAC CTT GGG GCT

(continued)

Table 4
(continued)

Primer	Location	Orientation	Sequence (5′–3′)
Sequencing primers (cloning vector)			
T7	Vector	Sense	TAA TAC GAC TCA CTA TAG GG
SP6	Vector	Antisense	ATT TAG GTG ACA CTA TAG AA
Housekeeping-gene primers			
Beta-actin fwd	Exon 4	Sense	GGCATCGTGATGGACTCCGGTG
Beta-actin rev	Exon 6	Antisense	CTGCTGGAAGGTGGACAGCGA

Primers can be used in individual *IGHV* family-specific PCRs or as pools of sense or antisense primers as described in this chapter. Where pools of primers are used, primers should be at equimolar concentration (20 μM). We have not optimized the use of pooled *IGHC*-specific antisense primers

the *IGHJ(1–6)* mix of antisense primers or the consensus *IGHJ* antisense primer. (b) Mix of framework region (FR) 1 sense primers and downstream *IGHC* or *IGHJ* mix/*IGHJ* consensus antisense primer. (c) If a low template yield is anticipated, nested PCR strategies can be employed, such as leader mix to *IGHC* (~100 bp in), followed by FR1 mix to *IGHC* (~10 bp in) or *IGHJ* mix/*IGHJ* consensus primer.

For DNA the same sense primer mix can be used as for cDNA, but use of leader-specific primers will amplify an approximately 100-bp intron, increasing the size of the PCR product for sequence analysis. Importantly when amplifying *IGHV* genes from DNA, the *IGHC* antisense primers will be too distant and no product will be amplified. Only the *IGHJ* mix or consensus antisense primers can be used for DNA.

1. Set up the PCR mixture as follows (a master mix of all reagents bar the template can be made and aliquoted prior to addition of template/H_2O for nontemplate control).

   37.25 μL H_2O.

   5 μL 5× Green GoTaq Flexi buffer.

   1.5 μL MgCl (25 mM).

   1 μL dNTP mix (10 mM).

   1 μL sense primer mix (each primer at 20 μM, gives final conc. 0.4 μM).

   1 μL antisense primer mix (each primer at 20 μM, final conc. 0.4 μM).

   0.25 μL GoTaq Hotstart taq (conc).

   3 μL cDNA/DNA.

2. PCR cycle conditions are 95 °C for 2 min, followed by 30 cycles of 94 °C for 30 s, 58 °C for 30 s, 72 °C for 1 min, then a final extension step of 72 °C for 5 min. For leader-specific sense primers the annealing temperature can be increased to 65 °C.
3. Run products on a 1–2% agarose gel containing 0.004% Web-Green Advance DNA stain (2 μL/50 mL gel), using a 1 kb ladder for sizing at 100–150 V for 30–45 min.
4. Visualize on a UV imager and excise the band of the expected product size (300–500 bp). Excised bands can be stored at −20 °C for later gel extraction if required (*see* **Note 7** for RT-PCR trouble shooting).
5. Extract amplified DNA from excised gel slices using QIAquick Gel Extraction Kit (*see* **Note 8**) in accordance with manufacturer's protocol, with elution in 30–50 μL buffer EB, depending on band intensity. This DNA can now be used for direct sequencing (go to Subheading 3.2.5) or cloned to assess intraclonal variation in *IGHV* gene sequence.

3.2.4 Analysis of Tumor Clonality

1. In advance prepare: LB agar plates containing 100 μg/mL ampicillin and LB.
2. Set up a ligation reaction using the pGEM-T vector System I: 2× Rapid Ligation Buffer (5 μL) + PCR DNA (3 μL) + pGEM-T Vector (0.5 μL) + T4 DNA Ligase (1 μL).
3. Incubate at RT for 3–4 h or overnight at 4 °C.
4. For transformation, remove an aliquot of *E. coli* XL-1 blue (*see* **Note 9**) competent cells from −80 °C storage and thaw on ice for 5–10 min.
5. Add ligated product (5 μL) to competent cells (50 μL) on ice.
6. Gently mix by shaking tube and incubate on ice for 20–30 min.
7. Heat shock cells by incubating at 42 °C for 45 s; return to ice for further 2–5 min.
8. Add LB (0.2 mL) and shake for 1 h at 37 °C.
9. Prepare an ampicillin-containing LB agar plate by coating with IPTG (0.1 M; 100 μL/plate) and X-gal (25 μL/plate).
10. Plate out all cells (250 μL vol) and allow to dry at 37 °C for approx. 30 min.
11. Invert plate and incubate overnight at 37 °C.
12. Incubate ~1 h at 4 °C to maximize blue/white colony selection.
13. Pick ten white colonies (*see* **Note 9** for cloning trouble shooting) and culture each in 2 mL LB containing 50 μg/mL ampicillin in a 20 mL universal tube or 15 mL falcon tube overnight in an orbital shaking 37 °C incubator.

14. Using ~1.5 mL of culture, extract plasmid DNA using QIAprep Spin Miniprep Kit according to manufacturer's instructions. Elute plasmids in 80 μL (*see* **Note 10**) Buffer EB.
15. Confirm inserts by digestion with restriction enzymes flanking the cloning site and running on an agarose gel.

3.2.5 DNA Sequence Analysis

The protocol described here is for the central DNA sequencing service available in our laboratory that uses an ABI 3130xl genetic analyzer. Countless commercial services are available if required.

1. Set up separate sense and antisense sequencing reactions; Big Dye Terminator mix v1.1 (2 μL) + 5X sequencing buffer (2 μL) + Sequencing primer (Table 4, *see* **Note 11**) at 1.6 μM (1 μL) + plasmid DNA or gel-extracted PCR DNA (5 μL).
2. PCR amplification is as follows: 25 cycles of 96 °C for 10 s, 50 °C for 5 s, 60 °C for 4 min.
3. Precipitate sequencing products; Combine 2 μL 3 M sodium acetate (pH 5.2) + 10 μL PCR sequence reaction mixture + 30 μL 100% EtOH and incubate on ice for 10 min.
4. Centrifuge at 12,000–15,000 × *g* at 4 °C for 30 min.
5. Decant supernatant and wash pellet with 175 μL room temperature 70% EtOH.
6. Centrifuge 12,000–15,000 × *g* for 10 min.
7. Remove supernatant (pipette or decant) and air dry pellet at 37 °C.
8. Resuspend pellet in 10 μL HiDi formamide and vortex mix for >15 s ready for loading on the ABI 3130xl genetic analyzer.
9. Resuspended pellets can be stored at −20 °C for ~7 days prior to sequencing.

3.2.6 Analysis and Interpretation of Data

1. Using MacVector, or similar alternative sequence analysis software, align each pair of sense/antisense sequences and correct any errors observed in the electropherogram files to create a consensus sequence.
2. Ensure that any errors introduced by the sequencing primers, particularly if performing direct sequencing of PCR product, are corrected or the area removed as these errors will be included in the subsequent calculation of *IGHV* gene homology to germline.
3. Submit the consensus sequence to IMGT V-Quest at http://www.imgt.org/IMGT_vquest/vquest?livret=0&Option=humanIg
4. Record the identified *IGHV* gene (and percent identity to germline), the *IGHD* gene and *IGHJ* gene, along with the CDR3 length and amino acid translation (Fig. 3).

Result summary:	Productive IGH rearranged sequence (no stop codon and in-frame junction)		
V-GENE and allele	Homsap IGHV3-23*01 F, or Homsap IGHV3-23D*01 F	score = 1345	identity = 96.53% (278/288 nt)
J-GENE and allele	Homsap IGHJ4*02 F	score = 195	identity = 89.58% (43/48 nt)
D-GENE and allele by IMGT/JunctionAnalysis	Homsap IGHD6-6*01 F	D-REGION is in reading frame 2	
FR-IMGT lengths, CDR-IMGT lengths and AA JUNCTION	[25.17.38.11]	[8.8.20]	CAKDNGPSRAARPNRYQFFAYW

CDR3 length

Fig. 3 IMGT/V-QUEST output. An example of a leukemia case subjected to conventional RT-PCR and Sanger sequencing as described using IGHV leader mix-IGHD. A consensus sequence was constructed in Macvector and submitted to IMGT/V-QUEST. The IGHV gene (with % identity to germline), IGHD and IGHJ genes are identified. The CDR3 length is *highlighted* and the translation of the junction provided

5. If assessment of tumor clonality is to be performed, once a single clone has been analyzed using IMGT V-Quest, MacVector can be used to align all clones of the same PCR product: (a) Prepare a germline sequence consisting of the IMGT-identified *IGHV* gene followed by the IMGT-identified *IGHJ* gene (up-to-date sequences can be downloaded from the IMGT website at http://www.imgt.org/vquest/refseqh.html#V-D-J-C-sets), separated by sufficient spaces (in lieu of an *IGHD* gene/CDR3 region) to allow alignment with the cloned sequence. (b) With this germline sequence perform a database alignment to a folder containing all the consensus cloned sequences from **step 1** for the originating PCR product (Fig. 4a). (c) For ease of viewing under the Options tab choose "Format Aligned Display" and set the line length to something between 150 and 200, blocking to 3, numbering to 10, and marking to 0. In the Characters section choose upper, one and single. Then, again under Options, choose "Format annotated Display," select score line (but not query line), leave the first three boxes empty, set the numbering to 10 and the vertical alignment to 0, vertical align character to |, Match character to "character" and Mismatch character to "upper." (d) Record both the frequency and positions of any intraclonal variation in the *IGHV* gene sequences (Fig. 4a). (e) Using a monospaced font in Word or similar clear alignment figures can be prepared for presentation or publication (Fig. 4b).

4 Notes

1. Published data suggests healthy donor PB typically yields 1.16×10^6 MNC/mL [15] with B-cells accounting for 6.4–16.5% of these [16]. We generally obtain ~1×10^6 MNC/mL with B-cells accounting for 1–10% of these.

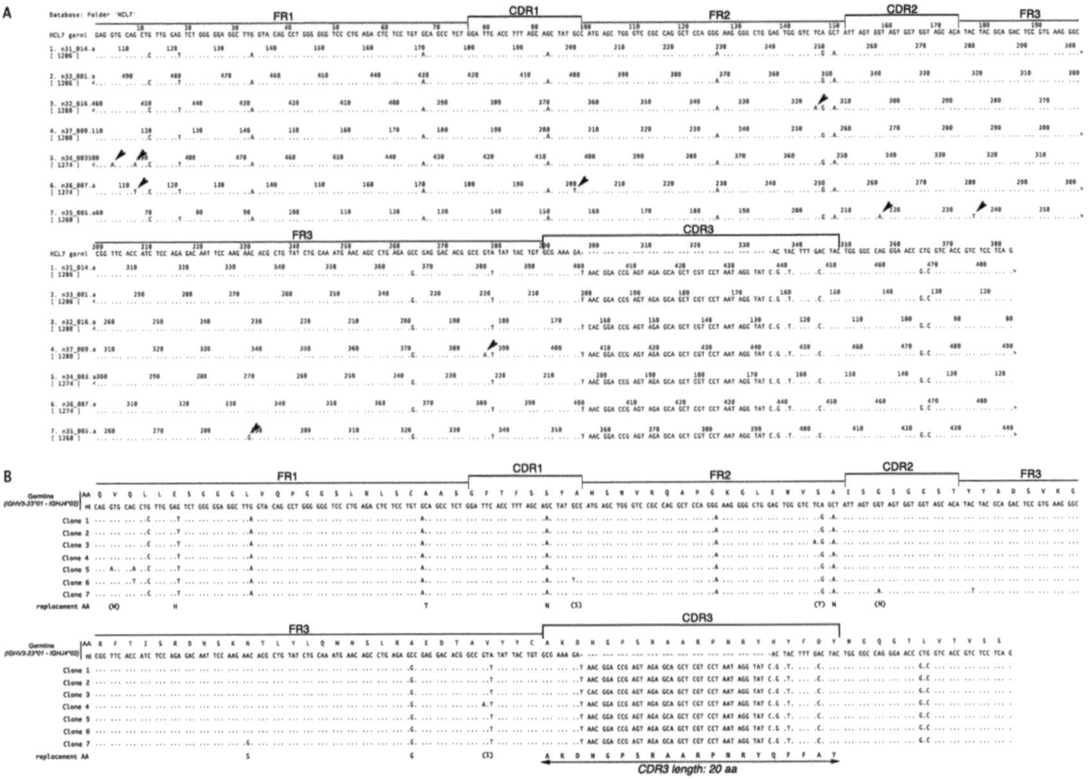

Fig. 4 Constructed germline sequence aligned to clones using database align tool in Macvector. (**a**) A leukemia tumor clone identified as described in legend to Fig. 3. Following identification of the *IGHV* and *IGHJ* genes using IMGT/V-QUEST, the germline *IGHV* gene sequence was imported into Macvector and aligned to a folder containing all leukemia-derived clonal sequences. The FRs and CDRs are marked, including the CDR3 showing identical sequence in all clones, and derivation from a monoclonal tumor. The IMGT translation of the junction is shown along with CDR3 length. Positions of intraclonal variation are highlighted with *black arrowheads*. (**b**) Data output edited to show alignment of tumor-derived clones with *IGHV* germline, identifying impact of SHM and ICV

In healthy donors BM aspirates typically yield 15–30 million MNC/mL, with plasma cells accounting for 0.5–1% of these [17] B-cell yield from normal BM aspirates averages 1.5×10^6 B-cells/mL [18]. Following collection into LH or EDTA tubes, dilute the sample 1:2 in RPMI 1640 before mononuclear cell isolation.

2. Variations of RNA preparation and cDNA synthesis protocols. This protocol describes repertoire sequencing from total RNA isolated from B-cell samples. Any efficient method of total RNA extraction may be used, such as Trizol reagent extraction or column-based purification (e.g., Zymo Research Directzol total RNA kit). Similarly, many cDNA synthesis methods based on MMLV reverse transcriptase have been shown to work, as well as the SuperScript III reverse transcriptase system (Thermo Fisher Scientific) using a reaction temperature of 50 °C.

3. PCR contamination—Good PCR practice is essential for the generation of meaningful results using this technique. Cross-contamination must be avoided by the use of spatially separated laboratory space for pre- and post-amplification steps. cDNA synthesis and PCR steps must use a PCR hood and dedicated equipment, including lab coats, pipettes, reagent storage freezer, and cDNA synthesis thermal block so that amplicons from previous reactions cannot contaminate pre-amplification samples.
4. mpNGS PCR Trouble Shooting—Poor performance of cDNA synthesis or PCR reactions may originate from a variety of sources and must be overcome by step-by-step investigation. RNA should be checked with a bioanalyzer for integrity—R.I. N. (RNA integrity number) scores of >6 typically indicate sufficient quality of RNA and the protocols described have been tested down to starting RNA quantities of 2 ng. Amplification of "housekeeping genes" from cDNA may also be useful in confirming cDNA integrity. Both cDNA synthesis and PCR reagents are sensitive to freeze–thaw cycles and should be aliquotted at first use. The Phusion flash master mix should be stored at −20 °C as directed and not −80 °C as repeated transitions between these temperatures can cause loss of activity.
5. These procedures can be performed successfully with lower cell numbers than described here. In this case care must be taken to keep samples/template as concentrated as possible. For RNA extraction the RNeasy Micro Kit (Qiagen) can be used and for DNA extraction, the elution volume can be reduced to 50 μL. A nested PCR approach can also be employed, as outlined (Subheading 3.2.3), to maximize amplification of *IGHV* gene product.
6. DNA-free RNA is not generally required for *IGHV* sequencing, as the primers used will either amplify a larger product or no product from gDNA. If, however, the RNA/cDNA is to be used for other amplifications where gDNA might interfere, such as quantitative-PCR or PCRs where primers cannot be designed to cross exon boundaries, the DNase step should be included.
7. If no PCR product is observed, first test the integrity of the cDNA by amplifying a housekeeping gene such as beta-actin or GAPDH. If this fails, check the originating RNA for concentration and purity using Nanodrop or similar. If the RNA is poor, consider repeating the extraction. If the RNA is good, then repeat the cDNA synthesis. However, if the cDNA does produce a strong band following amplification of a housekeeping gene, check that template-appropriate pairs of

sense and antisense primers were used for the *IGHV* PCR and consider lowering the annealing temperature. Finally separate *IGHV* family PCRs can provide a product where an IGHV mix has previously failed.

8. When using the QIAquick Gel Extraction Kit or QIAprep Spin Miniprep kit, U-Prep columns (Dutscher Scientific, Brentwood, UK) can be substituted for the supplier's miniprep columns in order to make use of excess buffers. U-Prep columns are not recommended for use with the RNeasy/DNeasy nucleic acid extraction kits.
9. *E. coli* XL-1 blue competent cells are used in preference over JM109 cells as XL-1 blue cells tend to produce a greater number of colonies (although these colonies tend to be smaller). JM109 cells can be used, but colony number may become a limiting factor. If no colonies are present, allow the plates to incubate longer at 37 °C initially. If colonies do not develop, repeat the transformation using the remaining ligated product or repeat the ligation and transformation. The pGEM-T Vector protocol recommends a 1-h incubation at RT. However, we find that >2 h improves yield and we most commonly incubate at RT for 3–4 h. If PCR product concentration might be limited, the pGEM-T Vector ligation can be performed in a larger volume to allow more PCR product to be included. Combine 2× Rapid Ligation Buffer (10 μL) + PCR DNA (7 μL) + pGEM-T Vector (1 μL) + T4 DNA Ligase (2 μL). Failure of the competent cells is a possibility, especially if there have been issues with temperature maintenance at −80 °C. If repeating the ligation/transformation does not help, try alternative stocks of competent cells.
10. Elution in 80 μL Buffer EB gives plasmid DNA at optimal concentration for the described downstream Sanger sequencing protocol. Alternative sequencing protocols may require more/less concentrated DNA.
11. The sequencing primer can be either T7/SP6 (Table 4) for plasmid DNA or the relevant sense/antisense PCR primer for direct sequencing of gel-extracted PCR DNA.

Acknowledgement

This work was funded by Bloodwise (UK).

References

1. Stevenson F, Sahota S, Zhu D, Ottensmeier C, Chapman C, Oscier D, Hamblin T (1998) Insight into the origin and clonal history of B-cell tumors as revealed by analysis of immunoglobulin variable region genes. Immunol Rev 162:247–259

2. Georgiou G, Ippolito GC, Beausang J, Busse CE, Wardemann H, Quake SR (2014) The promise and challenge of high-throughput sequencing of the antibody repertoire. Nat Biotechnol 32(2):158–168. doi:10.1038/nbt.2782
3. Tonegawa S (1983) Somatic generation of antibody diversity. Nature 302 (5909):575–581
4. Cook GP, Tomlinson IM (1995) The human immunoglobulin VH repertoire. Immunol Today 16(5):237–242. doi:10.1016/0167-5699(95)80166-9
5. MacLennan IC (1994) Germinal centers. Annu Rev Immunol 12:117–139. doi:10.1146/annurev.iy.12.040194.001001
6. Boyd SD, Gaeta BA, Jackson KJ, Fire AZ, Marshall EL, Merker JD, Maniar JM, Zhang LN, Sahaf B, Jones CD, Simen BB, Hanczaruk B, Nguyen KD, Nadeau KC, Egholm M, Miklos DB, Zehnder JL, Collins AM (2010) Individual variation in the germline Ig gene repertoire inferred from variable region gene rearrangements. J Immunol 184(12):6986–6992. doi:10.4049/jimmunol.1000445
7. Glanville J, Kuo TC, von Budingen HC, Guey L, Berka J, Sundar PD, Huerta G, Mehta GR, Oksenberg JR, Hauser SL, Cox DR, Rajpal A, Pons J (2011) Naive antibody gene-segment frequencies are heritable and unaltered by chronic lymphocyte ablation. Proc Natl Acad Sci U S A 108(50):20066–20071. doi:10.1073/pnas.1107498108
8. Briney BS, Willis JR, McKinney BA, Crowe JE Jr (2012) High-throughput antibody sequencing reveals genetic evidence of global regulation of the naive and memory repertoires that extends across individuals. Genes Immun 13 (6):469–473. doi:10.1038/gene.2012.20
9. Benichou J, Glanville J, Prak ET, Azran R, Kuo TC, Pons J, Desmarais C, Tsaban L, Louzoun Y (2013) The restricted DH gene reading frame usage in the expressed human antibody repertoire is selected based upon its amino acid content. J Immunol 190(11):5567–5577. doi:10.4049/jimmunol.1201929
10. Weston-Bell N, Townsend M, Di Genova G, Forconi F, Sahota SS (2009) Defining origins of malignant B cells: a new circulating normal human IgM(+)D(+) B-cell subset lacking CD27 expression and displaying somatically mutated IGHV genes as a relevant memory population. Leukemia 23(11):2075–2080. doi:10.1038/leu.2009.178
11. Cowan G, Weston-Bell NJ, Bryant D, Seckinger A, Hose D, Zojer N, Sahota SS (2015) Massive parallel IGHV gene sequencing reveals a germinal center pathway in origins of human multiple myeloma. Oncotarget 6(15): 13229–13240. doi:10.18632/oncotarget.3644
12. Larimore K, McCormick MW, Robins HS, Greenberg PD (2012) Shaping of human germline IgH repertoires revealed by deep sequencing. J Immunol 189(6):3221–3230. doi:10.4049/jimmunol.1201303
13. Shugay M, Britanova OV, Merzlyak EM et al (2014) Towards error-free profiling of immune repertoires. Nat Meth 11:653–655. doi:10.1038/nmeth.2960
14. Weston-Bell NJ, Tapper W, Gibson J, Bryant D, Moreno Y, John M, Ennis S, Kluin-Nelemans HC, Collins AR, Sahota SS (2016) Exome sequencing in classic hairy cell leukaemia reveals widespread variation in acquired somatic mutations between individual tumours apart from the signature BRAF V(600)E lesion. PLoS One 11(2):e0149162. doi:10.1371/journal.pone.0149162
15. Corkum CP, Ings DP, Burgess C, Karwowska S, Kroll W, Michalak TI (2015) Immune cell subsets and their gene expression profiles from human PBMC isolated by Vacutainer Cell Preparation Tube (CPT) and standard density gradient. BMC Immunol 16:48. doi:10.1186/s12865-015-0113-0
16. Autissier P, Soulas C, Burdo TH, Williams KC (2010) Evaluation of a 12-color flow cytometry panel to study lymphocyte, monocyte, and dendritic cell subsets in humans. Cytometry A 77(5):410–419. doi:10.1002/cyto.a.20859
17. Terstappen LW, Johnsen S, Segers-Nolten IM, Loken MR (1990) Identification and characterization of plasma cells in normal human bone marrow by high-resolution flow cytometry. Blood 76(9):1739–1747
18. Clark P, Normansell DE, Innes DJ, Hess CE (1986) Lymphocyte subsets in normal bone marrow. Blood 67(6):1600–1606

Chapter 22

How to Simulate a Germinal Center

Philippe A. Robert, Ananya Rastogi, Sebastian C. Binder, and Michael Meyer-Hermann

Abstract

Germinal centers host a mini-evolutionary environment where B cells can mutate their receptor and be selected depending on its affinity to target antigens in a process called affinity maturation. Starting from founder cells with a weak B cell receptor affinity, germinal centers release output cells as antibody-secreting cells or memory cells with a very high affinity, a property which is essential for pathogen clearance and immune memory. Therapeutic interventions on the germinal centers are tantalizing approaches to improve vaccines or to support rejection of chronic pathogens such as HIV. However, the complexity of the selection processes makes it very hard to make reliable predictions. Here, we present in detail how to build an agent-based model (hyphasma), accounting for the dynamics of the germinal center. It encompasses the core quantitative traits of affinity maturation, and allowed to make reliable predictions in previous studies.

Key words Germinal center, Affinity maturation, Agent-based modeling, Somatic hypermutation

1 Introduction: Spatial and Functional Structure of Germinal Centers

During the formation of a Germinal Center (GC), two zones emerge, the dark zone and the light zone (Fig. 1). While the steps leading to this characteristic structure are still poorly understood, the events happening in both of them have been extensively characterized and are critical for a proper affinity maturation of antibodies.

The dark zone contains stromal cells producing CXCL12 and proliferating B cells named centroblasts. The clonal expansion of centroblasts is accompanied by somatic hypermutation (SHM) specifically targeting the receptor (BCR), and consequently the generation of further lineages with diverse affinities towards the antigen. Cutting-edge single-cell sequencing methods now allow to characterize lineages generated from the same founder clone as harboring the same VDJ recombination pattern, together with their mutation history, shedding light on the diversity of mutations happening in one GC in vivo [1].

Dinis Pedro Calado (ed.), *Germinal Centers: Methods and Protocols*, Methods in Molecular Biology, vol. 1623, DOI 10.1007/978-1-4939-7095-7_22, © Springer Science+Business Media LLC 2017

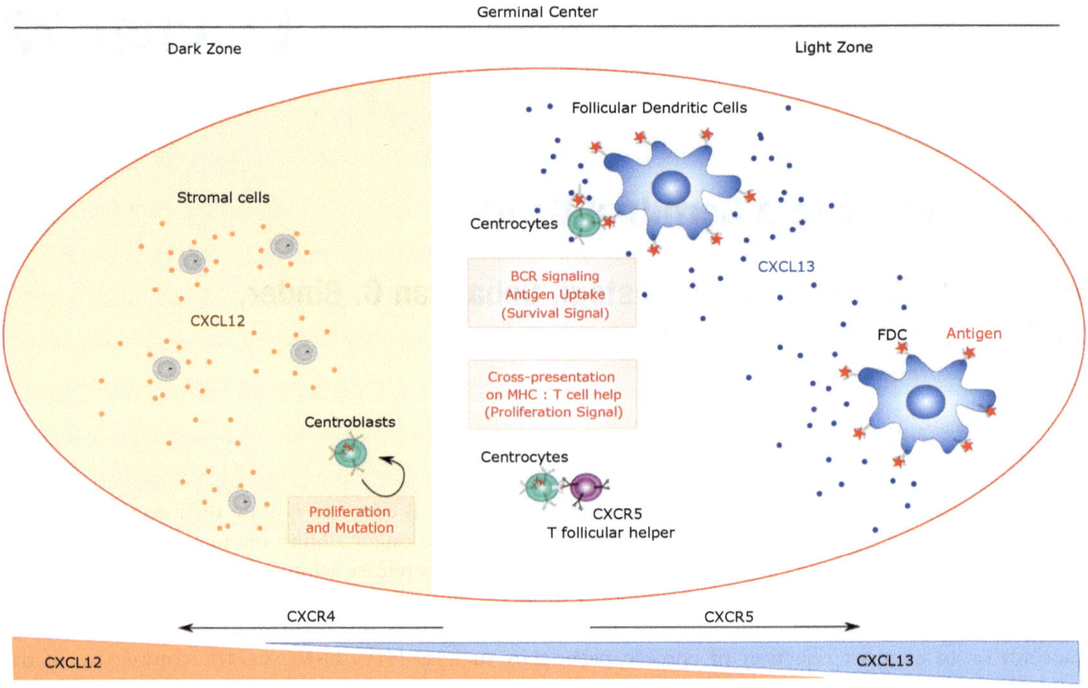

Fig. 1 Structure of a Germinal Center. B cells as centroblasts proliferate and mutate in the *dark Zone*, while centrocytes are screening the *light zone* for antigen and interactions with Tfhs. The network of FDCs and stromal cells generates CXCL12/CXCL13 gradients, allowing the cells to preferentially migrate on the *dark/light zone* depending on their expression of CXCR4/CXCR5, respectively

The light zone, however, is composed of nondividing B cells enduring high apoptosis rates and looking for survival and proliferation signals, hereby named centrocytes. It further contains a stromal network of Follicular Dendritic Cells (FDCs) producing CXCL13 and carrying antigens in their native structure on immune complexes bound via complement Fc receptors. Centrocytes with high affinity to an antigen can bind to it, receive BCR signaling necessary for their survival, and capture this antigen. Finally, centrocytes that successfully captured an antigen can process and further present it on their MHC class II, thereby acting as an Antigen Presenting Cell (APC) for the T helper cells present in the GC, called T Follicular Helper (Tfh) cells. The interaction between Tfhs and centrocytes further gives bidirectional signals, leading to the proliferation of the B cell together with the maintenance of the phenotype of the Tfh [2]. Interestingly, it has been shown that the antigen carried by B cells is not always equally distributed among daughter cells upon division [3], raising the possibility of B cell fate decision being based on the amount of carried antigen [4].

The GC is a very dynamic structure. B cells and T cells constantly migrate on the static network of stromal cells and FDCs. Centroblasts express CXCR4, allowing them to preferentially migrate and stay in the dark zone by following CXCL12 gradients [5]. They undergo a finite number of divisions, in a range of 2–6 [6, 7] divisions within 6–10 h per division, before they lose the expression of CXCR4 and become centrocytes. Centrocytes and Tfhs are targeted to the light zone in virtue of their expression of CXCR5. Once centrocytes complete the selection steps of finding antigen and T cell help, they further express CXCR4 and initiate proliferation and SHM again. The subsequent recycling process between the light zone and the dark zone allows for multiple rounds of mutation/selection in the GC (Fig. 2).

The stage at which cells leave the GC is still not fully understood. Here, we allow two hypotheses: either 1/centrocytes with high affinity after being selected by Tfhs or 2/centroblasts that could keep enough antigen after recycling and dividing, are authorized to become output cells and leave the GC. The latter assumption is named here "LEDA" (Leave through the Dark zone) and was predicted to be more advantageous than alternative hypotheses regarding cell output [4, 8].

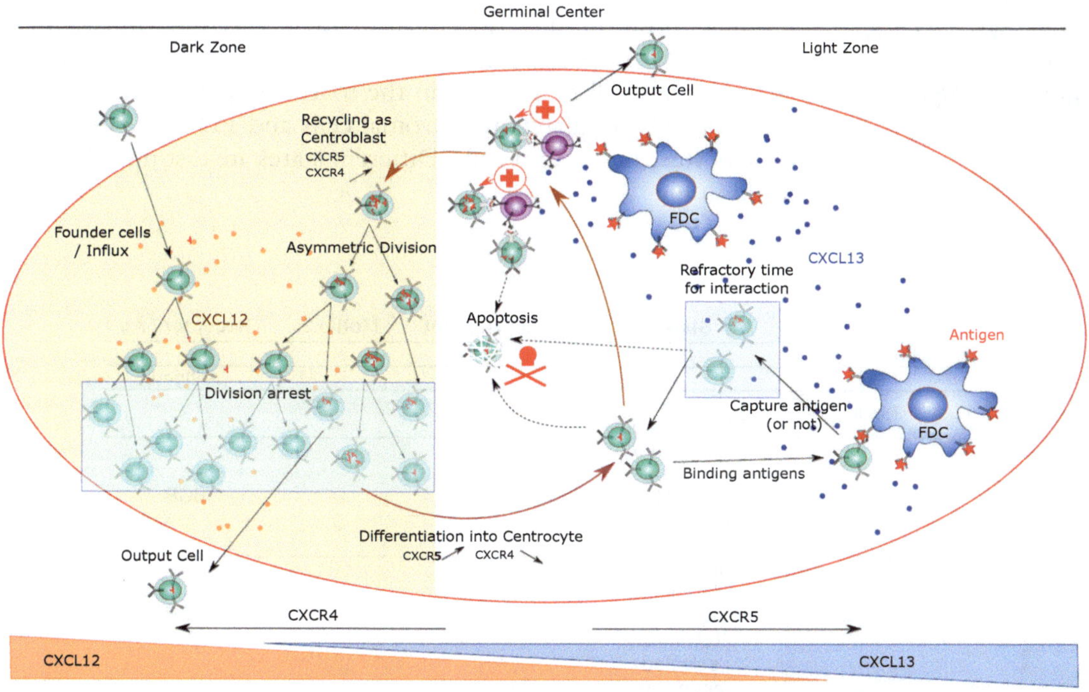

Fig. 2 The recycling processes and steps of selection incorporated in the simulation. Founder cells, initiated by long lasting T–B interactions, enter the GC as CXCR4+ centroblast and proliferate and mutate in the *dark zone*. When they stop dividing, they differentiate to centrocytes, become CXCR4−, and migrate to the *light zone* where they capture antigen at the surface of FDCs, cross-present it on Tfh cells upon encounter to get selected, and recycle as CXCR4+ centroblast. The recycled centroblasts potentially still carry antigens and can undergo asymmetric division. The centroblasts that still own a significant amount of antigen after dividing are suitable to become output cells (LEDA), or alternately a fraction of Tfh-selected B-cells

Here, we give a detailed description of our program (hyphasma), introduced in [9] to simulate the evolution of a GC. An algorithmic description is proposed, separately for each process, in order to be syntax independent while giving enough precisions to be reproduced. The main loop together with the parameter values are given at the end.

The selection pressure on B cells emerges from a combination of processes, among which are the abundance or frequency of antigens, the capacity to take an antigen depending on the BCR affinity, and the chance to find a Tfh. It is therefore not straightforward to disentangle the effect of the mechanisms underlying an efficient affinity maturation, and even harder to make predictions on how to improve it as a therapeutical purpose. Hyphasma incorporates all these effects, and allowed to investigate their relative contribution and to predict or support new findings.

2 Building the Agent-Based Model Step-by-Step

2.1 Representation of the System: Cells, Signals, Antigens, BCRs, and Affinities

A cell is represented as an agent, with its own properties. For instance, position, direction of movement, receptor (in case of B cell or T cell), or phase of cell cycle (in case of a centroblast). The list of properties will be detailed in the following parts. The different types of cells simulated are: stromal cells and FDCs, centroblasts, centrocytes, and output cells. Different states are distinguished and listed in Fig. 3.

	States	Multi-gridpoint	Motile ?	Receptor / Ag
Stromal cell		N	N	-
FDC		Y	N	Antigen
T cell	Tcnormal	N	Y	-
	TC-CCcontact		N	
Centroblasts,	cb_G1, cb_G0,cb_S, cb_G2	N	Y	BCR
	cb_M		N	
	stop_dividing cb_divide		Y	
Centrocytes,	unselected	N	Y	BCR
	FDCcontact		N	
	FDCselected		Y	
	Tccontact		N	
	selected		Y	
	apoptosis		N	
Out cells		N	Y	BCR

Fig. 3 The different types of cells considered and their states. FDCs occupy multiple sites while one B or T cell fills one position of the grid. B and T cells at an interacting state are not motile, together with stromal cells and FDCs. B cells carry their own BCR

The space is represented as a grid where a gridpoint has the critical size of a cell (5 μm), and the GC area is defined as a sphere of allowed positions in the grid. At each point of the GC is stored which cell is present and the concentration of each chemokine. Two cells cannot overlap, meaning that one gridpoint can only be taken by one cell, and a cell cannot move to an already occupied position.

The information needed to represent the state of a GC at any time can be stored and used in the following way.

First, as spatial data storages:

`Grid[x][y][z]`	`storing a cell type and an ID to access the cell at this position`
`CXCL12[x][y][z]`	`storing the amounts of CXCL12 and CXCL13 at this position boundary`
`CXCL13[x][y][z]`	`storing the amounts of CXCL12 and CXCL13 at the GC borders`

The agents are stored in separate lists: (and carry their own properties)

`CBlist[], CClist[], TClist[], OUTlist[], FDClist[], STROMAlist[].`

2.2 Representation of Antigens and Receptors

The distribution of antigens in the GC is still poorly understood. Following infection, antibodies are produced with high affinity towards specific epitopes of the pathogen's antigens, but it is still not clear whether each GC carries mainly one antigen and supports preferentially directed affinity maturation towards this antigen, or if each GC carries a large diversity of antigens. In our model, we take the assumption that only one antigen per GC, as a first approach.

Regarding the receptor specificity, another interesting property of the GC is the cross-checking of B cells by T cells, which is likely to avoid nonspecific activation of B cells that could lead to autoimmunity [10]. If a B cell response is initiated without the activation of T cells, no proper GC is formed, and only few antibodies are produced [11]. Further, if T cells are depleted during the course of a GC, the GC is shown to shrink and terminate [12], most likely because centrocytes do not receive a rescue signal then.

From a mechanistic point of view of this cross-checking, B cells need to internalize antigen, process it and present it to the Tfh, implying that the epitopes seen by the Tfh are different than the real 3D epitopes of the native antigen as it can be bound by the B cells. Therefore, even if a B cell is specific to a particular antigen epitope, it can most probably be rescued by several Tfh that can recognize different parts of the antigen, as presented on the MHC. Here, we assume that all the Tfh cells that could enter the GC recognize a sub-peptide of the antigen and therefore can support the selection of all B cells, provided the B cells present enough antigen. Another way to put it is that we only simulate Tfhs that can support B cells,

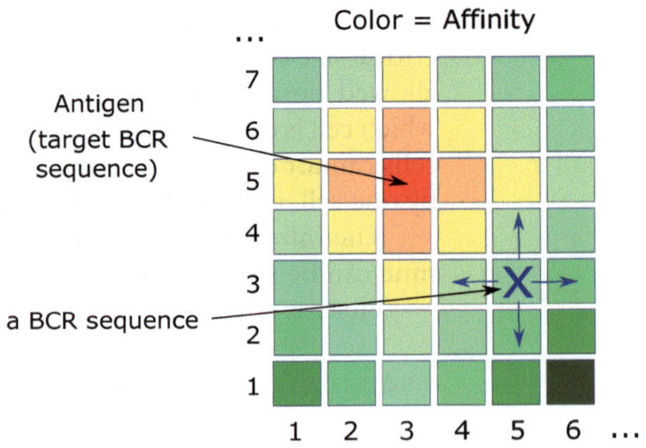

Fig. 4 Example of BCRs (53) and an Antigen (35) represented in a shape space with dimension $L = 2$. A mutation is similar to a movement in the shape space, and the affinity depends on the Hamming distance between the target antigen and the BCR. The affinity is highest if the BCR matches the antigen (*red*) and decreases with the number of mutations away

and that bystander Tfhs with a different specificity could be present but are ignored. This hypothesis allows to exclude the complexity of the TCR sequences from the model.

The representation of BCRs and antigens in the model can be incorporated with different degrees of detail. An optimal representation would be to give a real nucleotidic sequence to BCRs and an amino-acid sequence to the antigen. However, it is not yet possible to reliably simulate the binding affinity of any two proteins.

Therefore, a simplified representation of antigen and BCRs has to be taken. Here, a BCR and the target antigen are represented as a series of $L = 4$ numbers: ex 3594, and a mutation is represented as a $+/-1$ in one of the numbers (*see* Algorithm 1). This representation was first introduced in [13] and can be seen as a position in an L-dimensional grid called the "Shape Space". The (Hamming) distance between a TCR and the antigen is defined as the number of mutations remaining to be the exact same sequence, and the affinity is derived from a Gaussian of the distance, with a maximum affinity of 1 if the BCR matches the antigen (an example is given with a size of $L = 2$ in Fig. 4). The affinity between two positions relies on two parameters: $\eta = 2$ and Γ (the width of the Gaussian).

$$\text{Affinity}(\text{TCR}, \text{BCR}) = \exp\left(-\frac{\text{Hamming}(\text{TCR}, \text{BCR})^{\eta}}{\Gamma^{\eta}}\right)$$

The choice of the size L is critical, as by increasing L, the ratio of beneficial versus detrimental mutations reduces. For instance, from a position in the shape space, the number of neighbors with increased/decreased affinity is 1/1 for $L = 1$, 2/2 or 1/3 for

$L = 2$, etc. To recapitulate the dynamics of mutations/affinity maturation, it was shown that the optimal value of L is $L = 4$ (discussed in [14]).

Algorithm 1 Mutation

```
procedure MUTATE(Cell C, pMutation)
    if With probability pMutation then
        Take a random position i ∈ [1..L]
        Apply ± 1 to this position, if it is possible within [0 − 9]
    end if
end procedure
```

2.3 CXCRs Expression/Sensitivity to Chemokines

The characteristic structure of the GC relies on chemotaxis, and therefore the dynamic expression of chemokine receptors by T cells and B cells. Chemotaxis to CXCL12 and CXCL13 applies to cells expressing CXCR4 and CXCR5, respectively. While all cells in the GC express CXCR5 with a limited difference between centroblasts and centrocytes [15], the expression of CXCR4 is dynamic and restricted to centroblasts. We assume here that CXCR4 induced chemotaxis is dominant over CXCR5, and that, in dependence of CXCR4 expression, the cells will be sensitive to CXCL12 or CXCL13.

In the model, the chemotactic capacity of cells is represented as a Boolean property:

```
isResponsive2signals[CXCL12], meaning that it expresses CXCR4 (dominates over CXCR5
                              expression)
isResponsive2Signals[CXCL13], meaning it is CXCR4- and follows CXCL13 due to CXCR5
                              expression
```

Additionally, we include a dynamic modulation of the signaling: a high local dose of chemokine allows for the desensitization of the receptors, while resensitization can occur when the dose becomes low enough again. The decision of sensitivity to signals falls into two procedures: one which initiates the proper receptors when a cell is changing phenotype, and another function, called at each time-step, that updates the level of the receptor depending on the current level of chemokine (*see* Algorithm 2).

Algorithm 2 Dynamic updating of chemotaxis

```
procedure InitiateChemokineReceptors(Cell C, type T)
    if type = Centroblast then                                  ▷ Seeder, new-comer by influx, or recycling cell
        C.responsive2signal[CXCL12] ⟵ TRUE
        C.responsive2signal[CXCL13] ⟵ FALSE
    end if
    if type = Centrocyte then                                   ▷ From centroblast to centrocyte
        C.responsive2signal[CXCL12] ⟵ FALSE
        C.responsive2signal[CXCL13] ⟵ TRUE
    end if
    if type = Outcell then
        C.responsive2signal[CXCL13] ⟵ FALSE
        C.responsive2signal[CXCL12] ⟵ FALSE
    end if
end procedure

procedure UpdateChemokinesReceptors(Cell C)
    if C is a Centrocyte then
        if C.state = unselected or FDCselected then             ▷ i.e. if C is motile
            if CXCL13[C.x, C.y, C.z] > CXCL13crit then
                C.responsive2signal[CXCL13] ⟵ FALSE             ▷ Desensizitation
            end if
            if CXCL13[C.x, C.y, C.z] < CXCL13recrit then
                C.responsive2signal[CXCL13] ⟵ TRUE              ▷ Resensizitation
            end if
        end if
    else if C is a centroblast then
        if CXCL12[C.x, C.y, C.z] > CXCL12crit then
            C.responsive2signal[CXCL12] ⟵ FALSE                 ▷ Desensizitation
        end if
        if CXCL13[C.x, C.y, C.z] < CXCL12recrit then
            C.responsive2signal[CXCL12] ⟵ TRUE                  ▷ Resensizitation
        end if
    end if
end procedure
```

2.4 Movement of Cells

The movement of cells is performed by updating their position at each time-step (Algorithm 3). To avoid biases due to the discretization of space into a grid, each cell carries a polarity, defining a direction of movement, that can be kept during multiple steps. The choice of the next position in the grid is made in order to accurately follow the polarity direction after multiple steps. The polarity is not changed at each time-step but rather after a persistent time is completed, allowing to be independent of the chosen time-step of simulation. The properties required for movement in the GC (T cells, centroblasts, and centrocytes) are:

`polarity (3D vector)`	current intended direction of movement
`isResponsive2Signals[CXCL12, CXCL13]`	sensitivity to chemotaxis

Following a Polarity

First, each type of cell has a specific velocity (see parameter values in Subheading 22.2.12), which is translated into a probability of moving of a gridpoint size per time-step. Cells do not move at each time-step, but rather fast cells move more often, to recapitulate a correct observed speed.

To perform a move in the grid according to the (floating point) polarity, the new (floating point) wanted position is computed, and the list of free neighboring positions is sorted based on their distance with the target. One of the equally best positions is chosen randomly, provided it does not go against the direction of the polarity. It allows cells to follow a direction during multiple time-steps and to walk around obstacles if the best position is not available. If no suitable position is found, the cell stays where it was.

In the specific case that two neighboring cells intend to go in the opposite direction (towards each other), and therefore could not move, the movement procedure is extended by allowing swapping of two such cells. It is described in the main loop in Algorithm 10 and is performed by storing the list of cells that could not move. Each such cell is allowing to swap with a neighbor with a probability 50%. This 50% is used because the neighbor cell might also be in the list (and will also be considered), or alternately, if not in the list, then only the first cell should have moved. Swapping would move the other cell for free, increasing the total displacement of cells of 2. *dx*.

Defining a New Polarity

When a new polarity needs to be computed, at the end of the persistence time, it is computed based on a random movement defined by a turning angle from the previous polarity and is further influenced by chemotaxis. The turning angle distribution is directly taken from experimental measurements [16] while the chemotactic attraction to CXCL12 and CXCL13 is performed by adding a small contribution in the direction to the gradient of both signals. The weight between chemotaxis and random move is supposed to be dependent on the strength of the gradient. Details on how chemotaxis was included in the model can be found in [17]. Additionally, the preferential localization of Tfhs into the light zone is done by a preference to follow a direction called 'north' here, and pointing to it.

Algorithm 3 Updating the position and polarity of cells at each time-point

procedure MOVE(Cell C)
  if With probability $persistentLengthTime(C.type)$ **then** ▷ (Cell-type specific)
    get a new turning angle θ from the distribution observed in [16]
    and a random angle ϕ in $[0..2\pi]$ to generate a new polarity:
    $\overrightarrow{C.polarity} \longleftarrow TurnAngle(\overrightarrow{polarity}, \Theta, \Phi)$ ▷ Turns the polarity by an angle Θ in a random direction ϕ
    if C.isResponsive2Signals[CXCL12] **then**
      Computes the gradient vector for CXCL12 at the position of the cell:
      (If needed, concentrations outside the grid are taken from 'boundary' values)
      $\overrightarrow{Gradient_{CXCL12}} \longleftarrow \frac{1}{2.dx} \cdot \begin{bmatrix} CXCL12[x+1,y,z]-CXCL12[x-1,y,z] \\ CXCL12[x,y+1,z]-CXCL12[x,y-1,z] \\ CXCL12[x,y,z+1]-CXCL12[x,y,z-1] \end{bmatrix}$
      $\overrightarrow{C.polarity} \longleftarrow \overrightarrow{C.polarity} + \left(\frac{chemoMax}{1.0+\exp(chemoSteep*(chemoHalf-2.dx.\|\overrightarrow{Gradient_{CXCL12}}\|))}\right).\overrightarrow{Gradient_{CXCL12}})$
    end if
    if C.isResponsive2Signals[CXCL13] **then**
      Computes the gradient vector for CXCL13 (same formula)
      $\overrightarrow{C.polarity} \longleftarrow \overrightarrow{C.polarity} + \left(\frac{chemoMax}{1.0+\exp(chemoSteep*(chemoHalf-2.dx.\|\overrightarrow{Gradient_{CXCL13}}\|))}\right).\overrightarrow{Gradient_{CXCL13}}$
    end if
    if C is a TCell **then** ▷ $\overrightarrow{north}$ points to the light zone
      $\overrightarrow{C.polarity} \longleftarrow (1.0 - northWeight) * \overrightarrow{C.polarity} + northWeight * \overrightarrow{north}$
    end if
    Normalize $\overrightarrow{C.polarity}$
  end if
  $pDifu \longleftarrow Speed(C.type) * \frac{dt}{dx}$ ▷ probability to move from the speed, cell-type specific
  if With probability $pDifu$ **then**
    $\overrightarrow{WantedPosition} \longleftarrow \overrightarrow{oldPosition} + \overrightarrow{C.polarity}$
    $FreeNeighbors[] \longleftarrow$ list of neighbor positions without cells ▷ (i.e. $Space[x,y,z] = empty$)
    Sort FreeNeighbors according to the distance with $\overrightarrow{WantedPosition}$
    Move the cell to the best position (randomly chosen among the equally best ones):
    Grid[newposition].type = C.type
    Grid[newposition].ID = Grid[oldposition].ID
    Grid[oldposition].type = empty, .ID = -1
  end if
end procedure

procedure TURNANGLE($polarity$, Θ , Φ)
  Take a plane containing the current polarity, where $\overrightarrow{v}$ completes polarity to form an oriented base.
  Rotate $\overrightarrow{polarity}$ by the angle Θ following the axis $\overrightarrow{v}$.
  Rotate $\overrightarrow{polarity}$ by the angle Φ along the initial $\overrightarrow{polarity}$ vector.
end procedure

2.5 Proliferation and Mutation of Centroblasts

At the stage of centroblasts, several substates are distinguished in order to account for the cell cycle phase. Each phase of the cell cycle has an average duration which is sampled from a Gaussian distribution (see parameters in Subheading 22.2.12).

`states : cb_G1 , cb_S, cb_G2, cb_M, cb_divide, and cb_stop_dividing`

It is assumed here that the number of divisions a centroblast will perform is known in advance. Centroblasts arising from newly

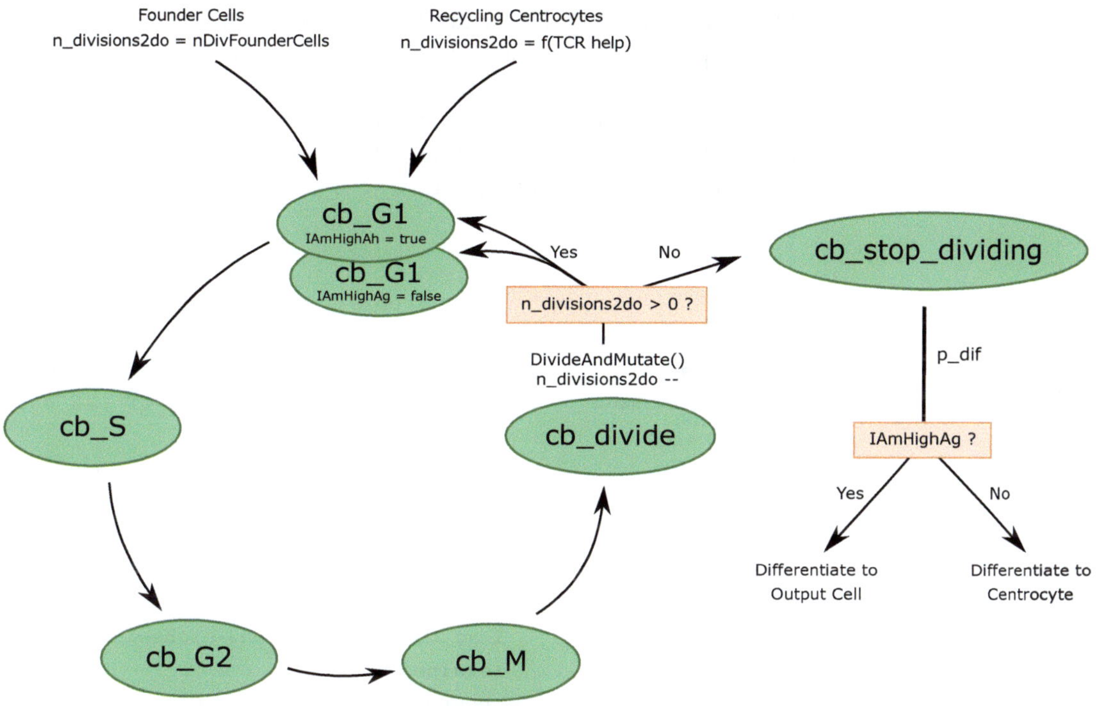

Fig. 5 Steps of updates at the centroblast stage. New centroblasts start in the G1 phase of the cell cycle. After completion of all phases, depending on the remaining number of divisions to do, the cells will divide or enter in the state *cb_stop_dividing* and wait for differentiating into centrocyte or output cell provided it has enough retained antigen

arrived founder cells will get a fixed number of divisions of (*nDiv FounderCells*). In the case of a recycled centrocyte, the number of divisions is supposed to be dependent on the strength of the T cell help received (*see* Subheading 22.2.8).

Finally, asymmetric division of the retained antigen is performed by using 72% of divisions being asymmetric [3], in which case a polarity index of the antigens in daughter cells is sampled by a Gaussian distribution around a mean value of 88% [3].

To implement the hypothesis that cells retaining the highest amount of antigen preferentially leave the GC as output cell, a property *IAmHighAg* is given to centroblasts. When a cell recycles from centrocyte to centroblast, it is marked positive for this property (and it has retained antigen). However, at each division, only one cell, with the highest amount of retained antigen, keeps this property. In case of symmetric division, none of the cells keeps it. The decision to become an output cell is then restricted to cells that are still *IAmHighAg* positive (LEDA), or alternately done on a probabilistic basis (LEDAX model in [8]).

It was proposed with our model [8] and later confirmed [7] that centroblasts that are recycling with a high amount of antigens harbor a lower mutation rate. We implement this Dynamic Somatic

Hypermutation (DSH) by giving a mutation rate to each cell, and by computing this probability at recycling (*see* Subheading 22.2.8).

The rules for updating the state of a cell are shown in Fig. 5 and can be separated into two procedures: one to initiate a cycle, and another one to update the state (Algorithm 4).

Properties of a centroblast:

state	cb_G1 / cb_S / cb_G2 / cb_M / cb_divide / cb_stop_dividing
endOfThisPhase	duration of the current cell-cycle phase
cycleStateTime	time since the beginning of the current phase
retainAg	remaining amount of retained antigen
pMutation	probability of mutation per division for this cell
nDivisions2do	remaining number of divisions to perform as centroblast
IAmHighAg	I am the cell with the highest amount of retained antigen (compared with the progeny of the same recycled centrocyte)

Algorithm 4 Updating events at the centroblast Stage

```
procedure InitiateCycle(Cell C, divisions)
    if divisions = 0 then
        C.state ⟵ cb_stop_dividing
    else
        C.state ⟵ cb_G1
        C.cycleStateTime ⟵ 0
        C.endOfThisPhase ⟵ getDuration(cb_g1)      ▷ Taken from a normal distribution. See parameters in 2.12
        C.nDivisions2do ⟵ divisions
    end if
end procedure

procedure Progress Cycle(Cell C)
    C.cycleStateTime += dt
    if C.cycleStateTime ≥ C.endOfThisPhase then
        Progress : C.state ⟵ (cb_G1 → cb_S → cb_G2 → cb_M → cb_divide)
        C.endOfThisPhase ⟵ getDuration(state)
        C.cycleStateTime ⟵ 0
    end if
end procedure

procedure DivideAndMutate(Cell C)
    if With probability pNow then
        Find a free position(x,y,z) in the neighborhood
        if Not found then
            end procedure                    ▷ it will be called later as long as state = cb_divide
        end if
        Creates a new cell C' with the same properties of C, in CBlist[] at index newID
        C.IAmHighAg ⟵ false
        C'.IAmHighAg ⟵ false
        Grid[x,y,z].ID ⟵ newId
        Grid[x,y,z].type ⟵ Centroblast
        InitiateCycle(C, nDivisions2do-1)
        InitiateCycle(C', nDivisions2do-1)
        if time ≥ mutationStartTime then
            mutate(C, pMutation)
            mutate(C', pMutation)
        end if
        if With probability divideAgAsymmetric then          ▷ If the division is asymmetric
            if C.retainedAg = 0 then
                C'.retainedAg ⟵ 0
            else
                sep ⟵ random number taken with a gaussian distribution around polarityIndex
                Repeat until sep ∈ [0..1]
                C'.retainedAg ⟵ sep . C. retainedAg
                C.retainedAg ⟵ (1- sep). C.retainedAg
                if sep > 0.5 then
                    C'.IAmHighAg ⟵ true
                else
                    C.IAmHighAg ⟵ true
                end if
            end if
        else                                                  ▷ symmetric division
            C'.retainedAg ⟵ C.retainedAg / 2
            C.retainedAg ⟵ C.retainedAg / 2
        end if
    end if
end procedure
```

2.6 Selection Steps at the Centrocyte Stage

Antigen Collection

Centrocytes start at the state of unselected and are looking for interactions with FDCs and the antigens they carry, while they have a high probability to die by apoptosis. We assume here that the antigen collection phase has a predefined time-window, and that, at each interaction with an FDC position, the antigen is taken with a probability directly depending on the BCR affinity. In this way, centrocytes with a high-affine BCR can internalize a huge amount of antigen compared to low affinity ones. As a first step of selection, B cells that could not capture any antigen during the collection period are killed. Additionally, we assume that cells leaving an interaction with an FDC (successful or not) are not able to interact again and look for antigen for a small refractory time *testDelay* [18]. This avoids repeatedly testing on the same FDC and limits the maximum amount of antigen that can be collected.

Properties needed by a centrocyte during FDC selection (unselected FDCcontact stage) (Fig. 6)

`state`	`unselected / FDCcontact / FDCselected / TCcontact / selected / apoptosis`
`selectedClock`	`the time since a cell became an unselected centrocyte.` `FDC selection finishes when selected_clock > collectFDCperiod`
`selectable`	`Boolean to say if the cell can make interactions with FDCs`
`clock`	`time since last interaction.` `the cell becomes selectable again when clock > testDelay`
`nFDCcontacts`	`number of antigens captured up to now (successful contacts)`

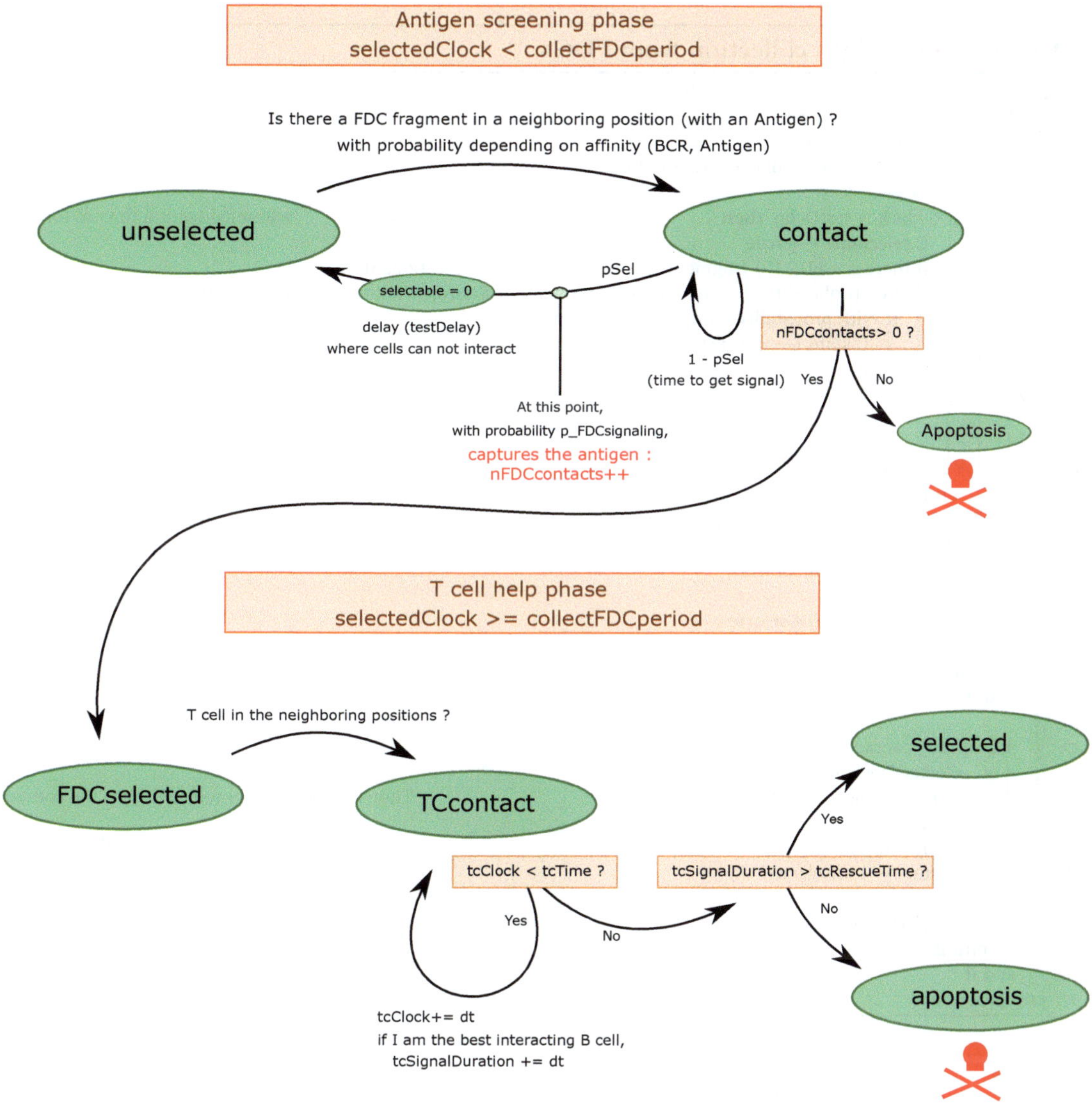

Fig. 6 Selection steps at the centrocyte stage. The first phase of antigen collection depends on the BCR affinity and is time-limited with the same duration for all B cells, while the T cell selection phase relies on receiving prolonged signal from a T cell which is only helping the best interacting B cell around. After a certain time without T cell rescue, the B cell dies, generating an additional selection bottleneck

Algorithm 5 Antigen collection from FDCs

```
procedure PROGRESSFDCSELECTION(Cell C)
    if C.state = unselected then
        C.selectedClock += dt
        if C.selectedClock <= collectFDCperiod then
            C.clock++
            if C.clock > testDelay then                                  ▷ Resensitization time to interact
                C.selectable ⟵ true
                if there is a FDC (F) fragment (f) in the neighboring positions then
                    bindprobability ⟵ affinity(BCR, antigen) . ( F.antigenAmount[f] / antigenSaturation)    ▷ See § 2.9
                    if With probability bindprobability then
                        C.state ⟵ contact
                    else                                                  ▷ desensitizes interaction capacity
                        C.clock ⟵ 0
                        C.selectable ⟵ false
                    end if
                end if
            end if
        else if C.selectedClock > collectFDCperiod then                  ▷ when antigen collection time is over
            if C.nFDCcontacts = 0 then
                C.state ⟵ apoptosis
            else
                C.state ⟵ FDCselected                                     ▷ proceed to next state
            end if
        end if
    end if

    if C.state = contact then
        C.selectedClock += dt
        if With probability pSel then                                    ▷ Time to be positively selected from the contact
            C.nFDCcontacts++
            F.antigenAmount[f] - -                                        ▷ Captures the antigen !
            C.state ⟵ unselected
            C.clock ⟵ 0
            C.selectable ⟵ false                                          ▷ refractory time for next interactions
        end if
    end if
end procedure
```

T-cell Help

When the antigen collection period is finished, centrocytes carry a number of internalized antigens (nFDCcontacts) and need to find a Tfh to receive a proliferation signal. When a B cell finds a neighbor Tfh, we assume that it requires sustained signaling for a long time in order to get selected. This is performed by counting for how long the B cell interacts with T cells and by integrating the received signal. However, since a Tfh cell might have different centrocyte neighbors, we assume that only the neighbor with the highest amount of antigen will receive the signal. Finally, if the T cell is not rescued by Tfh help in a defined time-window, it dies by apoptosis. The differentiation decision (to recycle as centroblast or to become an output) is delayed by a differentiation time called *individualDiffDelay* here. It is important to maintain a proper light zone to dark zone ratio of cells in simulations.

The following properties are needed in the T cell selection stages (FDCselected, TCcontact, and selected):

`tcClock`	`time since the start of interaction with a T cell`
`tcSignalDuration`	`accumulated signal from the current interacting T cell`
`selectedClock`	`time since the cell has been selected`
`individualDifDelay`	`time staying as selected, before recycling to centroblast`

Algorithm 6 Screening for T cell help at the centrocyte stage

```
procedure PROGRESSTCELLSELECTION(Cell C)
    if C.state = FDCselected then
        if there is a T cell TC in a neighboring position P then
            C.state ⟵ TCcontact
            C.tcClock ⟵ 0
            C.tcSignalDuration ⟵ 0
            UpdateTCell(TC, C)        ▹ See below § 2.7 - This function puts the T cell in state TC-CCcontact
        end if
    end if

    if C.state = TCcontact then
        C.tcClock += dt
        if the T cell has no other neighbor B cells with equal or more antigens then
            C.tcSignalDuration += dt        ▹ Accumulates signal from the Tfh
        end if
        if C.tcSignalDuration ≥ tcRescueTime then
            C.state ⟵ selected
            C.selectedClock ⟵ 0
            C.individualDifDelay ⟵ random number taken from a gaussian around 6 hours    ▹ See parameters in 2.12
            LiberateTCell(TC, C)
        else if C.tcClock ≥ tcTime then
            C.state ⟵ Apoptosis
            LiberateTCell(TC, C)
        end if
    end if

    if C.state = selected then
        selectedClock += dt
        if selectedClock > individualDifDelay then
            if With probability pDif then
                if With probability pDif2out(time) then        ▹ Alternative to LEDA
                    Differ2Out(C)
                else
                    Differ2CB(C)        ▹ Recycling as centroblast
                end if
            end if
        end if
    end if
end procedure
```

2.7 Update of T Cells

In this model, the T cells do not bear any antigen specificity and provide help to interacting B cells. However, when a T cell is close to multiple B cells, it tends to repolarize its synapse towards the cell providing the highest signaling [19]. Here, each T cell maintains a list of interacting B cells in the vicinity, thanks to the functions UpdateTCells, when a B cell enters in contact with a T cell and LiberateTCell when it dies or when it leaves the TCcontact state. Finally, at each time-step, a T cell will only provide help to the interacting B cell with the highest amount of retained antigen (Algorithm 7).

The properties required for a T cell are, on top of the motile cells fields (polarity and isResponsive2signals),

`state`	`TCnormal (default) / TC-CCcontact`
`ListInteractingB[]`	`The list of IDs of B cells currently in interaction`

Algorithm 7 Updating the T cells according to B cells interactions

```
procedure UPDATETCELL(Cell TC, Cell B)
    Add B to TC.ListInteractingB[]
    TC.state ⟵ TC-CCcontact
end procedure
procedure LIBERATETCELL(Cell TC, Cell B)
    Remove B from TC.ListInteractingB[]
    if TC.ListInteractingB[] is empty then
        TC.state ⟵ TCnormal
    end if
end procedure
```

2.8 Differentiation, Recycling, Output, and Inflow

The transitions between cell stages require a few adjustments (Algorithm 8).

First, when a cell becomes an output cell, we suppose it becomes insensitive to chemokines. Optionally, the "move" procedure can be extended to give a "south polarity" (as opposite to Tfh), to target them preferentially to the dark zone before leaving the GC.

When a centrocyte recycles to centroblasts, we took the hypothesis of a dynamic mutation rate according to the affinity of a B cell (quantified as the amount of antigen it could internalize in the light zone), which was also observed in [7]. Since the mechanisms involved are not fully quantified, this dependency is simulated by a Hill function.

Additionally, it is believed that the strength of the T cell–B cell interactions impacts on the number of divisions that a B cell will later perform in the dark zone after recycling. The mechanisms responsible for this tuning of proliferation are still poorly understood, and therefore we took a Hill function as a general function to convert the affinity of the BCR into a number of divisions to perform.

Finally, in case a cell kept antigen in the dark zone, even after cycles of divisions, we let the possibility to keep it for the next phase of antigen collection in the light zone (Option *DeleteAgInFreshCC* = *false*). In this case, the cell memorizes previous successful FDC contacts even before starting antigen collection.

Algorithm 8 Transition between centroblasts, centrocytes, and output cells

```
procedure DIFFER2OUT(Cell C)
    Move the cell to OUTlist[]
    InitiateChemokineReceptors(C, Out cell)
    NbOutCells[C.BCR] ++                         ▷ (Storing the number of output cells per BCR sequence)
    Grid[C.x, C.y, C.z].type ⟵ Out cell
    Grid[C.x, C.y, C.z].ID ⟵ index in OUTlist[]
end procedure

procedure DIFFER2CB(Cell C)
    Move the cell to CBlist[]
    InitiateChemokineReceptors(C, Centroblast)
    Grid[C.x, C.y, C.z].type ⟵ Centroblast
    Grid[C.x, C.y, C.z].ID ⟵ index in CBlist[]

    Attributing a mutation rate dynamically tuned depending on the amount of captured antigen
    C.pMutation ⟵ pMut + (pMutAfterSelection − pMut).Affinity(C.BCR)^pMutAffinityExponent

    Computing the number of division depending on the amount of antigen captured:
    agFactor ⟵ nFDCcontacts^pMHCdep_nHill
    ndiv ⟵ pMHCdepPmin + (pMHCdepPmax − pMHCdepPmin).agFactor/(agFactor+pMHCdepK^pMHCdep_nHill)
    InitiateCycle(nDiv)

    C.retainedAg ⟵ nbFDCcontacts
    C.IAmHighAg ⟵ true
end procedure

procedure DIFFER2CC(Cell C)
    Move the cell to CClist[]
    InitiateChemokineReceptors(C, Centrocyte)
    Grid[C.x, C.y, C.z].type = Centrocyte
    Grid[C.x, C.y, C.z].ID = index in CClist[]
    C.state ⟵ unselected
    if DeleteAgInFreshCC then                    ▷ (Optional)
        nFDCcontacts ⟵ 0
    else
        nFDCcontacts ⟵ ⌊C.retainedAg + 0.5⌋
        antigen from recycling is further conserved
    end if
end procedure
```

2.9 Model Initialization

To start a simulation, the location of the cells and the initial distribution of signals have to be defined. The 3D grid is initialized to be of size $N \times N \times N$. The points which are not in a central sphere of radius $N/2$ representing the GC are considered as unavailable for cells (outside). All positions within this sphere are initially marked with $type = empty$ and $ID = -1$.

The initial cells are put randomly in free positions. The dark zone is defined as positions with $Grid[x = 1..N, y = 1..N, z > N/2]$, and a predefined number of stromal cells are distributed in the dark zone. Similarly, FDCs are randomly distributed in the light zone.

The FDCs are multi-fragment cells, meaning that they possess branches that can cover multiple positions in the grid. Here, they are designed as straight branches in each direction, of length *DendriteLength*, and each position occupied by an FDC is called a fragment. The molecules of the antigen are dispatched from the beginning and equally distributed inside the FDCs fragments. The branches are transparent in the sense that other cells may occupy a position of the branch. This is performed by storing an additional information in the grid, namely *FDClisti*, meaning the index of the FDC that has a branch at that point.

Recently, the number of GC founder cells was determined and was found in the range of 100 clones [1]. Therefore, two variants of initialization are implemented in hyphasma: Either 3 founder clones are predefined in the beginning and divide and mutate 12 times in the expansion phase of the GC. In an alternative model, the GC simulation starts from 0 founder clones and founder cells continuously enter the GC reaction for a few days (Following Algorithm 9, lines 36–47 [1]), each of them dividing six times. In both variants, the quality, i.e., the initial affinity, of the founder cells can be restricted and adapted to the described experiment. For example, in some transgenic experiments, the quality of the founder cells is substantially higher than in experiments based on wildtype germline B cell stimulations. In the latter case, a realistic starting quality is in the range of a Hamming distance of 4–8.

Antibody Feedback

Additionally, the possibility of antibody feedback, a process by which antibodies can diffuse back to a GC and cover the antigens at the surface of the FDCs, can be incorporated [20]. It consists into allowing antibodies to bind antigens on fragments of FDCs, which then become immune complexes, not accessible from B cells (the binding probably depending only on the "free antigens").

Additional arrays can be used to keep track of the cells secreting antibody (still in the GC or already out).

```
NbOutCells[n]
NbOutCellsProduce[n]
AntibodyPerSeq[n]
```

It does not require to keep the cells as agents, but just to remember how many cells are producing antibodies for each antibody sequence n. Similarly, the concentration of each antibody type can be stored. Note that this concentration needs to take into account the effect of dilution of antibodies in the blood and in the peripheral organ (*abProdFactor*).

We suppose that the amount of antibodies is much higher than antigens, and therefore the consumption of antibodies by making immune complexes is neglected. The binding of antibodies to antigens on FDCs depends on their affinity. The *kon* and *koff* values of antibodies are derived from the affinity (see parameters Subheading 22.2.12). Optionally, to save computational time, antibodies can further be classified into bins of affinity, and the formation of immune complexes then only depends on the amount of antibodies for each bin (not shown).

The properties of an FDC are summarized as follows:

```
volume               number of fragments
fragments[].x,y,z    position of each fragment
antigenAmount[]      amount of antigens available at each fragment
icAmount[]           amount of antigens covered (hidden) by
                     feedback antibodies
```

Algorithm 9 Initialization

```
procedure INITIALIZECELLS
    for NStromalCells times do
        Generate a random position ∈ (x ∈ [0..N], y ∈ [0..N], z ∈ [N/2..N]), in the border of the GC ▷ In the dark zone
        Repeat until this position is empty
        newID = position of adding a Stromal Cell in STROMAlist[]
        Grid[x,y,z].ID = newID
        Grid[x,y,z].type = stromal
    end for

    for FDCnumber times do
        Generate a random position within (x ∈ [0..N], y ∈ [0..N], z ∈ [0..0.5N])        ▷ In the Light Zone
        Repeat until this position is empty
        Create a new FDC F
        newID = position of adding F in FDClist[]
        Grid[x,y,z].ID = newID
        Grid[x,y,z].type = FDC
        F.fragments[0] = (x,y,z)
        for i = 0 to DendriteLength do
            F.fragments[6i+1] = (x+i,y,z)
            F.fragments[6i+2] = (x-i,y,z)
            F.fragments[6i+3] = (x,y+i,z)
            F.fragments[6i+4] = (x,y-i,z)
            F.fragments[6i+5] = (x,y,z+i)
            F.fragments[6i+6] = (x,y,z-i)
            remove fragments outside of the GC or overlapping with existing FDC branches.
        end for
        F.Volume ⟵ number of fragments
        agPerFrag ⟵ antigenAmountPerFDC/F.volume
        for i = 0 to Volume-1 do
            Grid[fragment[i].(x,y,z)].FDClisti ⟵ newID
            F.antigenAmount[i] ⟵ agPerFrag
        end for
    end for

    for SeederNumber times do                ▷ Can be also performed to add seeders cells later
        Generate a random position within (x ∈ [0..N], y ∈ [0..N], z ∈ [0..0.5N])        ▷ In the Light Zone
        Repeat until this position is empty
        Create a new CentroBlast B
        B.BCR ⟵ get new BCR, randomly or from a predefined repertoire
        B.pMutation ⟵ pMut(time)
        InitiateCycle(B,nDivFounderCells)
        InitiateChemokineReceptors(B, Centroblast)
        newID = position of adding B in CBlist[]
        Grid[x,y,z].ID = newID
        Grid[x,y,z].type = CentroBlast
    end for

    for TCnumber times do
        Generate a random position (x ∈ [0..N], y ∈ [0..N], z ∈ [0..0.5N])        ▷ In the Light Zone
        Repeat until this position is empty
        Create a new T cell T
        newID = position of adding T in TClist[]
        Grid[x,y,z].ID = newID
        Grid[x,y,z].type = Tcell
    end for
end procedure
```

2.10 Main Simulation

Now that each process and event has been described separately, together with the data structure suitable for describing a GC state, we depict the time-evolution of our GC in silico in Algorithm 10.

At each time-point, each kind of cell is updated according to its properties.

Stromal cells and FDCs do not move but secrete chemokines. Antibodies produced by output producers can bind antigens on the FDCs and create immune complexes, allowing for a negative antibody feedback. Output cells move. The chemokines are finally diffused according to the time-step and the diffusion constants (algorithm not shown, see parameter values in Subheading 22.2.12). Since the chemokine producing cells are not moving, it is possible to simulate beforehand the steady-state amount of chemokines in space and perform simulations with this fixed spatial distribution of chemokines.

The state of centroblasts and centrocytes is updated and can lead to the decision of division, death, differentiation, or recycling.

The function Move is called only for cells in a motile state, and at the end of the time-step, swapping cells that intended to go to reverse directions (exchange their position) are allowed, to reduce the effect of the discretization of the space into a grid.

Algorithm 10 Hyphasma: Simulation of a germinal center

```
procedure HYPHASMA
    for time = tmin to tmax by steps of dt do
        for each stromal cell S in STROMAlist[] do
            signalSecretion(S, CXCL12, p_mkCXCL12)
        end for
        for each cell F in FDClist[] taken in a random order do
            signalSecretion(F, CXCL13, p_mkCXCL13)
            for each fragment f of the cell F do                                ▷ Antibody Feedback
                for each antibody sequence n do
                    d_ic ⟵ d_t.(k_on(n). F.antigenAmount[f] . antibodyPerSeq[n] - k_off(n).F.icAmount[f])
                    F.antigenAmount[f] − = d_ic
                    F.icAmount[f] + = d_ic
                end for
            end for
        end for
        DiffuseSignal(CXCL12, CXCL13)
        for each BCR sequence n do                                               ▷ Output cells becoming producers
            transfert ⟵ NbOutCells[n]. pmDifferentiationRate . dt
            NbOutCells[n] -= transfert
            NbOutCellsProduce[n] += transfert
        end for
        for each BCR sequence n do                                               ▷ Amount of antibody produced
            AntibodyPerSeq[n] + = NbOutCellsProduce[n] . abProdFactor - antibodyDegradation . AntibodyPerSeq[n]
        end for                              ▷ The degradation rate is taken to be the same for all antibodies
        for each cell C in OUTlist[] taken in a random order do
            Move(C)
            if C is in a border of the GC then
                remove the cell from the simulation                              ▷ This cell is already counted in nbOutCells[]
            end if
        end for
        for each cell C in CBlist[] taken in a random order do
            UpdateChemokinesReceptors(C)
            ProgressCellCycle(C)
            if C.state = cb_Divide then
                DivideAndMutate(C)
            end if
            if C.state = cb_stop_dividing then
                if With probability pDif then
                    if C.IamHighAg then                                          ▷ LEDA hypothesis
                        differ2OUT(C)
                    else
                        differ2CC(C)
                    end if
                end if
            end if
            if C.state ≠ cb_M then
                Move(C)
            end if
        end for
        for each cell C in CClist[] taken in a random order do
            UpdateChemokinesReceptors(Cell C)
            ProgressFDCSelection(C)
            ProgressTCellSelection(C)
            if C.state = apoptosis then
                Remove cell from simulation
                End procedure
            end if
            if C.state ≠ (FDCcontact or TCcontact) then
                Move(C)
            end if
        end for
        for each cell T in TClist[] taken in a random order do
            if T.state = TCnormal then
                Move(T)
            end if                         ▷ The state of T cells is updated from ProgressTCellSelection()
        end for
        Add newcoming seeder B cells if desired                ▷ Follow same steps as Algorithm 2.9, lines 36-47
        for each cell C whose movement could not be performed do
            if another neighbor cell N intended to exchange but couldn't move then
                Swap the position of the two cells with probability 0.5
            end if
        end for
    end for
end procedure
```

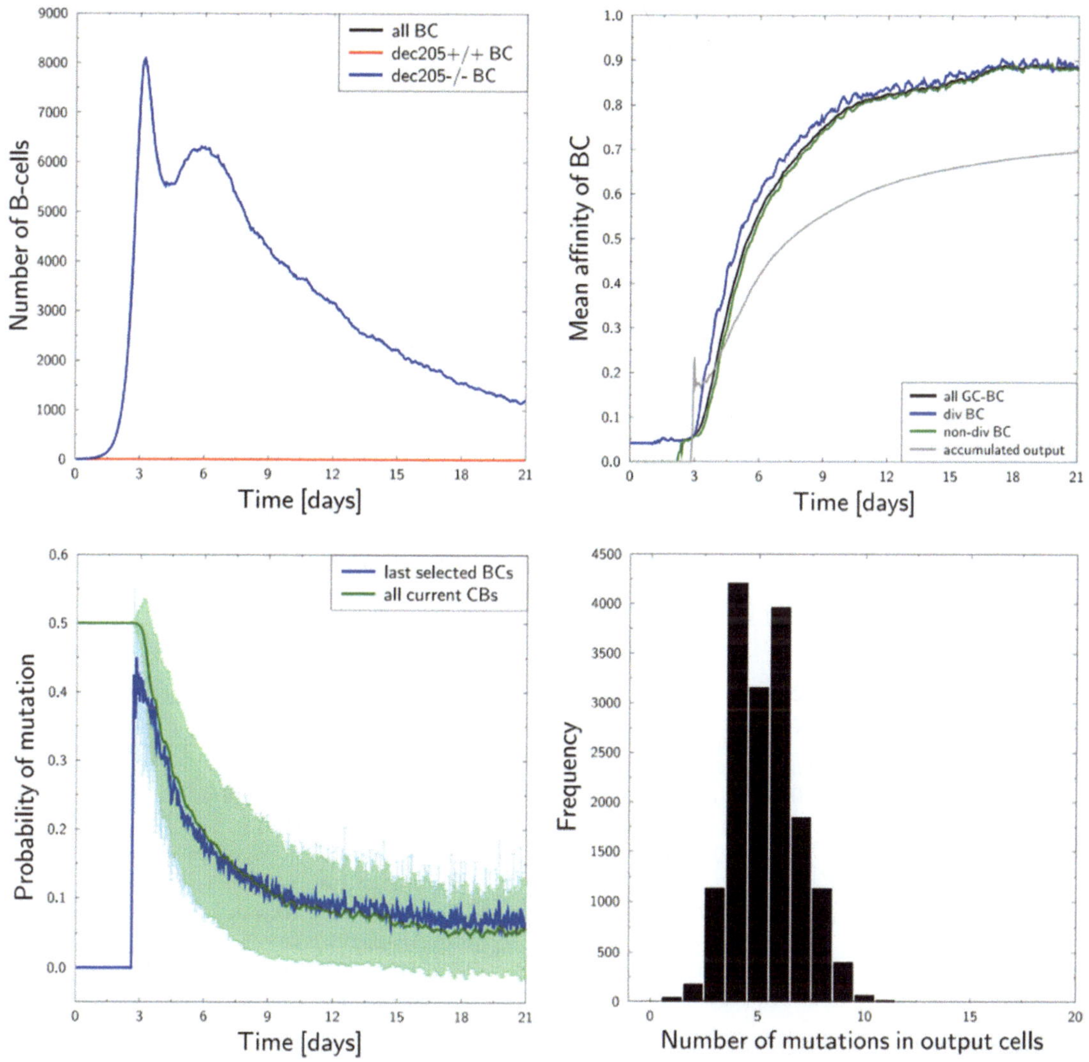

Fig. 7 A representative simulation in the LEDA model. The evolution over time of the number of B cells, the average affinity, the mutation rate (DSH), and the distribution of mutations in output cells is depicted for one simulation

2.11 Example of Simulation

A typical simulation is shown in Fig. 7. The number of B cells over time often follows two peaks, followed by a slow phase of shrinking. The affinity of B cells increases to reach high affinity values, whereas the mutation rate is significantly inhibited due to DSH.

2.12 List of Parameters

All the parameters needed to run a simulation, and mentioned in the algorithms, are shown in Tables 1 and 2.

The transformation between abstract affinity ($\in [0..1]$) in the model and real affinity is given by the following formula (with $ExpMin = 5.5$ and $ExpMax = 9.5$). Binding/unbinding rates follow:

Table 1
List of parameters and their values, part 1

Parameter name	Value	Unit	Description
Space discretization			
dt	0.002	h	Time-steps
tmin	0.0	h	Beginning time
tmax	504.0	h	End time
dx	5	μM	Lattice constant of space grid
–	160	μM	Radius of GC
N	64		Nb of gridpoints in each dimension: 2*Radius/dx
Initialization			
NStromalCells	300		Number of initial stromal cells
FDCnumber	200		Total number of FDCs
–	40	μM	Length of FDC arms (microm)
DendriteLength	8	points	(Length of arms/dx)
AntigenAmountPerFDC	3000		Presented Ag per FDC in units of threshold
seederNumber	3		Total Number of initial B Cells
Tcnumber	250		Number of initial TC
Affinity			
Gamma	2.8		Width of Gaussian affinity weight function
Eta	2		Exponent of the hamming distance
Dynamic update of chemokine receptors			
CXCL13crit	0.8e−10	mol	Critical CXCL13 concentration for desensitization [mol]
CXCL13recrit	0.6e−10	mol	Critical CXCL13 concentration for resensitization [mol, − 1 for none]
CXCL12crit	60.0e−10	mol	Critical CXCL12 concentration for desensitization [mol]
CXCL12recrit	40.0e−10	mol	Critical CXCL12 concentration for resensitization [mol, − 1 for none]
Chemotaxis			
Turning angles Thetas		rad	Distribution directly taken from Allen 2004
chemoMax	10		Maximum weight of chemokine to random polarity
chemoSteep	1.00E+10		Steepness of weight reduction with chemokine gradient l/mol
chemoHalf	2.00E−11	mol/l	Chemokine gradient of half weight mol/l
northweight	0.1		Tendency for TC to stay in the LZ (weight in 0..1)
persistentLengthTime(Cell Type)			
persistentLengthTime (Centrocyte)	1.5	min	Persistence of CC polarity
persistentLengthTime (Centroblast)	1.5	min	Persistence of CB polarity
persistentLengthTime(T cell)	1.7	min	Persistence of polarity of TC
persistentLengthTime (Out Cell)	0.75	min	Persistence of OUT polarity

(continued)

Table 1
(continued)

Parameter name	Value	Unit	Description
Speed(Cell Type)			
Speed(Centrocyte)	7.5	μM/min	Mean CC velocity
Speed(Centroblast)	7.5	μM/min	Mean cell velocity
Speed(T cell)	10	μM/min	Velocity of TC
Speed(Out Cell)	3.0	μM/min	Mean OUT velocity
Duration of cell cycle phases (according to Gaussian distribution with parameters)			
getDuration(c_G1)	2.0	h	Duration of CB cell cycle phase G1
getDuration(c_S)	1.0	h	Duration of CB cell cycle phase S
getDuration(c_G2)	2.5	h	Duration of CB cell cycle phase G2
getDuration(c_M)	0.5	h	Duration of CB cell cycle phase M
All with a standard deviation of	1.0	h	Width of Gaussian variation of phases
Differentiation rates			
StartDifferentiation	72.0	h	Start Differentiation
pDif (CB to CC)	dt * 0.1	proba per dt	Rate for differentiation of centroblasts to centrocytes (h)
p_dif2out_target (from CC)	0.0	[= 0 in the LEDA case]	Output probability towards Antibody production
width	12.0		Width of smooth onset of differentiation to output (hours)
weight(time)	time dep		= 1.0 + exp((start_differentiate - time)/ smooth_differentiation_time)
pDif2out(time)	time dep	[= 0 in the LEDA case]	= p_dif2out_target/weight(time)
pDif(CC to CB)	dt * 0.1		dt * Rate of selected CC-differentiation (h)
dif_delay	6.0	h	Delay of selected CC-differentiation
getNewDiffDelay()	proba. law		= dif_delay * (1 + 0.1 * log((1. − x)/x) where x is random in [0..1]
agDeletedInFreshCC	true		Retained antigen is deleted in fresh CC

$$
\begin{aligned}
RealAffinity(BCR, Antigen) &= 10^{expMin+Affinity(BCR,Antigen).(expMax-expMin)}(1/\text{mol})\\
k_{\text{on}}(BCR) &= Constant = 1e6(1/\text{mols}) = 3.6e9(1/\text{molh})\\
k_{\text{off}}(BCR, Antigen) &= \frac{k_{\text{on}}(BCR)}{RealAffinity(BCR, Antigen)}(1/\text{h})
\end{aligned}
$$

3 Conclusions

Simulations of complex biological systems are a powerful tool for the testing of hypotheses. The presented code *hyphasma* was employed in GC research in three different ways, for which a few examples of applications of the code are provided:

Table 2
List of parameters and their values, part 2

Parameter name	Value	Unit	Description
Dynamic number of divisions			
nDivFounderCells	12		(StartDifferentiation - tmin)/ total_cell_cycle_duration
	6		Note: in case influx of cells is allowed, this value is more physiological
pMHCdepHill	1		pMHC-dependent division number Hill: Hill-coefficient (n_P)
pMHCdepMin	1		pMHC-dependent division number Hill: minimum (P_min)
pMHCdepMax	6		pMHC-dependent division number Hill: maximum (P_max)
pMHCdepK	6		pMHC-dependent division number Hill: half (K_P)
Mutation and division			
–	24	h	Start Mutation
pmut(time <StartMutation)	0	proba/BCR/div	Mutation probability
pmut(time >StartMutation)	0.5	proba/BCR/div	Mutation probability
pmutafterselection	0.0	proba/BCR/div	Mutation probability after TC selection [0..1] (target value of DSH)
pmutaffinityexponent	1		Affinity-dependent mutation upon TC contact (affinity-exponent)
pNow (proba divide)	dt * 9.0	proba in dt	dt * Proliferation rate (h) [speed of processing to division]
divideAgSymmetric	0.72	proba per div	Divide antigen asymmetrically on daughter BC (1: 100%; 0: no)
polarityIndex	0.88		Distribute antigen asymmetric (0.5 <polarity index <1.0)
width	0.04		Coefficient of variation around the polarity index
Selection steps			
testDelay	0.02	h	Time gap between affinity tests (– 1=move, in h)
collectionFDCperiod	0.7	h	Duration of CC collection of antigen by serial FDC encounters (h)
antigenSaturation	20		Ag saturation per FDC-fragment in units of threshold (1: constant binding probability):
pSel	dt * 0.05	proba in dt	dt * Rate of positive selection at FDCs (h)
tcTime	0.6	h	Duration of TC-CC-interaction
tcRescueTime	0.5	h	Minimum duration of TC-CC-polarization for CC-rescue in hours

(continued)

Table 2
(continued)

Parameter name	Value	Unit	Description
Production/Diffusion rates			
	1000	$\mu M^2/h$	Diffusion constant of chemotaxis signal [CXCL13]
	1000	$\mu M^2/h$	Diffusion constant of CXCL12
	1.00E−08	mol/(h FDC)	Rate of CXCL13 production
pMKCXCL13	~ dt * 0.75	molecules/dt/point	rate * dt * dx^3 * Navogadro
	4.00E−07	mol/(h cell)	Rate of CXCL12 production
pMKCXCL12	~ dt * 30	molecules/dt/point	rate * dt * dx^3 * Navogadro
boundary[CXCL13]	0	mol	Boundary for CXCL13
Antibody			
pmDifferentiationRate	24	h	Half time of plasma cell differentiation to antibody producing cell
antibodies_production	1.00E−17	mol/(h cell)	Production rate of systemic antibodies by output population
V_blood	1.00E−02	*L*	Blood volume (order of magnitude, mice)
N_GC	1000		Number of GC generating antibody producing plasma cells
abProdFactor			= antibodies_production * dt * N_GC * V_blood * 1e15
antibodyDegradation	30		Antibody degradation half time (days):
k_{on}(seq n)	1E+6	1/(mol s)	k_{on} for building immune complex
k_{off}(seq n)	seq. dep.	1/s	$k_{off} = f$(BCR affinity)

Data interpretation

Using quantitative data in order to determine how a particular mechanism works.

Example: In 2007, three groups measured B cell migration in GCs [16, 21, 22]. The data were consistent with each other but the interpretation was controversial. The surprise in the data was that B cells basically were measured to do random walk, which was in conflict with the belief that cells would do chemotaxis in order to switch between both zones. In addition, transmigration of B cells between the zones was rare. Further, more cells moved from the dark to the light zone than vice versa, setting in question the recycling hypothesis that B cells do affinity maturation in repeated rounds of mutation and selection. GC simulations with hyphasma were proving that the random walk measurement was not in contradiction with chemotaxis of a few cells, that the amount of transzone migration was exactly what has to be expected in the recycling model, and that the asymmetry of transzone migration was expected due to selection in the light zone. A quantitative analysis of the two-photon data predicted a *transient chemotaxis model* according to which B cells would sensitize and desensitize of the chemokine in a concentration-dependent way [17].

Theory prediction

Using the current knowledge in order to develop a consistent theory of GC reactions.

Example: It was generally believed that B cells compete for antigen. The code presented here was used in order to distinguish eight different selection mechanisms [18]. It was shown that competition for antigen is not robust and less efficient for affinity maturation than other mechanisms. In hyphasma, the most efficient mechanism was competition for T cell help. This prediction was supported 4 years later by the finding that B cells got signals for recycling to the dark zone phenotype and division [15]. A quantitative analysis of the B cell expansion found in this experiment induced the prediction that the number of divisions is regulated by Tfh in an affinity-dependent manner between 1 and 6 divisions [8]. This prediction was proven true 2 years later [7] and further implications could be drawn from this mechanism [6].

Theory distinction

Testing of various hypotheses for consistency with a particular data set.

Example: Experiments that suppress or knowout a particular part of the GC reaction may lead to a complex phenotype involving several readouts and may also include subtle differences. Such a complex phenotype does not allow to determine the impact point of the suppressed molecule because the interdependency of the different modulated read-outs is unclear. For example, the suppression of the microRNA-155 generated such a complex phenotype with changed B cell population dynamics, affinity maturation, fraction of apoptotic cells, and amount of generated output cells. Hyphasma was used to test six different models of how microRNA-155 would impact on the GC reaction assuming that the whole phenotype stems from a single impact point. In this way, two of the six tested hypotheses were found compatible with the phenotype, namely a defect in BCR signaling as well as increased apoptosis of positively selected B cells [23]. This result focused further experimentation onto these two mechanisms. Similar applications were done with hyphasma in the context of the impact of TLR-4 on GCs [24] and of sequential switching to the IgE isotype [25].

While the development of a code for a particular biological system can take several years, once it is setup, it has a great potential of interpreting experiments and generating novel theories. We believe that the fundamental reason that it worked out in the case of hyphasma and the GC reaction is a strict mechanistic implementation on the cellular scale. Subcellular processes were implemented in a phenomenological way and nothing but the boundary conditions were implemented on the GC scale. Only then, the appearance of the GC in silico can be tracked back to the mechanisms that made this appearance emerge. It is this principle of scale integrity that makes the difference between a reproductive simulation and a predictive theory.

References

1. Tas JM et al (2016) Visualizing antibody affinity maturation in germinal centers. Science. doi:10.1126/science.aad3439
2. Vinuesa CG, Linterman MA, Goodnow CC, Randall KL (2010) T cells and follicular dendritic cells in germinal center B-cell formation and selection. Immunol Rev 237:72–89
3. Thaunat O et al (2012) Asymmetric segregation of polarized antigen on B cell division shapes presentation capacity. Science 335:475–479
4. Dustin ML, Meyer-Hermann M (2012) Antigen feast or famine. Science 335:408–409
5. Allen CD et al (2004) Germinal center dark and light zone organization is mediated by CXCR4 and CXCR5. Nat Immunol 5:943–952
6. Meyer-Hermann M (2014) Overcoming the dichotomy of quantity and quality in antibody responses. J Immunol 193:5414–5419
7. Gitlin AD, Shulman Z, Nussenzweig MC (2014) Clonal selection in the germinal centre by regulated proliferation and hypermutation. Nature 509:637–640
8. Meyer-Hermann M et al (2012) A theory of germinal center B cell selection, division, and exit. Cell Rep 2:162–174
9. Meyer-Hermann ME, Maini PK (2005) Interpreting two-photon imaging data of lymphocyte motility. Phys Rev E 71:061912
10. Linterman MA et al (2009) Follicular helper T cells are required for systemic autoimmunity. J Exp Med 206:561–576
11. De Vinuesa CG et al (2000) Germinal centers without T cells. J Exp Med 191:485–494
12. Baumjohann D et al (2013) Persistent antigen and germinal center B cells sustain T follicular helper cell responses and phenotype. Immunity 38:596–605
13. Perelson AS, Oster GF (1979) Theoretical studies of clonal selection: minimal antibody repertoire size and reliability of self-non-self discrimination. J Theor Biol 81:645–670
14. Meyer-Hermann M, Deutsch A, Or-Guil M (2001) Recycling probability and dynamical properties of germinal center reactions. J Theor Biol 210:265–285
15. Victora GD et al (2010) Germinal center dynamics revealed by multiphoton microscopy with a photoactivatable fluorescent reporter. Cell 143:592–605

16. Allen CD, Okada T, Tang HL, Cyster JG (2007) Imaging of germinal center selection events during affinity maturation. Science 315:528–531
17. Figge MT et al (2008) Deriving a germinal center lymphocyte migration model from two-photon data. J Exp Med 205:3019–3029
18. Meyer-Hermann ME, Maini PK, Iber D (2006) An analysis of B cell selection mechanisms in germinal centers. Math Med Biol 23:255–277
19. Duchez S, Rodrigues M, Bertrand F, Valitutti S (2011) Reciprocal polarization of T and B cells at the immunological synapse. J Immunol 187:4571–4580
20. Zhang Y et al (2013) Germinal center B cells govern their own fate via antibody feedback. J Exp Med 210:457–464
21. Schwickert TA et al (2007) In vivo imaging of germinal centres reveals a dynamic open structure. Nature 446:83–87
22. Hauser AE et al (2007) Definition of germinal-center B cell migration in vivo reveals predominant intrazonal circulation patterns. Immunity 26:655–667
23. Nakagawa R et al (2015) MicroRNA-155 controls affinity-based selection by protecting c-MYC+ B cells from apoptosis. J Clin Invest 126
24. Garin A et al (2010) Toll-like receptor 4 signaling by follicular dendritic cells is pivotal for germinal center onset and affinity maturation. Immunity 33:84–95
25. He J-S et al (2013) The distinctive germinal center phase of IgE+ B lymphocytes limits their contribution to the classical memory response. J Exp Med 210:2755–2771

Index

Dinis Pedro Calado (ed.), *Germinal Centers: Methods and Protocols*, Methods in Molecular Biology, vol. 1623,
DOI 10.1007/978-1-4939-7095-7, © Springer Science+Business Media LLC 2017

Q

R

S

T

U

V

FSC
www.fsc.org
MIX
Papier aus verantwortungsvollen Quellen
Paper from responsible sources
FSC® C105338

If you have any concerns about our products,
you can contact us on
ProductSafety@springernature.com

In case Publisher is established outside the EU,
the EU authorized representative is:
Springer Nature Customer Service Center GmbH
Europaplatz 3, 69115 Heidelberg, Germany

Printed by Libri Plureos GmbH
in Hamburg, Germany